T0254520

CONTEMPORARY ISSUES IN ESTUARINE PHYSICS

Estuaries are areas of high socioeconomic importance, with 22 of the 32 largest cities in the world being located on river estuaries. Estuaries bring together fluxes of fresh and saline water, as well as fluvial and marine sediments, and contain many biological niches and high biological diversity. Increasing sophistication of field observation technology and numerical modeling have led to significant advances in our understanding of the physics of these systems over the last decade.

This book introduces a classification for estuaries before presenting the basic physics and hydrodynamics of estuarine circulation and the various factors that modify it in time and space. It then covers special topics at the forefront of research, such as turbulence, fronts in estuaries and continental shelves, low inflow estuaries, and implications of estuarine transport for water quality.

With contributions from some of the world's leading authorities on estuarine and lagoon hydrodynamics, this volume provides a concise foundation for academic researchers, advanced students and coastal resource managers.

Arnoldo Valle-Levinson received a PhD from the State University of New York at Stony Brook in 1992 before going on to work at Old Dominion University (Norfolk, VA). He joined the University of Florida (Gainsville, FL) in 2005, where he is now a Professor in the Department of Civil and Coastal Engineering. His research focuses on bathymetric effects on the hydrodynamics of estuaries, fjords and coastal lagoons. Professor Valle-Levinson is the recipient of a CAREER award from the US National Science Foundation, a Fulbright Fellowship for research in Chile, and a Gledden Fellowship from the University of Western Australia. He has worked extensively in several Latin-American countries, where he also teaches courses on estuarine and coastal hydrodynamics. He is also an associate editor for the journals *Continental Shelf Research* and *Ciencias Marinas*.

CONTEMPORARY ISSUES IN ESTUARINE PHYSICS

Edited by

A. VALLE-LEVINSON
University of Florida

CAMBRIDGE
UNIVERSITY PRESS

University Printing House, Cambridge CB2 8BS, United Kingdom

One Liberty Plaza, 20th Floor, New York, NY 10006, USA

477 Williamstown Road, Port Melbourne, VIC 3207, Australia

4843/24, 2nd Floor, Ansari Road, Daryaganj, Delhi - 110002, India

79 Anson Road, #06-04/06, Singapore 079906

Cambridge University Press is part of the University of Cambridge.

It furthers the University's mission by disseminating knowledge in the pursuit of education, learning and research at the highest international levels of excellence.

www.cambridge.org
Information on this title: www.cambridge.org/9781108447003

First published 2010
First paperback edition 2017

A catalogue record for this publication is available from the British Library

ISBN 978-0-521-89967-3 Hardback
ISBN 978-1-108-44700-3 Paperback

Contents

List of contributors

Robert J. Chant
IMCS Rutgers University
71 Dudley Road
New Brunswick, NJ 08901
USA

Carl T. Friedrichs
Virginia Institute of Marine Science
P.O. Box 1346
Gloucester Point, VA 23062-1346
USA

W. Rockwell Geyer
Woods Hole Oceanographic Institution
Applied Ocean Physics and Engineering
98 Water Street
Mail Stop 12
Woods Hole, MA 02543
USA

David A. Jay
Portland State University
Department of Civil and Environmental Engineering
P.O. Box 751
Portland, OR 97207-0751
USA

John Largier
Bodega Marine Lab
University of California, Davis

P.O. Box 247
Bodega Bay, CA 94923
USA

Lisa V. Lucas
US Geological Survey
345 Middlefield Road, MS #496
Menlo Park, CA 94025
USA

Stephen G. Monismith
Department of Civil and Environmental Engineering
Stanford University
Stanford, CA 94305-4020
USA

James O'Donnell
University of Connecticut
1084 Shennecossett Road
Groton, CT 06340
USA

Arnoldo Valle-Levinson
Department of Civil and Coastal Engineering
University of Florida
365 Weil Hall, P.O. Box 116580
Gainesville, FL 32611
USA

Clint Winant
Scripps Institution of Oceanography, UCSD
9500 Gilman Drive
La Jolla, CA 92093-0209
USA

Preface

This book resulted from the lectures of a PanAmerican Advanced Studies Institute (PASI) funded by the United States National Science Foundation and the Department of Energy. The topic of the PASI was "Contemporary Issues in Estuarine Physics, Transport and Water Quality" and was held from July 31 to August 13, 2007 at the Unidad Académica Puerto Morelos of the Mexican National University (UNAM). One of the requirements was that the PASI had to involve lecturers and students from the Americas, with most from the United States. The institute was restricted to advanced graduate students and postdoctoral participants. Because of the requirements, this book includes authors who work in the United States but tries to be comprehensive in including aspects of estuarine systems in different parts of the world. The book, however, reflects regional experiences of the authors and obviously does not include exhaustive illustrations throughout the world. Nonetheless, it is expected to motivate studies, in diverse regions, that address problems outlined herein.

This book should be appropriate for advanced undergraduate or graduate courses on estuarine and lagoon hydrodynamics. It should also serve as a reference for the professional or environmental manager in this field. The sequence of chapters is designed in such a way that the topic is introduced in terms of estuaries classification (Chapter 1). This is followed by the basic hydrodynamics that drive the typically conceived estuarine circulation consisting of fresher water moving near the surface toward the ocean and saltier water moving below in opposite direction (Chapter 2). This chapter also presents the implications of estuarine circulation on salinity stratification. The chapter sequence then deals with processes that modify the basic circulation pattern, such as tides. The theoretical framework for tides in different systems is treated in Chapter 3. The effect of tides on estuarine circulation is presented at intratidal and subtidal time scales in Chapter 4. Chapters 5 and 6 deal with effects of bathymetry on estuarine hydrodynamics. The effects of lateral bathymetry and lateral circulation on estuarine circulation are explored in

Chapter 5. Chapter 6 depicts the circulation driven by tides and winds under varying bathymetry, to compare with the results of Chapter 3 for tides. The rest of the chapters deal with selected topics related to estuarine physics: turbulence is studied in Chapter 7; fronts in estuaries and continental shelves are covered in Chapter 8; processes in low-inflow estuaries are discussed in Chapter 9; and water quality implications are presented in Chapter 10.

The effort of putting this book together was made possible by the interest and dedication of the chapter authors, whose gathering at the PASI was supported by funding from the United States National Science Foundation, under project IOISE-0614418. Special recognition to David Salas de León and Adela Monreal, from the National University of Mexico (UNAM), for their tremendous contributions and original ideas in the organization of the PASI. Mario Cáceres, David Salas Monreal, Gilberto Expósito and Miguel Angel Díaz were extremely helpful with the logistics during the PASI. Particular gratitude to the Academic Unit of UNAM in Puerto Morelos, Brigitta Van Tussembroek, Director of the Unit at the time of the PASI, for allowing the use of their facilities for this activity.

1

Definition and classification of estuaries

ARNOLDO VALLE-LEVINSON
University of Florida

This chapter discusses definitions and classification of estuaries. It presents both the classical and more flexible definitions of estuaries. Then it discusses separate classifications of estuaries based on water balance, geomorphology, water column stratification, and the stratification–circulation diagram – Hansen–Rattray approach and the Ekman–Kelvin numbers parameter space.

The most widely accepted definition of an estuary was proposed by Cameron and Pritchard (1963). According to their definition, an estuary is (a) a semienclosed and coastal body of water, (b) with free communication to the ocean, and (c) within which ocean water is diluted by freshwater derived from land. Freshwater entering a semienclosed basin establishes longitudinal density gradients that result in long-term surface outflow and net inflow underneath. In classical estuaries, freshwater input is the main driver of the long-term (order of months) circulation through the addition of buoyancy. The above definition of an estuary applies to temperate (classical) estuaries but is irrelevant for arid, tropical and subtropical basins. Arid basins and those forced intermittently by freshwater exhibit hydrodynamics that are consistent with those of classical estuaries and yet have little or no freshwater influence. The loss of freshwater through evaporation is the primary forcing agent in some arid systems, and causes the development of longitudinal density gradients, in analogy to temperate estuaries. Most of this book deals with temperate estuaries, but low-inflow estuaries are discussed in detail in Chapter 9.

1.1. Classification of estuaries on the basis of water balance

On the basis of the definitions above, and in terms of their water balance, estuaries can be classified as three types: positive, inverse and low-inflow estuaries (Fig. 1.1). Positive estuaries are those in which freshwater additions from river discharge, rain and ice melting exceed freshwater losses from evaporation or

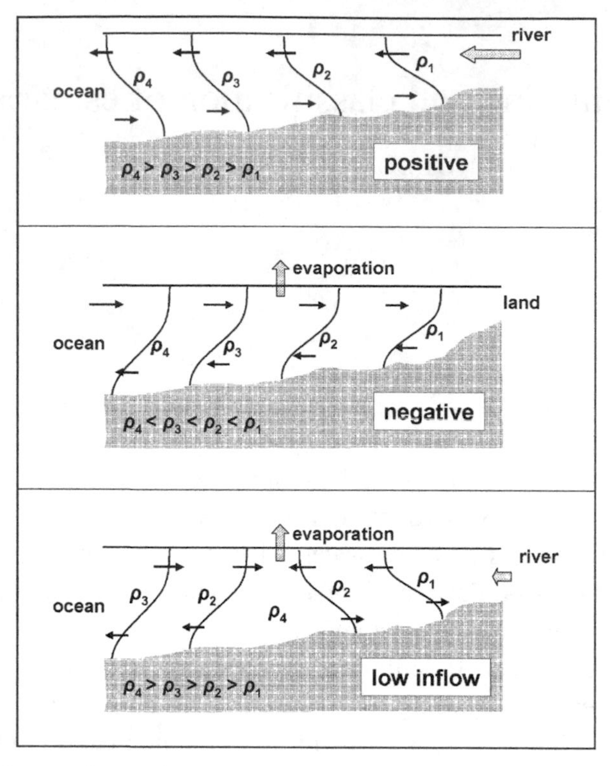

Figure 1.1. Types of estuaries on the basis of water balance. Low-inflow estuaries exhibit a "salt plug".

freezing and establish a longitudinal density gradient. In positive estuaries, the longitudinal density gradient drives a net volume outflow to the ocean, as denoted by stronger surface outflow than near-bottom inflow, in response to the supplementary freshwater. The circulation induced by the volume of fresh water added to the basin is widely known as "estuarine" or "gravitational" circulation.

Inverse estuaries are typically found in arid regions where freshwater losses from evaporation exceed freshwater additions from precipitation. There is no or scant river discharge into these systems. They are called inverse, or negative, because the longitudinal density gradient has the opposite sign to that in positive estuaries, i.e., water density increases landward. Inverse estuaries exhibit net volume inflows associated with stronger surface inflows than near-bottom outflows. Water losses related to inverse estuaries make their flushing more sluggish than positive estuaries. Because of their relatively sluggish flushing, negative estuaries are likely more prone to water quality problems than positive estuaries.

Low-inflow estuaries also occur in regions of high evaporation rates but with a small (on the order of a few m^3/s) influence from river discharge. During the dry and hot season, evaporation processes may cause a salinity maximum zone (sometimes referred to as a salt plug, e.g., Wolanski, 1986) within these low-inflow estuaries. Seaward of this salinity maximum, the water density decreases, as in an inverse estuary. Landward of this salinity maximum, the water density decreases, as in a positive estuary. Therefore, the zone of maximum salinity acts as a barrier that precludes the seaward flushing of riverine waters and the landward intrusion of ocean waters. Because of their weak flushing in the region landward of the salinity maximum, low-inflow estuaries are also prone to water quality problems.

1.2. Classification of estuaries on the basis of geomorphology

Estuaries may be classified according to their geomorphology as coastal plain, fjord, bar-built and tectonic (Fig. 1.2; Pritchard, 1952). Coastal plain estuaries, also called drowned river valleys, are those that were formed as a result of the Pleistocene increase in sea level, starting ~15,000 years ago. Originally rivers, these estuaries formed during flooding over several millennia by rising sea levels. Their shape resembles that of present-day rivers, although much wider. They are typically wide (on the order of several kilometers) and shallow (on the order of 10 m), with large width/depth aspect ratios. Examples of these systems are Chesapeake Bay and Delaware Bay on the eastern coast of the United States.

Fjords are associated with high latitudes where glacial activity is intense. They are characterized by an elongated, deep channel with a sill. The sill is related to

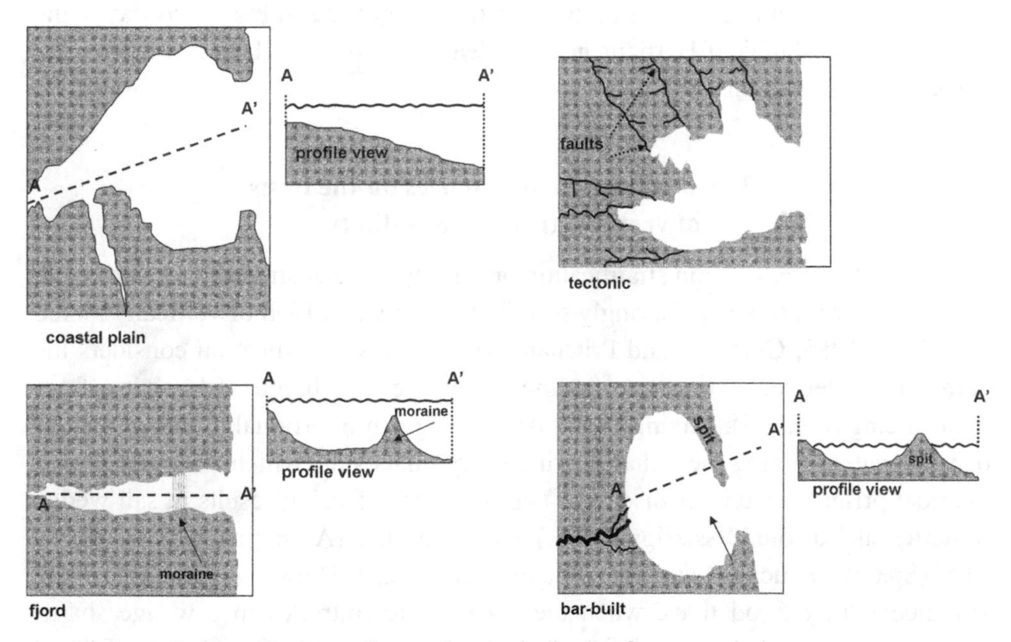

Figure 1.2. Classification of estuaries on the basis of geomorphology.

a moraine of either a currently active glacier or an extinct glacier. In the sense of the glacier activity, it could be said that there are riverine and glacial fjords. Riverine fjords are related to extinct glaciers and their main source of buoyancy comes from river inputs. They are usually found equatorward of glacial fjords. Glacial fjords are found in high latitudes, poleward of riverine fjords. They are related to active glaciers and their main source of buoyancy is derived from melting of the glacier and of snow and ice in mountains nearby. Fjords are deep (several hundreds of meters) and narrow (several hundreds of meters) and have low width/depth aspect ratios with steep side walls. Fjords are found in Greenland, Alaska, British Columbia, Scandinavia, New Zealand, Antarctica and Chile.

Bar-built estuaries, originally embayments, became semienclosed because of littoral drift causing the formation of a sand bar or spit between the coast and the ocean. Some of these bars are joined to one of the headlands of a former embayment and display one small inlet (on the order of a few hundred meters) where the estuary communicates with the ocean. Some other sand bars may be detached from the coast and represent islands that result in two or more inlets that allow communication between the estuary and the ocean. In some additional cases, sand bars were formed by rising sea level. Examples of bar-built estuaries abound in subtropical regions of the Americas (e.g., North Carolina, Florida, northern Mexico) and southern Portugal.

Tectonic estuaries were formed by earthquakes or by fractures of the Earth's crust, and creases that generated faults in regions adjacent to the ocean. Faults cause part of the crust to sink, forming a hollow basin. An estuary is formed when the basin is filled by the ocean. Examples of this type of estuary are San Francisco Bay in the United States, Manukau Harbour in New Zealand, Guaymas Bay in Mexico and some Rias in NwSpain.

1.3. Classification of estuaries on the basis of vertical structure of salinity

According to water column stratification or salinity vertical structure, estuaries can be classified as salt wedge, strongly stratified, weakly stratified or vertically mixed (Pritchard, 1955; Cameron and Pritchard, 1963). This classification considers the competition between buoyancy forcing from river discharge and mixing from tidal forcing (Fig. 1.3). Mixing from tidal forcing is proportional to the volume of oceanic water entering the estuary during every tidal cycle, which is also known as the tidal prism. Large river discharge and weak tidal forcing results in salt wedge estuaries such as the Mississippi (USA), Rio de la Plata (Argentina), Vellar (India), Ebro (Spain), Pánuco (Mexico), and Itajaí-Açu (Brazil). These systems are strongly stratified during flood tides, when the ocean water intrudes in a wedge shape. Some of these systems lose their salt wedge nature during dry periods. Typical

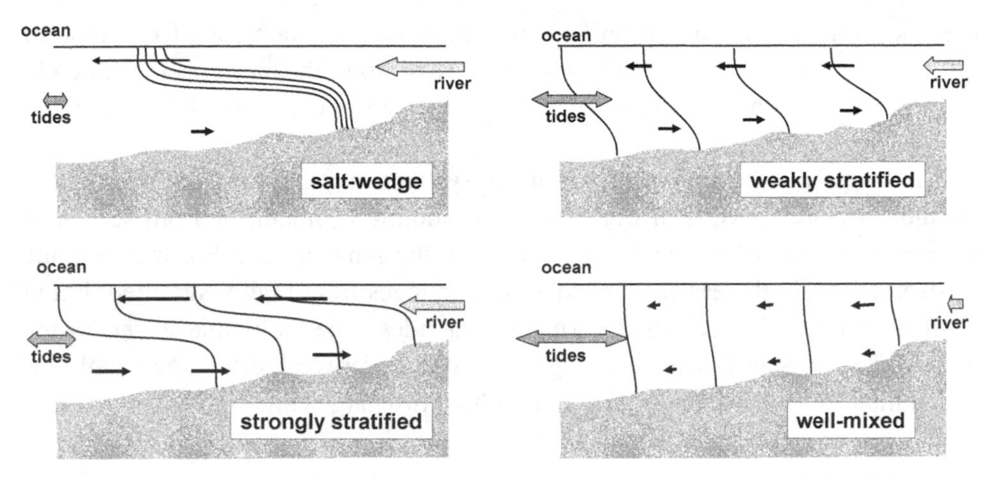

Figure 1.3. Classification of estuaries on the basis of vertical structure of salinity.

tidally averaged salinity profiles exhibit a sharp pycnocline (or halocline), with mean flows dominated by outflow throughout most of the water column and weak inflow in a near-bottom layer. The mean flow pattern results from relatively weak mixing between the inflowing ocean water and the river water.

Moderate to large river discharge and weak to moderate tidal forcing result in strongly stratified estuaries (Fig. 1.3). These estuaries have similar stratification to salt wedge estuaries, but the stratification remains strong throughout the tidal cycle as in fjords and other deep (typically >20 m deep) estuaries. The tidally averaged salinity profiles have a well-developed pycnocline with weak vertical variations above and below the pycnocline. The mean flow exhibits well-established outflows and inflows, but the inflows are weak because of weak mixing with freshwater and weak horizontal density gradients.

Weakly stratified or partially mixed estuaries result from moderate to strong tidal forcing and weak to moderate river discharge. Many temperate estuaries, such as Chesapeake Bay, Delware Bay and James River (all in the eastern United States) fit into this category. The mean salinity profile either has a weak pycnocline or continuous stratification from surface to bottom, except near the bottom mixed layer. The mean exchange flow is most vigorous (when compared to other types of estuaries) because of the mixing between riverine and oceanic waters.

Strong tidal forcing and weak river discharge result in vertically mixed estuaries. Mean salinity profiles in mixed estuaries are practically uniform and mean flows are unidirectional with depth. In wide (and shallow) estuaries, inflows may develop on one side across the estuary and outflow on the other side, especially during the dry season. Parts of the lower Chesapeake Bay may exhibit this behavior in early autumn. In narrow well-mixed estuaries, inflow of salinity may only occur during the flood tide because the mean flow will be seaward. Examples of this type of estuary are scarce because, under well-mixed conditions, the mean (as in the tidally averaged sense) flow will most likely be driven by wind or tidal forcing (e.g., Chapter 6).

It is essential to keep in mind that many systems may change from one type to another in consecutive tidal cycles, or from month to month, or from season to season, or from one location to another inside the same estuary. For instance, the Hudson River, in the eastern United States, changes from highly stratified during neap tides to weakly stratified during spring tides. The Columbia River, in the western United States, may be strongly stratified under weak discharge conditions and similar to a salt-wedge estuary during high discharge conditions.

1.4. Classification of estuaries on the basis of hydrodynamics

A widely accepted classification of estuaries was proposed by Hansen and Rattray (1966) on the basis of estuarine hydrodynamics. It is best to review this classification after acquiring a basic understanding of estuarine hydrodynamics, e.g., after Chapter 6 of this book. This classification is anchored in two hydrodynamic non-dimensional parameters: (a) the circulation parameter and (b) the stratification parameter. These parameters refer to tidally averaged and cross-sectionally averaged variables. The circulation parameter is the ratio of near-surface flow speed u_s to sectionally averaged flow U_f. The near-surface flow speed is typically related to the river discharge and, for the sake of argument, on the order of 0.1 m/s. The depth-averaged flow U_f is typically very small, tending to zero, in estuaries of vigorous water exchange because there will be as much net outflow as net inflow. In estuaries with weak net inflow, such as well-mixed and salt-wedge systems, the depth-averaged flow will be similar in magnitude to the surface outflow. Therefore, the circulation parameter is >10 in estuaries with vigorous gravitational circulation and close to 1 in estuaries with unidirectional net outflow. In general, the greater the circulation parameter, the stronger the gravitational circulation.

The other non-dimensional parameter, the stratification parameter, is the ratio of the top-to-bottom salinity difference ∂S to the mean salinity over an estuarine cross-section S_0. A ratio of 1 indicates that the salinity stratification (or top-to-bottom difference) is as large as the sectional mean salinity. For instance, if an estuary shows a sectional mean salinity of 20, for it to exhibit a stratification parameter of 1 it must have a very large stratification (on the order of 20). In general, estuaries will most often have stratification parameters <1. The weaker the water column stratification, the smaller the stratification parameter will be.

The two parameters described above can be used to characterize the nature of salt transport in estuaries. The contribution by the diffusive portion (vs the advective

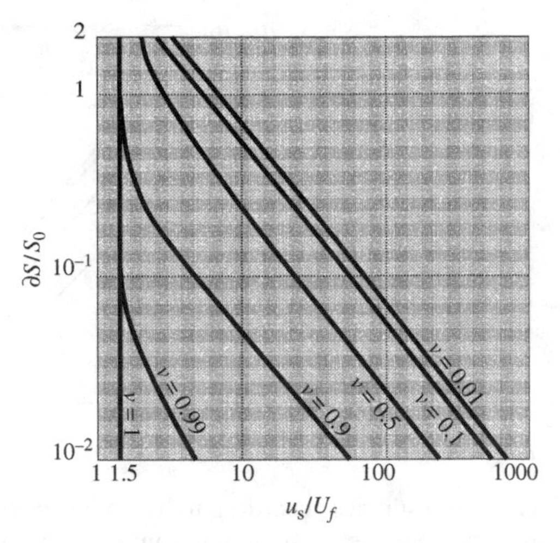

Figure 1.4. Diffusive salt flux fraction in the stratification/circulation parameter space (redrawn from Hansen and Rattray, 1966).

portion) of the total salt flux into the estuary can be called ν. The parameter ν may oscillate between 0 and 1. When ν is close to 0, up-estuary salt transport is dominated by advection, i.e., by the gravitational circulation. In this case, mixing processes are weak, as in a highly stratified estuary (fjord). When ν approaches 1, the total salt transport is dominated by diffusive processes (e.g., tidal mixing), as in unidirectional net flows. The parameter ν may be portrayed in terms of the stratification and circulation parameters (Fig. 1.4). This diagram shows that salt transport is dominated by advective processes under high gravitational circulation or strongly stratified conditions. It also shows that diffusive processes dominate the salt flux at low circulation parameter (unidirectional net flows) regardless of the stratification parameter. Between those two extremes, the salt transport has contributions from both advective and diffusive processes. The more robust the stratification and circulation parameters, the stronger the contribution from advective processes to the total salt flux will be.

These concepts can be used to place estuaries in the parameter space of the circulation and stratification parameters. The lower-left corner of the parameter space (Fig. 1.5A) describes well-mixed estuaries with unidirectional net outflows, i.e., seaward flows with no vertical structure or type 1 estuaries. These are well-mixed estuaries, type 1a, implying strong tidal forcing and weak river discharge (or large tidal prisms relative to freshwater volumes). There are also estuaries with depth-independent seaward flow but with highly stratified conditions. These type 1b estuaries have large river discharge compared to tidal forcing. In type 1 estuaries, in general, the upstream transport of salt is overwhelmingly dominated by diffusive processes ($\nu \approx 1$, Fig. 1.5B).

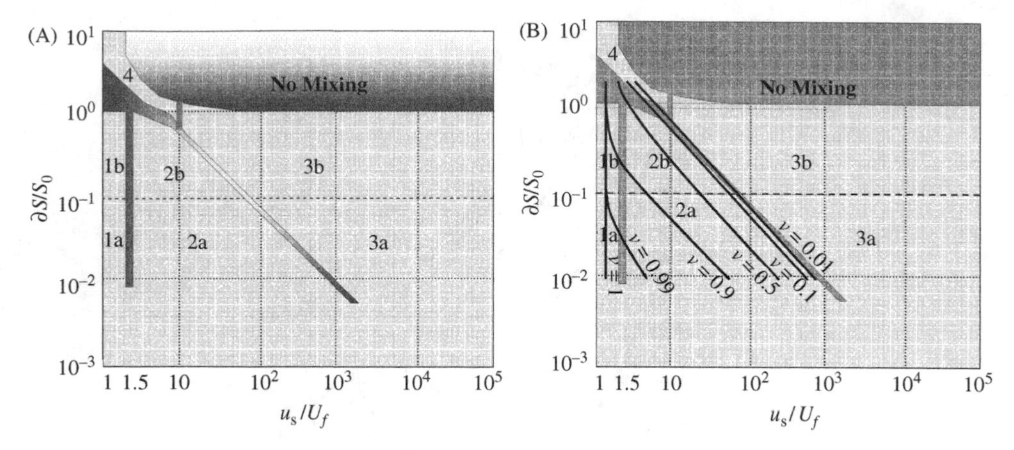

Figure 1.5. Classification of estuaries according to hydrodynamics, in terms of the circulation and stratification parameters (redrawn from Hansen and Rattray, 1966). (A) Type 1 estuaries show no vertical structure in net flows; in type 2 estuaries, the net flows reverse with depth; type 3 estuaries exhibit strong gravitational circulation; and type 4 estuaries are salt wedge. (B) Includes lines of diffusive salt flux showing the dominance of advective salt flux for type 3 estuaries and diffusive flux for type 1.

Type 2 (Fig. 1.5B) estuaries are those where flow reverses at depth, and include most temperate estuaries. These systems feature well-developed gravitational circulation and exhibit contributions from advective and diffusive processes to the upstream salt transport ($0.1 < \nu < 0.9$). Type 2a estuaries are well mixed or weakly stratified and type 2b estuaries are strongly stratified. Note that strongly stratified and weakly stratified estuaries of type 2 may exhibit similar features in terms of the relative contribution from diffusive processes to the upstream salt transport (Fig. 1.5B).

Type 3 estuaries are associated with fjords, where gravitational circulation is well established: strong surface outflow and very small depth-averaged flows, typical of deep basins. This flow pattern results in large values (>100–1000) of the circulation parameter (Fig. 1.5A). Type 3a estuaries are moderately stratified and type 3b are highly stratified. The peculiarity about these systems is that the upstream transport of salt is carried out exclusively by advective processes ($\nu < 0.01$, Fig. 1.5B).

Finally, type 4 estuaries exhibit seaward flows with weak vertical structure and highly stratified conditions as in a salt-wedge estuary. In type 4 estuaries, the diffusive fraction ν lines tend to converge, which indicates that in type 4 estuaries, salt transport is produced by both advective and diffusive processes. In the Hansen–Rattray diagram, it is noteworthy that some systems will occupy different positions in the parameter space as stratification and circulation parameters change from spring to neap tides, from dry to wet seasons, and from year to year.

Analogous to the classification of estuaries in terms of the two non-dimensional parameters discussed above, estuarine systems can also be classified in terms of the lateral structure of their net exchange flows. The lateral structure may be strongly influenced by bathymetric variations and may exhibit vertically sheared net exchange flows, i.e., net outflows at the surface and near-bottom inflows (e.g., Pritchard, 1956), or laterally sheared exchange flows with outflows over shallow parts of a cross-section and inflows in the channel (e.g., Wong, 1994). The lateral structure of exchange flows may ultimately depend on the competition between Earth's rotation (Coriolis) and frictional effects (Kasai *et al.*, 2000), as characterized by the vertical Ekman number (*Ek*). But the lateral structure of exchange flows may also depend on the Kelvin number (*Ke*), which is the ratio of the width of the estuary to the internal radius of deformation.

The Ekman number is a non-dimensional dynamical depth of the system. Low values of *Ek* imply that frictional effects are restricted to a thin bottom boundary layer (weak frictional, nearly geostrophic conditions), while high values of *Ek* indicate that friction affects the entire water column. The lateral structure of density-driven exchange flows may be described in terms of whether the flows are vertically sheared or unidirectional in the deepest part of the cross-section (Valle-Levinson, 2008). Under low *Ek* (< 0.001, i.e., < -3 in the abscissa of Fig. 1.6), the lateral structure of exchange flows depends on the dynamic width of the system (Fig. 1.6). In wide systems ($Ke > 2$, i.e., > 0.3 in the ordinate of Fig. 1.6), outflows and inflows are separated laterally according to Earth's rotation, i.e., the exchange flow is laterally sheared. In narrow systems ($Ke < 1$, i.e., < 0 in the ordinate of Fig. 1.6) and low *Ek* (still < 0.001, i.e., < -3 in the abscissa of Fig. 1.6), exchange flows are vertically sheared. In contrast, under high *Ek* (> 0.3, i.e., > -0.5 in the abscissa of Fig. 1.6) and for all *Ke*, the density-driven exchange is laterally sheared independently of the width of the system. Finally, under intermediate *Ek* ($0.01 < Ek < 0.1$, i.e., between -2 and -1 in the abscissa of Fig. 1.6), the exchange flow is preferentially vertically sheared but exhibiting lateral variations.

The main message is that under weak friction ($Ek < 0.01$), both depth and width are important to determine whether the density-driven exchange is vertically or horizontally sheared. This is illustrated by the fact that the contour values in the entire region of $Ek < 0.01$ (i.e., < -2 in the abscissa) in Fig. 1.6 vary with both *Ek* and *Ke*. In contrast, under $Ek > 0.01$ the depth is the main determinant as to whether the exchange is vertically or horizontally sheared. This is shown by the fact that the contour values in Fig. 1.6 vary mostly with *Ek* but very little with *Ke*. A future challenge of this approach is to determine the variability of a particular system in the *Ek* vs *Ke* parameter space. It is likely that an estuary will describe an ellipse of variability in this plane from spring to neaps and from wet to dry seasons, or from year to year.

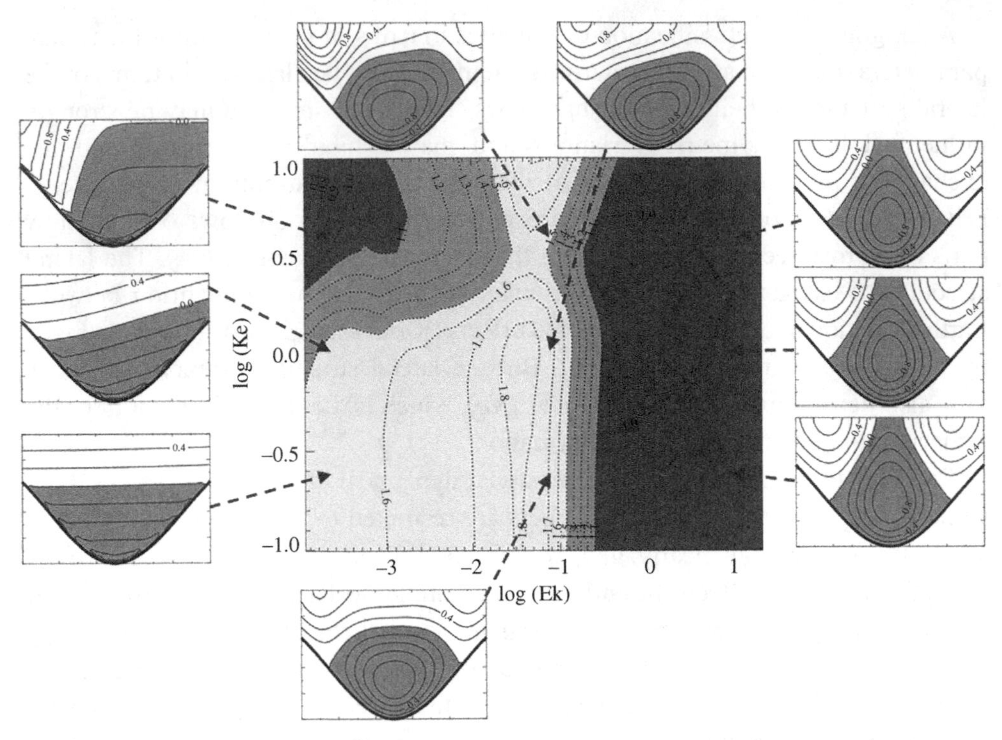

Figure 1.6. Classification of estuarine exchange on the basis of *Ek* and *Ke*. The subpanels appearing around the central figure denote cross-sections, looking into the estuary, of exchange flows normalized by the maximum inflow. Inflow contours are negative and shaded. The vertical axis is non-dimensional depth from 0 to 1 and the horizontal axis is non-dimensional width, also from 0 to 1. The central figure illustrates contours of the difference between maximum outflow and maximum inflow over the deepest part of the channel and for different values of *Ek* and *Ke*. Note that the abscissa and ordinate represent the logarithm of *Ek* and *Ke*. Dark-shaded contour regions denote net inflow throughout the channel, i.e., laterally sheared exchange flow as portrayed by the subpanels whose arrows point to the corresponding *Ek* and *Ke* in the dark-shaded regions. Light contour regions illustrate vertically sheared exchange in the channel as portrayed by the subpanels whose arrows point to the corresponding *Ek* and *Ke* in the light-shaded regions. Intermediate-shaded regions represent vertically and horizontally sheared exchange flow, similar to the second subpanel on the left, for $\log(Ke) = 0$ and $\log(Ek) \sim -3.7$.

All of the above classifications depend on diagnostic parameters that require substantial information about the estuary, i.e., on dependent variables. In addition, they do not take into account the effects of advective accelerations, related to lateral circulation, that may be of the same order of magnitude as frictional effects (e.g., Lerczak and Geyer, 2004). Some of these nuances are discussed further in Chapter 5 of this book. Future schemes will require taking those advective effects into account. In the following chapter, in addition to presenting the basic hydrodynamics in estuaries, another approach for classifying estuaries based on different dynamical properties is discussed. Such an approach, consistent with that of Prandle (2009), uses only the river discharge velocity and the tidal current velocity as the parameters needed to classify estuaries.

References

Cameron, W. M. and D. W. Pritchard (1963) Estuaries. In M. N. Hill (ed.), *The Sea*, Vol. 2. John Wiley & Sons, New York, pp. 306–324.

Hansen, D. V. and M. Rattray, Jr. (1966) New dimensions in estuary classification. *Limnol. Oceanogr.* **11**, 319–326.

Kasai, A., A. E. Hill, T. Fujiwara and J. H. Simpson (2000) Effect of the Earth's rotation on the circulation in regions of freshwater influence. *J. Geophys. Res.* **105**(C7), 16,961–16,969.

Lerczak, J. A. and W. R. Geyer (2004) Modeling the lateral circulation in straight, stratified estuaries. *J. Phys. Oceanogr.* **34**, 1410–1428.

Prandle, D. (2009) *Estuaries: Dynamics, Mixing, Sedimentation and Morphology*. Cambridge University Press, India, 246pp.

Pritchard, D. W. (1952) Estuarine hydrography. *Adv. Geophys.* **1**, 243–280.

Pritchard, D. W. (1955) Estuarine circulation patterns. *Proc. Am. Soc. Civil Eng.* **81**(717), 1–11.

Pritchard, D. W. (1956) The dynamic structure of a coastal plain estuary. *J. Mar. Res.* **15**, 33–42.

Valle-Levinson, A. (2008) Density-driven exchange flow in terms of the Kelvin and Ekman numbers. *J. Geophys. Res.* **113**, C04001, doi:10.1029/2007JC004144.

Wolanski, E. (1986) An evaporation-driven salinity maximum zone in Australian tropical estuaries. *Est. Coast. Shelf Sci.* **22**, 415–424.

Wong, K.-C. (1994) On the nature of transverse variability in a coastal plain estuary. *J. Geophys. Res.* **99**(C7), 14,209–14,222.

2

Estuarine salinity structure and circulation

W. R. GEYER

Woods Hole Oceanographic Institution

2.1 The horizontal salinity gradient

Estuaries show a great diversity of size, shape, depth, and forcing characteristics, but a general characteristic of estuaries is the presence of a horizontal salinity gradient (Fig. 2.1). Normally the salinity decreases from the ocean toward the head of the estuary due to freshwater input; in the case of inverse estuaries, the salinity increases in the landward direction due to excess evaporation (see Chapters 1 and 9). This salinity gradient is the key dynamical variable that makes estuaries different from any other marine or lacustrine environment. The horizontal salinity gradient is the key driving force for the estuarine circulation, which in turn plays a key role in maintaining salinity stratification in estuaries. The combined influence of the estuarine circulation and stratification determines the fluxes of salt and freshwater within the estuary, and their intensity varies with the strength of the freshwater inflow. Because of these dynamics, estuaries are often the most strongly stratified aquatic environments, but they also tend to have vigorous water and salt exchange, due to the estuarine circulation.

This chapter explores the coupled equations involving the estuarine circulation, stratification, salt flux and freshwater inflow. A major outcome of this analysis is to reveal the essential importance of horizontal salinity gradient in the estuarine circulation, stratification, and salt balance, but also to find that the horizontal salinity gradient ultimately depends on the strength of the freshwater outflow and the intensity of mixing by tidal currents.

2.2. The estuarine circulation

The first known reference to the estuarine circulation was contributed by Pliny the Elder in reference to the vertically varying flow in the Strait of Bosphorus, which connects the high-salinity Mediterranean to the more brackish Black Sea. He found that as fishermen lowered their nets, they would tug in the eastward direction once

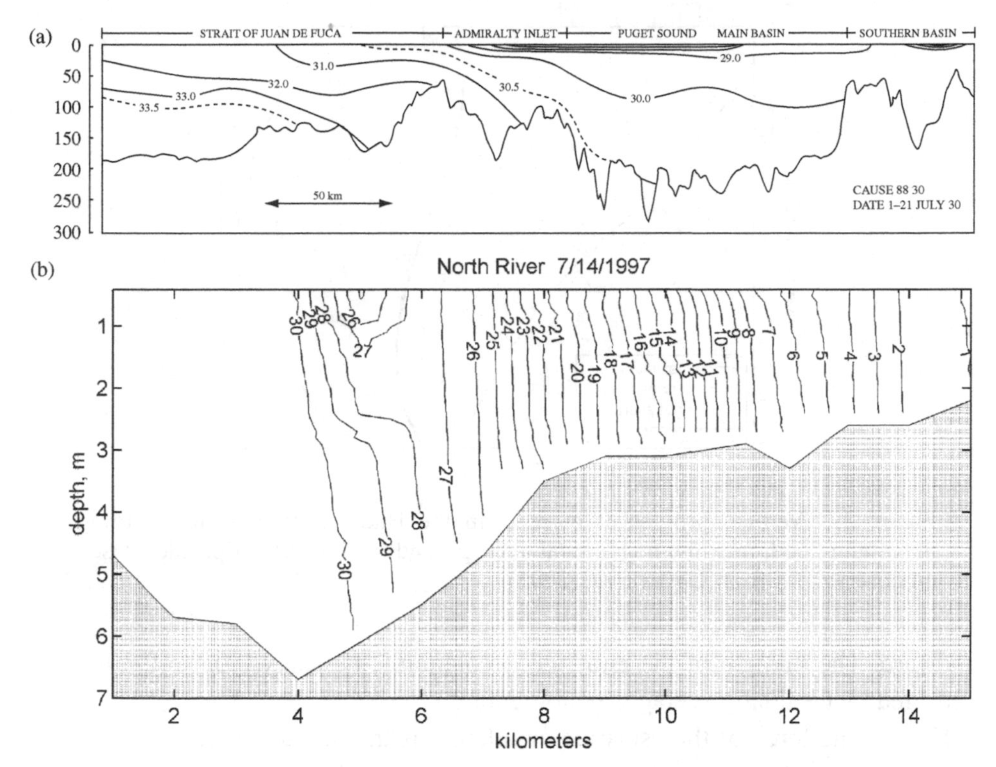

Figure 2.1. Two estuarine cross-sections showing salinity distributions. The upper panel is Puget Sound (Washington State, from Collias *et al.*, 1974), one of the largest estuaries in North America. The lower panel is the North River (Marshfield, MA), a very small estuary. In both cases there is a horizontal salinity gradient (saltier toward the ocean) that provides the driving force for the estuarine circulation. The extreme difference in the strength of the salinity gradient between the two regimes is related to the extreme difference in depth as well as the relative strength of freshwater inflow.

they reached a certain depth, at which they were being swept by the deep, high-salinity inflow into the Black Sea. Nineteen hundred years later, Knudsen noted a similar bidirectional flow in the entrance to the Baltic Sea. His name has been immortalized for his quantification of the salt balance, which will be discussed in Section 2.4. In 1952, Pritchard was the first researcher to link the estuarine circulation to the forcing by the horizontal density gradient, using observations in the James River estuary to demonstrate the mechanism.

Pritchard pointed out that the tidal currents are typically much stronger than the estuarine circulation, but if the vertically varying horizontal currents are measured through the course of the tidal cycle and then averaged, the "residual" or estuarine circulation would be revealed, as shown in Fig. 2.2. In a "normal" or positive estuary (see Chapter 1), i.e., one with excess freshwater input, the near-bottom flow is

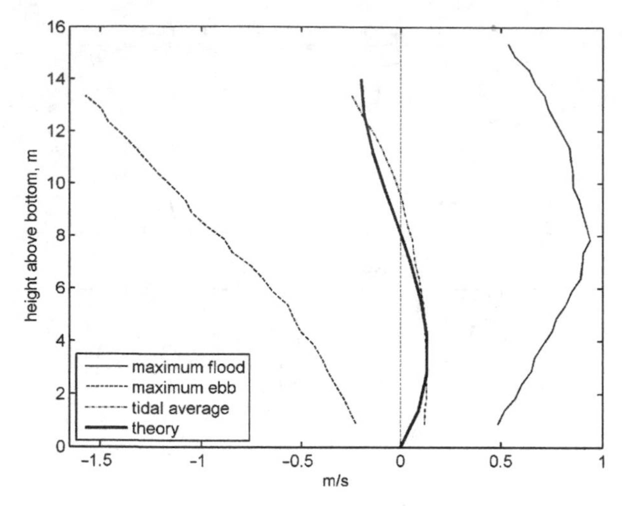

Figure 2.2. Vertical profiles of currents in the Hudson River estuary during maximum flood, maximum ebb, tidal average, and the theoretical profile. Based on observations from Geyer *et al.* (2001).

landward. The strength of the estuarine circulation is typically 0.05 to 0.3 m/s (as measured by the tidally averaged bottom inflow).

The driving force of the estuarine circulation is the horizontal salinity gradient $\partial s/\partial x$, which induces a vertically varying pressure gradient. The pressure gradient $\partial p/\partial x$ can be expressed as the combined influence of the surface slope $\partial \eta/\partial x$ and $\partial s/\partial x$:

$$\frac{1}{\rho}\frac{\partial p}{\partial x} = g\frac{\partial \eta}{\partial x} + \beta g \frac{\partial s}{\partial x}(h - z), \tag{2.1}$$

where ρ is the density of water (dominated by salinity), β is the coefficient of saline contraction, g is the acceleration due to gravity, h is the water depth, and z is the vertical coordinate measured upward from the bottom. Note that the second term on the right-hand side of equation (2.1), the $\partial s/\partial x$ term, is zero at the surface and maximal at the bottom, oriented in the direction that accelerates the bottom water into the estuary. The tidally averaged surface slope tilts in the other direction from the salinity gradient, with a magnitude large enough, relative to the salinity gradient, that the pressure gradient reverses somewhere close to the middle of the water column. Thus the surface water is driven seaward and the bottom water landward.

If the flow were starting from rest (imagine the estuary was frozen and it suddenly melted), the pressure gradient would result in acceleration of the surface and bottom waters in opposite directions. This acceleration would continue until some other force balanced the pressure gradient. The force that is most important for balancing the pressure gradient is the internal stress (or momentum flux) acting on the estuarine shear flow. Pritchard represented the turbulent stress in terms of an eddy viscosity A_z,

$$\tau = \rho A_z \frac{\partial u}{\partial z}, \tag{2.2}$$

where τ is the stress (in units of force/area or Pascals) and $\partial u/\partial z$ is the vertical shear of the horizontal flow. The units of A_z are length^2/time, just like the molecular viscosity, although it is several orders of magnitude larger. A practical way of thinking about the eddy viscosity is as a product of a turbulent velocity scale and a turbulent length scale. The magnitude of the eddy viscosity is set by the intensity of the tidal flow and the stratification; values range from 10^{-4} to 10^{-2} m^2/s in estuaries. The eddy viscosity varies significantly in space and time due to variations in forcing, but the momentum balance can be approximated using a constant value that represents an "effective" tidal average (although the estimate of its value is best obtained a posteriori).

The equation relating the pressure-gradient forcing to the estuarine circulation is a greatly simplified, tidally averaged representation of the horizontal momentum equation:

$$g\frac{\partial \eta}{\partial x} + \beta g \frac{\partial s}{\partial x}(h - z) = \frac{1}{\rho}\frac{\partial \tau}{\partial z} = A_z \frac{\partial^2 u}{\partial z^2}, \tag{2.3}$$

where the eddy viscosity has been assumed to be constant in the vertical to make the parameter dependence more clear. The magnitude of the surface slope is constrained by the mean outflow; if the river flow is small compared to the estuarine circulation (generally a valid approximation except for salt wedge or highly stratified estuaries), then

$$g\frac{\partial \eta}{\partial x} = -\frac{3}{8}\beta g \frac{\partial s}{\partial x} h \tag{2.4}$$

and the solution for the velocity is

$$u(z) = \frac{1}{48}\frac{\beta g \frac{\partial s}{\partial x} h^3}{A_z}\left(8\varsigma^3 - 15\varsigma^2 + 6\varsigma\right), \tag{2.5}$$

where $\varsigma = z/h$ is a non-dimensional depth varying from 0 at the bottom to 1 at the surface. The shape of the velocity profile is shown in Fig. 2.2. This should be considered a qualitative solution – the vertical structure of the velocity in actual estuaries differs due to the spatial and temporal variations of the eddy viscosity as well as other factors such as lateral advection (note comparison with actual observed profiles in Fig. 2.2). Nevertheless, equation (2.5) yields the general shape and magnitude of the estuarine circulation, and it is a useful approximation for analyzing the influence of the estuarine circulation on the stratification and salt balance.

In practice it is difficult to assign an appropriate value to A_z without actually solving the time-dependent equations with the inclusion of an advanced turbulence closure. A simple alternative to equation (2.5) can be obtained by replacing A_z with an equivalent expression involving the tidal velocity magnitude U_t, the depth h, and a bottom drag coefficient C_d:

$$A_z = \frac{1}{48\, a_o} C_d U_T h, \tag{2.6}$$

where the drag coefficient equates the magnitude of the bottom stress to the tidal velocity as $\tau_b = \rho C_d U_t^2$ (with typical values of $C_d \sim 3\times10^{-3}$) and a_o is a dimensionless constant yet to be determined related to the effectiveness of turbulent momentum flux. Note that this is an "effective" viscosity based on the tidally averaged flow, and its value is significantly smaller than the maximum eddy viscosity that would occur within a tidal cycle. Substituting this into equation (2.5), the magnitude of the estuarine circulation can be determined as

$$U_e = a_o \frac{\beta g \frac{\partial s}{\partial x} h^2}{C_d U_t}. \tag{2.7}$$

This formulation indicates the parameter dependence of the estuarine circulation relative to the key variables: linearly with $\partial s/\partial x$, inversely with tidal velocity, and quadratically with depth. Comparison with data from the Hudson River estuary verifies that equation (2.7) provides a reasonable estimate of the variability, with a value of a_o of about 0.3 (Fig. 2.2).

2.3. The stratification

The vertical variation of salinity, or stratification, is one of the more conspicuous characteristics of estuaries. Estuarine stratification varies considerably from one estuary to another, and one way of classifying estuaries (Chapter 1) is based on the strength of stratification: well mixed, partially mixed, highly stratified, and salt wedge. As a classification scheme this is not very reliable, however, as one estuary may vary from well mixed to highly stratified depending on the forcing conditions. Estuarine stratification is important for a number of reasons – it inhibits vertical mixing, which affects the dynamics and may lead to hypoxia in the subpycnocline waters. The stratification also plays a fundamental role in the salt balance, which will be discussed in Section 2.4.

What controls the stratification in estuaries? What makes stratification vary so much from one estuary to another, and for one estuary to vary from well mixed to highly stratified? The key to the variability of stratification is embodied in the equation for local salt conservation, which can be simplified for steady-state conditions and modest along-estuary variations in depth to

$$u(z)\frac{\partial s}{\partial x} = K_z\frac{\partial^2 s}{\partial z^2}, \tag{2.8}$$

where $u(z)$ is the vertically varying, tidally averaged velocity, and K_z is the eddy diffusivity of salt, with the same dimensions and similar magnitude to A_z. It is not immediately obvious why this equation would dictate the stratification. One way to illustrate the balance more clearly is to integrate equation (2.8) from the middle of the water column to the surface (neglecting vertical variations in $\partial s/\partial x$), obtaining

$$\alpha U_e\frac{\partial s}{\partial x} = -K_z\frac{\partial s}{\partial z_{mid}}, \tag{2.9}$$

where α is a constant of integration approximately equal to 0.3, and $\partial s/\partial z_{mid}$ is the vertical gradient of salinity in the middle of the water column. This equation indicates that mean advection of salt by the estuarine circulation is balanced by vertical mixing. The salinity of the upper layer would decrease due to seaward advection of lower-salinity water, but vertical mixing carries higher-salinity water up, balancing the effect of advection (Fig. 2.3). In the lower layer, advection would cause the salinity to increase, but vertical mixing transports low-salinity water downward to maintain a steady state.

As with the eddy viscosity, the eddy diffusivity can be represented in terms of tidal velocity and depth [cf. equation (2.6)]. Substituting this into equation (2.9) and noting that $\partial s/\partial z \approx \Delta s/h$, where Δs is the top-to-bottom salinity difference, we obtain

$$\Delta s = a_1\frac{U_e\frac{\partial s}{\partial x}h}{C_D U_t}, \tag{2.10}$$

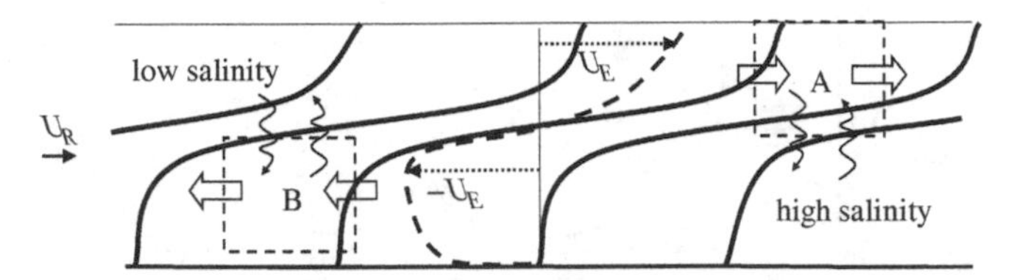

Figure 2.3. Schematic cross-section of an estuary showing the influence of advection and vertical mixing on the local salt balance. The thick lines are isohalines, and the thin lines and arrows indicate the estuarine circulation. In Box A (upper layer), horizontal advection causes a reduction of salinity, but vertical mixing compensates by replacing low-salinity water with underlying high-salinity water. The relative roles of advection and mixing are reversed in Box B (lower layer).

where a_1 is a constant that incorporates α and also depends on the shape of the salinity and velocity profiles. The value of a_1 was estimated from Hudson River data to be approximately 50. Combining equation (2.10) with equation (2.7) for the estuarine circulation,

$$\Delta s = a_o a_1 \frac{\beta g \left(\frac{\partial s}{\partial x}\right)^2 h^3}{\left(C_D U_t\right)^2}. \tag{2.11}$$

This equation indicates that the stratification depends quadratically on the salinity gradient and inversely on the square of the tidal velocity – i.e., the stratification is a lot more sensitive to changes in these forcing variables than is the estuarine circulation.

The sensitive dependence of stratification on the strength of the tides has been revealed in a number of studies of the spring–neap variations in stratification. During spring tides, tidal mixing is maximal, and equation (2.11) predicts that the stratification should reach a minimum, and vice versa. Haas (1977) first documented the spring–neap variations of stratification due to variations in tidal mixing in the subestuaries of Chesapeake Bay.

Data from the Hudson estuary are used to test equation (2.11), as shown in Fig. 2.4 (lower panel). These data indicate that the actual stratification is more sensitive to the spring–neap cycle than equation (2.11) predicts. The reason for this is that the

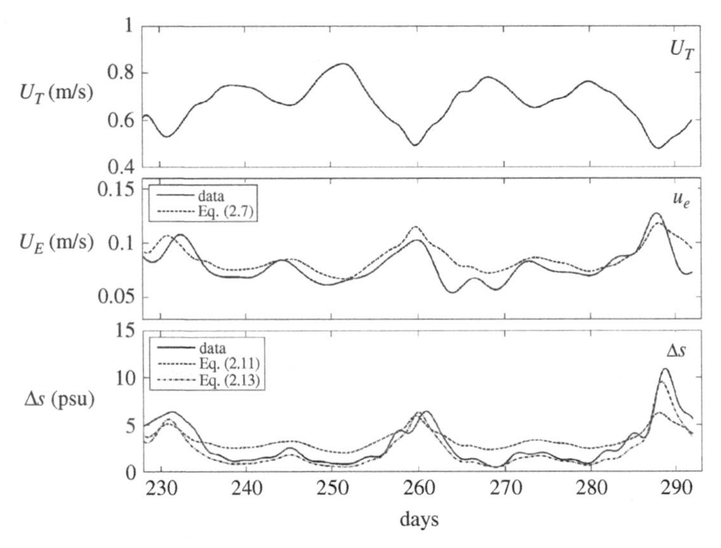

Figure 2.4. Time series observations from the Hudson River in the fall of 1995 showing fortnightly and monthly variations in tidal velocity amplitude (upper panel), estuarine velocity (middle panel), and stratification (lower panel). The theoretical predictions for estuarine velocity and stratification are also shown, based on observed values of U_T and $\partial s/\partial x$.

intensity of mixing does not just depend on the intensity of the tides; it also depends on the ambient stratification. This is a complex topic, which is treated in more detail in Chapter 7. Briefly, the parameter that best quantifies the importance of stratification on mixing is the gradient Richardson number Ri – the ratio of the stratification to the shear. This can be approximated for scaling purposes by a non-dimensional ratio of the vertical salinity difference to the tidal velocity:

$$Ri_T = \frac{\beta g \Delta s h}{{U_T}^2}. \tag{2.12}$$

For $Ri_T > 1$, mixing is strongly suppressed by the stratification, and for $Ri_T < 0.25$, mixing is relatively unaffected by stratification. Models of mixing in stratified boundary layers (e.g., Trowbridge, 1992) suggest that the mixing rate should depend on $Ri^{-1/2}$. If equation (2.11) is modified by the addition of a factor to account for the Ri-dependence,

$$\Delta s = a_o a_1' {Ri_T}^{1/2} \frac{\beta g \left(\frac{\partial s}{\partial x}\right)^2 h^3}{(C_D U_t)^2}, \tag{2.13}$$

the fit with the Hudson River data is significantly improved (Fig. 2.4). An explicit form for Δs that includes the stratification effect on turbulence is obtained by combining equations (2.12) and (2.13):

$$\Delta s = {a_o}^2 {a_1}'^2 \frac{(\beta g)^3 \left(\frac{\partial s}{\partial x}\right)^4 h^7}{{C_D}^4 {U_T}^6}. \tag{2.14}$$

Note the extreme sensitivity of the stratification to the tidal mixing according to this formulation, as well as the sensitivity to the horizontal salinity gradient and depth. The Hudson data provide some support for the sensitivity to tidal velocity, but equation (2.14) appears to overemphasize the dependence on horizontal salinity gradient, based on observations and model results. The consequences of this stratification-dependence on the overall estuarine balance are discussed in Section 2.5.

2.4. The salt balance

If an estuary is in steady state, then the amount of salt being transported past any cross-section has to be zero (except in the highly unusual case that there are significant sources of salt in the watershed). Considering a cross-section within the estuary with some tidally averaged value of salinity s_o, the freshwater outflow Q_r (volume per time) will cause a seaward transport $Q_{salt} = Q_r s_o$. In order to maintain a steady state, there must be mechanisms that transport salt into the estuary, to

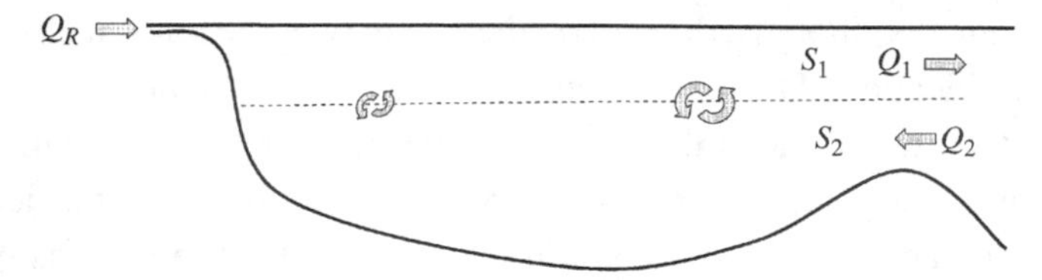

Figure 2.5. Schematic cross-section of an estuarine basin to illustrate Knudsen's relation for the salt balance. The salinity is assumed to be in steady state, which may only be valid for long time scales (weeks to months) for larger estuaries.

compensate for the advective loss due to the river outflow. Knudsen's analysis of the estuarine circulation of the Baltic Sea provides the essence of the estuarine salt balance. Knudsen's relation considers the integrated conservation of volume and salt in a basin that has a riverine source of freshwater and exchange flow with the ocean (Fig. 2.5). Volume and salt conservation yield

$$\begin{aligned} Q_1 &= Q_2 + Q_r, \\ Q_1 s_1 &= Q_2 s_2, \end{aligned} \tag{2.15}$$

where Q_1 and Q_2 are the volume transports in the upper and lower layers (the upper layer being directed seaward and the lower layer landward). Combining the two leads to Knudsen's relation:

$$Q_r s_1 = Q_2 \Delta s. \tag{2.16}$$

If $Q_2 \gg Q_r$ (which is generally the case for partially mixed and well-mixed estuaries), then $s_1 \approx s_2 \approx s_o$, and Knudsen's relation can be restated using U_e:

$$U_r s_o = a_2 U_e \Delta s, \tag{2.17}$$

where $u_r = Q_r / A_{cs}$ (the outflow velocity associated with the river discharge, A_{cs} being the local cross-sectional area of the estuary), and a_2 is a constant equal to approximately 0.5. Equation (2.17) indicates that the tendency for salt to be carried out of the estuary by the freshwater outflow is balanced by the net input of salt due to the estuarine circulation. The stronger the estuarine circulation, and the larger the vertical salinity difference, the more salt is transported into the estuary.

Equation (2.17) represents the steady-state balance, and it only considers the influence of the estuarine circulation on the salt balance. The salt content of estuaries is not actually constant in time, because a number of factors cause the left-hand and right-hand sides of equation (2.16) to change. The magnitude and causes of those variations will be discussed in Section 2.6. This equation also leaves out another contributor to the horizontal salt flux: the horizontal dispersion of salt (tidal dispersion), which is due mainly to tidal stirring. Tidal dispersion is particularly important in short estuaries, in which the length of the salt intrusion is comparable to the tidal excursion distance (5–15 km), and also in regions of abrupt changes in estuarine cross-section. But in large estuaries (>30 km long), and away from major changes in cross-sectional geometry, the salt transport induced by estuarine circulation generally dominates over tidal dispersion, and equation (2.17) is a good representation of the time-averaged salt balance.

2.5. The coupled equations

The key to the dynamics of estuaries is that the global salt balance has to be satisfied at the same time that the local momentum and salinity equations are in balance. This combination of equations for estuarine circulation, stratification, and salt balance leads to a constraint on the horizontal salinity gradient, which is in essence the master variable controlling the estuarine dynamics. First we will consider the formulation in which the influence of stratification on mixing is neglected, as it leads to a result that is often noted in the literature. Combining equations (2.7), (2.11) and (2.17), we obtain

$$U_r s_o = a_o{}^2 a_1 a_2 \frac{(\beta g)^2 \left(\frac{\partial s}{\partial x}\right)^3 h^5}{(C_d U_t)^3}. \tag{2.18}$$

Then, solving for $\partial s/\partial x$, we obtain

$$\frac{\partial s}{\partial x} = \frac{1}{(a_o{}^2 a_1 a_2)^{1/3}} \frac{C_d U_t U_r{}^{1/3} s_o{}^{1/3}}{(\beta g)^{2/3} h^{5/3}}. \tag{2.19}$$

Equation (2.19) indicates that for steady-state conditions, the horizontal salinity gradient depends on the one-third power of the river flow and the first power of the tidal velocity. These scaling relations are consistent with the Hansen and Rattray (1965) solution for the advection-dominated limit, the Chatwin (1976) solution for partially mixed estuaries, and the MacCready (1999) solution for advection-dominated estuaries. The essential finding is that the salinity gradient is relatively insensitive to variations in river flow, or alternatively that the estuarine salt flux is particularly sensitive to the salinity gradient.

Now, considering the solution for the horizontal salinity gradient when the Richardson number dependence on vertical mixing is included, we start with equation (2.14) for Δs in combination with equations (2.7) and (2.17) and obtain

$$\frac{\partial s}{\partial x} = \frac{1}{(a_o{}^3 a_1{}'^2 a_2)^{1/5}} \frac{C_d U_t{}^{7/5} U_r{}^{1/5} s_o{}^{1/5}}{(\beta g)^{4/5} h^{9/5}}. \tag{2.20}$$

When the influence of vertical mixing is included, the salinity gradient is *even less* sensitive to variability of river flow, i.e., the estuary is "stiff" (think of a steel spring) with respect to variations in river flow. Monismith *et al.* (2002) found even greater "stiffness" in the salinity gradient of northern San Francisco Bay, obtaining a $U_r^{1/7}$ dependence. This may be due in part to variations in geometry along the estuary, but the influence of stratification on mixing is likely an important contributor to the estuarine response to variations in river flow.

The expression for $\partial s/\partial x$ can be substituted back into the equations for U_e and Δs to obtain expressions for the estuarine circulation and stratification in terms of the forcing variables U_t and U_r, first without considering the Richardson number effect:

$$U_e = \left(\frac{a_o}{a_1 a_2}\right)^{1/3} (\beta g s_o h)^{1/3} U_r^{1/3}, \tag{2.21}$$

$$\frac{\Delta s}{s_o} = \left(\frac{a_1}{a_o a_2^2}\right)^{1/3} \frac{U_r^{2/3}}{(\beta g s_o h)^{1/3}}. \tag{2.22}$$

These equations reveal the surprising result that for steady-state balances, the estuarine circulation and stratification *do not* depend on the tidal velocity, even though equations (2.7) and (2.11) clearly indicate the inverse dependence on tidal mixing. This paradoxical result is due to the variation of horizontal salinity gradient with tidal velocity [equation (2.19)], which exactly compensates for the variations in tidal mixing. Real estuaries do show large variations due to changes in tidal mixing – this is because the steady-state assumption is violated in the spring–neap cycle – more on this in the next section.

For the case in which Richardson number effects are considered, we get slightly different equations for U_e and Δs:

$$U_e = \left(\frac{a_o^2}{a_1^2 a_2}\right)^{1/5} (\beta g s_o h)^{1/5} U_r^{1/5} U_T^{2/5}, \tag{2.23}$$

$$\frac{\Delta s}{s_o} = \left(\frac{a_1}{a_o a_2}\right)^{2/5} \frac{U_r^{4/5}}{U_T^{2/5} (\beta g s_o h)^{1/5}}. \tag{2.24}$$

These equations indicate that the estuarine velocity U_e increases with increasing river flow (as expected), and also increases with increasing tidal velocity (not expected). Stratification is found to be slightly more sensitive to river flow than in the case without the Richardson number dependence, and also it is found to vary inversely with tidal velocity.

Although the physics would suggest that equations (2.23) and (2.24) provide a more realistic representation of the variability of estuarine circulation and

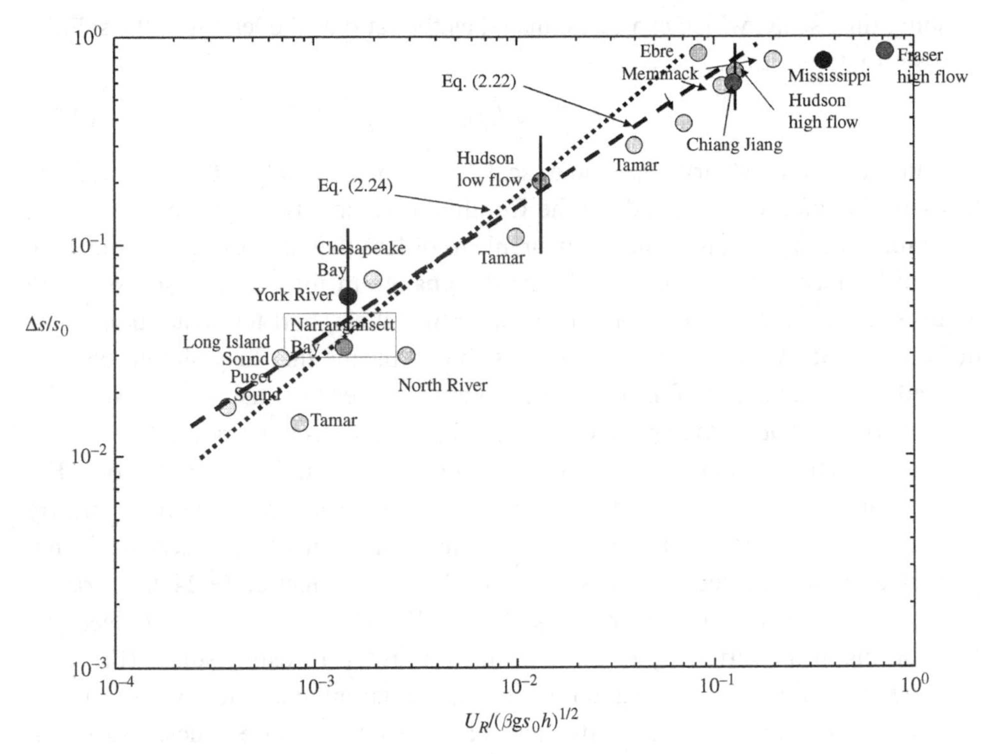

Figure 2.6. Estuarine stratification plotted as a function of freshwater velocity for a variety of different estuaries. Equation (2.22) provides a reasonable fit to most of these observations. At high values of U_r, the stratification asymptotically approaches a value of 1, consistent with the "salt-wedge" regime.

stratification, there have not been adequate analyses of these quantities among different estuaries to provide a definitive assessment. An analysis of stratification among a wide variety of estuaries (Fig. 2.6) indicates that the simple power law prediction of equation (2.22) does a better job than equation (2.24) of predicting the wide range of stratification observed in estuaries, even though it has a more simplified theoretical basis. There are other factors, such as estuarine geometry and temporal variability, that are not included in the theory but contribute to the actual variability among estuaries. These equations should be regarded at this time as providing guidance for the interpretation of variability among estuaries and within a particular estuary.

2.6. Temporal variability of the estuarine salinity structure and circulation

The combined equations presented in the previous section are based on a steady-state salt balance, meaning that the estuary neither gains nor loses salt. An estimate of the time scale for which that assumption is valid can be approximated by the flushing time scale, which can be estimated as the ratio of the length of the salinity intrusion L to the estuarine velocity:

$$T_F = L/U_E. \tag{2.25}$$

For the Hudson estuary, that time scale is approximately 10 days; for the Chesapeake, closer to 30; and for the Columbia River estuary, close to 1 day. If the forcing occurs at time scales comparable to or less than the residence time, then the salt balance will not "keep up" with the change in forcing, and so $\partial s/\partial x$ will adjust to the average forcing conditions rather than their short-term variations. This does not invalidate the coupled equations, but it means that they should be considered a representation of the conditions averaged over the flushing time scale.

A major contributor to temporal variability in estuaries is the fortnightly variation of tidal amplitude, mentioned earlier in the context of stratification variations. For larger estuaries such as the Hudson or the Chesapeake, $\partial s/\partial x$ remains nearly constant through the fortnightly cycle, so the variation of stratification is not predicted by the coupled equations [equation (2.22) or equation (2.24)], but rather by the "local" stratification balance [equation (2.11) or equation (2.14)]. Indeed, the large spring–neap variability of the stratification in the Hudson estuary (Fig. 2.4) indicates a significant amplification of the tidal variation, consistent with equation (2.24). In estuaries that are shorter and have faster response times, $\partial s/\partial x$ will increase during spring tides, partially offsetting the influence of increased mixing.

Other factors, such as changes in river flow and wind forcing, also contribute to temporal variability in estuaries, with the same caveat that they may occur at time scales faster or slower than the response time of the estuary, and their influence on the estuarine regime will depend on the relative time scales. The influence of temporal variability (including tides as well as lower-frequency processes) on the time-averaged estuarine regime is one of the most important topics in estuarine physics. Chapters 4, 5 and 9 examine these time-dependent processes.

2.7. Estuarine classification

Hansen and Rattray (1966) developed a scheme for estuarine classification called a "stratification–circulation" diagram, which has been the most commonly used approach over the last four decades. The Hansen–Rattray approach is intended as a *diagnostic* tool – given the observed stratification and circulation, what are the processes responsible for salt flux? Another approach to classification is the *prognostic* approach, in which the estuary is classified based on forcing variables, and the purpose of the classification is to predict the estuarine regime based on those forcing conditions. Due to the complexity of the processes and variability within and among estuaries, the prognostic approach could at best provide a rough estimate of the

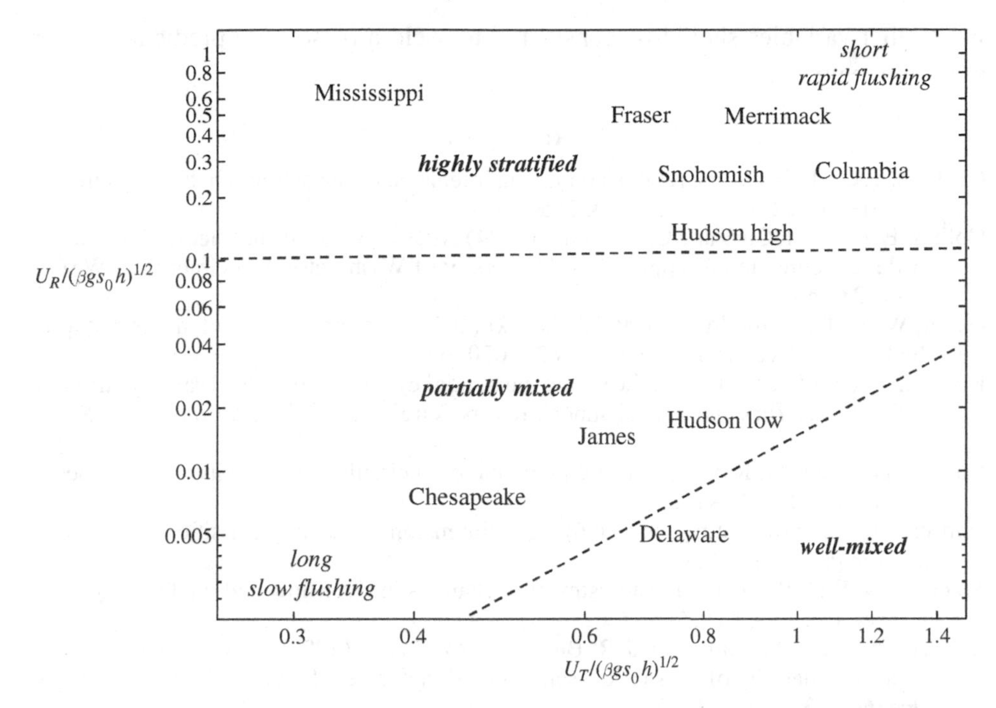

Figure 2.7. A framework for prognostic estuarine classification. The axes are the forcing variables: tidal velocity and "freshwater velocity", non-dimensionalized by a densimetric velocity scale. Stratification variations are mainly represented by the vertical position on the diagram. The length and flushing times depend on both parameters. Additional research may lead to a more quantitative approach using this framework.

conditions of a particular estuary – the quantitative prediction of estuarine processes is difficult even with a high-resolution numerical model. Nevertheless, the coupled equations provide a starting point for a prognostic classification of estuaries. This analysis suggests that the freshwater velocity U_R and tidal velocity U_T are "master variables" that may provide the framework for such a prognostic approach, as illustrated in Fig. 2.7. More strongly stratified estuaries appear on the upper part of the diagram, and weakly stratified on the lower part. The length of the estuary and flushing time scale depend inversely on $\partial s/\partial x$ [equation (2.19) or equation (2.20)], so the lower left corner of the diagram indicates long, slow-flushing regimes, whereas the upper right corner indicates short, rapidly flushing systems. Estuaries at similar points in the diagram would be expected to have similar dynamics, at least in a general sense. This framework does not account for variations in estuarine geometry – e.g., very deep vs very shallow, or very wide vs very narrow systems. Also, the issue of time-dependence is not represented in this diagram. More research is required to assess the general applicability of this framework, and to determine what other variables should be considered to yield a reasonable prediction of the estuarine regime.

References

Chatwin, P. C. (1976) Some remarks on the maintenance of the salinity distribution in estuaries. *Est. Coast. Mar. Sci.* **4**, 555–566.

Collias, E. E., N. McGary and C. A. Barnes (1974) Atlas of physical and chemical properties of Puget Sound and its approaches. University of Washington Sea Grant Pub. WSG 74–1, 285 pp.

Geyer, W. R., J. D. Woodruff and P. Traykovski (2001) Sediment transport and trapping in the Hudson River estuary. *Estuaries* **24**, 670–679.

Haas, L. W. (1977) The effect of the spring–neap tidal cycle on the vertical salinity structure of the James, York and Rappahannock Rivers, Virginia, U.S.A. *Est. Coast. Mar. Sci.* **5**, 485–496.

Hansen, D. V. and M. Rattray, Jr. (1965) Gravitational circulation in straits and estuaries. *J. Mar. Res.* **23**, 104–122.

Hansen, D. V. and M. Rattray, Jr. (1966) New dimensions in estuary classification. *Limnol. Oceanogr.* **11**, 319–325.

MacCready, P. (1999) Estuarine adjustment to changes in river flow and tidal mixing. *J. Phys. Oceanogr.* **29**, 708–726.

Monismith, S. G., W. Kimmerer, J. R. Burau and M. Stacey (2002) Structure and flow-induced variability of the subtidal salinity field in northern San Francisco Bay. *J. Phys. Oceanogr.* **32**, 3003–3019.

Pritchard, D. V. (1952) Salinity distribution and circulation in the Chesapeake Bay estuarine system. *J. Mar. Res.* **15**, 33–42.

Trowbridge, J. H. (1992) A simple description of the deepening and structure of a stably stratified flow driven by a surface stress. *J. Geophys. Res.* **97**, 15,529–15,543.

3

Barotropic tides in channelized estuaries

CARL T. FRIEDRICHS

Virginia Institute of Marine Science, College of William and Mary

3.1. Introduction

This chapter addresses the dynamics of cross-sectionally averaged tidal currents and elevation in channelized estuaries. The tides considered here are further assumed to be entirely barotropic and externally forced. Given these constraints, a series of estuarine geometries are examined which attempt to encompass generic, reasonably realistic scenarios found in nature. Typically, the goal in each case is to determine the lowest-order physical balances governing barotropic tides for a realistically relevant geometry, derive an analytical expression for the speed of tidal phase propagation, solve for the amplitude and phase of tidal velocity relative to that of elevation, and determine the lowest-order variation in tidal amplitude with distance along the estuary. A few less realistic, but classically studied cases (e.g., those involving an intermediate length, constant width channel) are also considered for completeness. For most cases, examples of estuaries that reasonably represent and justify the simplified dynamics are discussed.

In this chapter we generally assume that tidal elevation (η) and cross-sectionally averaged tidal velocity (u) along an estuary can be described at lowest order by

$$\eta(x, t) = a(x)\cos(\omega t - kx), \quad u(x, t) = U(x)\cos(\omega t - kx - \phi), \qquad (3.1\text{a,b})$$

where x is along-channel distance (positive into the estuary), t is time, a and U are tidal elevation and velocity amplitude, ω and k are tidal frequency and wavenumber, and ϕ is the relative phase between tidal elevation and velocity. Tidal frequency and wavenumber, in turn, are defined as $\omega = T/(2\pi)$ and $k = \lambda/(2\pi)$, where T and λ are the tidal period and wavelength. Note that equation (3.1) represents the tide as a forward-propagating, variable-amplitude wave without the explicit presence of a reflected wave. This approach is consistent with the manner in which observations of spatial variations in tidal phase, kx, are typically reported with distance along real estuaries. By using a formulation more-or-less consistent with commonly reported observations, observed variations in amplitude and phase can more easily be used to directly scale the governing equations.

The next three sections of this chapter present and then scale the governing equations of mass and momentum conservation. An essential step in this process is the identification of key length scales, corresponding inverse length scales (or "spatial rates of change"), and dimensionless ratios to be used subsequently to determine when and where to keep or neglect various dynamic or kinematic terms. Sections 3.5 through 3.14 then examine specific estuary geometries (e.g., short vs long, shallow vs deep, funnel-shaped vs non-convergent) that allow simplifications to be made. These simplifications are particularly relevant to better understanding fundamental physics in real tidal estuaries. The order of presentation proceeds from cases that are dynamically the most simple, yet still observationally useful, toward somewhat more complicated but naturally common "equilibrium" and "near-equilibrium" estuaries. Along the way, controls on tidal asymmetries are specifically considered in the context of short and/or shallow estuaries. For completeness, non-equilibrium channels and reflected wave cases are also briefly considered. Channels with lateral variations in bathymetry are considered in Chapter 6.

Themes emphasized in this chapter include the following: (i) it is valuable to use observations to scale the governing equations applicable to a tidal estuary in order to better understand the dominant physics and, in the process, avoid applying an overly complex and/or incorrect analytical framework; (ii) a unidirectional, up-estuary propagating waveform is applicable to a wide variety of long-channel geometries; (iii) the relative phase between tidal velocity and elevation can be shown to be close to "standing" (~ 90°) in many short and/or shallow estuaries without the need to explicitly include a classical reflected wave; (iv) there is a morphodynamic tendency for the amplitude of tidal velocity to vary along-estuary less dramatically than width and/or tidal phase; (v) variation in tidal amplitude and phase speed along such systems results when the net effects of width convergence and bottom friction do not quite balance; (vi) the total width of a tidal estuary (including tidal storage in marsh, shoals, and/or tidal tributaries) can play an important role in slowing the rate of up-estuary tidal wave propagation; and (vii) tidal variations in system width or channel depth can result in a shorter high- or low-water slack, shorter falling or rising tidal elevation, and stronger ebb or flood currents.

Several papers over the last two decades have presented generalized analytical solutions for cross-sectionally averaged barotropic tides in a wide range of estuarine geometries that include channel convergence (Jay, 1991; Friedrichs and Aubrey, 1994; Friedrichs *et al.*, 1998; Lanzoni and Seminara, 1998; Prandle, 2003, 2004; Savenije and Veling, 2005; Savenije *et al.*, 2008). These series of papers demonstrate the growing awareness among theoreticians that tidal amplitude, phase speed, and the relative phase between tidal velocity and elevation in most river valley estuaries are largely controlled by a competition between bottom friction and channel convergence. In addition, variations in tidal velocity amplitude along morphologically active, tidally energetic estuaries tend to be relatively small, and prominent reflected waves are uncommon. However, these advances in theoretical understanding have remained somewhat hidden from non-specialists by the formal mathematical analysis and presentation style of some authors.

This chapter, which is based on class lecture notes designed for non-specialists, aims to present a similar story as the above references, but in the most straightforward manner possible – while still retaining essential analytical components. For example, equations and variables are not non-dimensionalized here, and intuitive symbols are chosen wherever possible (e.g., L for length scales, A for areas, w for widths, etc.). Step-by-step approaches for scaling the governing equations and deriving analytical solutions are explicitly presented, and the series of cases considered starts with the simplest first. (More formal analyses often start out with more complete but possibly opaque solutions, followed by specific asymptotes.) Furthermore, the majority of the theoretical cases considered here are motivated and justified by clear observational analogues in order to ground theoretical insights in applied reality.

3.2. Governing equations

The generic estuarine geometry of interest to this chapter is shown in Fig. 3.1. The estuary is assumed to be channelized such that $w(x)$ is the width of the portion within which all along-estuary transport of mass and momentum is assumed to reside (e.g., Friedrichs and Aubrey, 1988). The tidally varying (but width-averaged) depth of the channelized portion is given by $h(x,t)$; $b(x,t) \geq w(x)$ is the tidally varying width of the estuary as a whole (including fringing marsh and shoals, intertidal flats and, potentially, smaller tidal tributaries); and x follows the possibly curving axis of flow along the main channel.

The cross-sectionally integrated conservation of mass (or "continuity") equation for barotropic tides is then given by

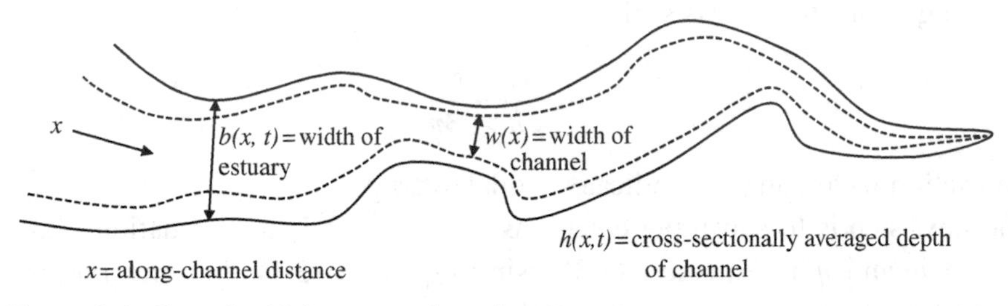

Figure 3.1. Generic tidal estuary plan-view geometry.

$$b\frac{\partial \eta}{\partial t} = -\frac{\partial}{\partial x}(uwh) \tag{3.2}$$

(e.g., Ippen and Harleman, 1966). The left-hand side (l.h.s.) of equation (3.2) represents the time rate of change in the wetted cross-sectional area of the estuary (including intertidal storage areas), while the right-hand side (r.h.s.) represents along-channel convergence in volume flux within the channel alone. It is assumed that there is negligible along-channel flow over areas outside the channelized portion of the estuary.

The cross-sectionally averaged tidal momentum equation is given approximately by

$$\frac{\partial u}{\partial t} + u\frac{\partial u}{\partial x} + g\frac{\partial \eta}{\partial x} + \frac{\tau_b}{\rho h} = 0 \tag{3.3}$$

(e.g., Ippen and Harleman, 1966), where g is gravitational acceleration, τ_b is bottom stress averaged over the channel perimeter, and ρ is water density. The terms in equation (3.3), from left to right, are local acceleration, advective acceleration, along-channel pressure gradient, and bottom friction. The above formulation neglects minor errors in $u\,\partial u/\partial x$ associated with averaging over the cross-section. However, it will soon be shown that the entire $u\,\partial u/\partial x$ term does not affect lowest-order solutions for a, U, or tidal propagation in the channelized systems of most interest to this chapter.

3.3. Scaling momentum

Our first step in simplifying and scaling the momentum equation is to linearize the friction term as follows:

$$\frac{\tau_b}{\rho h} = \frac{1}{\rho h}(\rho c_d u|u|) = ru\left[1 \pm O\left(\frac{a}{<h>} \pm \frac{8}{15\pi}\right)\right], \tag{3.4}$$

where "$O(\;)$" indicates "order of", and the two leading terms in the Fourier expansion arise from tidal variations in h and from linearizing $u|u|$ (Parker, 1991). In equation (3.4), the bottom drag coefficient c_d is on the order of 10^{-2} to 10^{-3} (depending on the roughness scale),

$$r = \frac{c_d}{<h>}\frac{8U}{3\pi} \tag{3.5}$$

is the friction factor, and $<>$ indicates a tidal average.

The next step is to substitute equations (3.4) and (3.5) into equation (3.3) and eliminate u and η in equation (3.3) using equation (3.1). For the purpose of evaluating the order of magnitude of the various terms in equation (3.3), after performing differentiations, we set sin( ) and cos( ) equal to 1. The momentum equation in terms of scales then becomes:

$$\pm O(\omega U) \pm O\left[U\left(L_U{}^{-1}U \pm kU\right)\right] \pm O\left[g\left(L_a{}^{-1}a \pm ka\right)\right] \\ \pm O\left[rU\left(1 \pm \frac{a}{<h>} \pm \frac{8}{15\pi}\right)\right] = 0. \tag{3.6}$$

Note that $\partial u/\partial x$ and $\partial \eta/\partial x$ in equation (3.3) each produce two terms in equation (3.6) because u and η are potentially dependent on x in two distinct ways: (i) x-dependence of the tidal amplitudes (via $\partial U/\partial x$ and $\partial a/\partial x$); (ii) phase-dependence of u and η (via $\partial(kx)/\partial x$). $L_U{}^{-1}$ and $L_a{}^{-1}$ in equation (3.6) are scales (with units of 1/km) for the spatial rates of change of tidal velocity and elevation along the channel – i.e., $L_U{}^{-1}$ and $L_a{}^{-1}$ scale the size of $\partial/\partial x$ in $\partial U/\partial x$ and $\partial a/\partial x$, respectively. Said another way, their inverses, L_U and L_a, are the characteristic along-channel distances over which U and a vary. $L_U{}^{-1}$ and $L_a{}^{-1}$ are formally defined as

$$L_U{}^{-1} = \frac{\partial U/\partial x}{U}, \quad L_a{}^{-1} = \frac{\partial a/\partial x}{a}. \tag{3.7a, b}$$

At first, the definitions in equation (3.7) may appear to be somewhat circular. But we will see that the magnitudes $L_U{}^{-1}$ and $L_a{}^{-1}$ can often be evaluated a priori directly from data (via observed e-folding lengths, for example), which aids in the process of appropriately simplifying equations (3.2) and (3.3) for application to real estuaries.

Dividing the terms in equation (3.6) by ωU, we see that the maximum size of the friction term relative to local acceleration (to leading order) is r/ω, and the maximum size of the advective term relative to $\partial u/\partial t$ is given by $(U/c)(1 + (kL_U)^{-1})$, where $c = \omega/k$ is the phase velocity of the tidal waveform. In mildly non-linear systems, we anticipate that U/c will usually be much less than 1. Thus we expect that advection will only be important where tidal velocity amplitude U changes dramatically over very short along-channel distances (i.e., if $L_U{}^{-1}$ is very large).

3.4. Scaling continuity

In some shallow estuaries of interest, the total estuary width b varies significantly in time due to the presence of extensive intertidal flats and/or marsh (Fig. 3.2). Under these conditions, scaling the l.h.s. of equation (3.2) yields:

$$b\frac{\partial \eta}{\partial t} = <b>\left[1 \pm O\left(\frac{\Delta b}{<b>}\right)\right]\frac{\partial \eta}{\partial t}, \tag{3.8}$$

where Δb is the amplitude of estuary width variation over the tidal cycle.

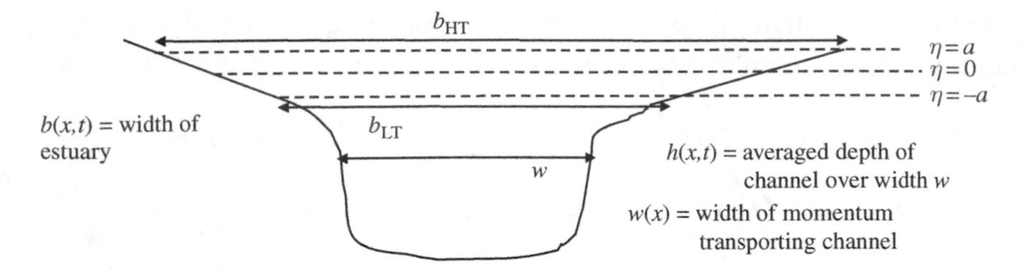

Figure 3.2. Generic tidal estuary cross-section geometry.

Expanding $\partial/\partial x$ on the r.h.s. of equation (3.2) while accounting for tidal variations in channel depth yields:

$$-\frac{\partial}{\partial x}(uwh) = -uw{<}h{>}\left(\frac{1}{u}\frac{\partial u}{\partial x} + \frac{1}{w}\frac{\partial w}{\partial x} + \frac{1}{{<}h{>}}\frac{\partial {<}h{>}}{\partial x}\right)\left[1 \pm O\left(\frac{a}{{<}h{>}}\right)\right]. \tag{3.9}$$

Balancing equations (3.8) and (3.9), applying the definitions of η and u, differentiating, and further scaling then gives:

$$\begin{aligned}&\omega a{<}b{>}\left[1 \pm O\left(\frac{\Delta b}{{<}b{>}}\right) \pm O\left(\frac{a}{{<}h{>}}\right)\right] \\ &= uw{<}h{>}k\ O\left[1 \pm \frac{L_U{}^{-1}}{k} \pm \frac{L_w{}^{-1}}{k} \pm \frac{L_h{}^{-1}}{k}\right],\end{aligned} \tag{3.10}$$

where

$$L_w{}^{-1} = \frac{\partial w/\partial x}{w}, \quad L_h{}^{-1} = \frac{\partial {<}h{>}/\partial x}{{<}h{>}} \tag{3.11a,b}$$

are the spatial rates of change in channel width and depth.

The mass balance represented by equation (3.10) is between the time rate of change of the wetted cross-sectional area on the l.h.s. and along-channel convergence of volume flux on the r.h.s. At leading order, flux convergence on the r.h.s. is due to the sum of (i) the along-channel rate of change in the phase of tidal velocity ($\sim\partial(kx)/\partial x \approx k$), (ii) the along-channel rate of change in the amplitude of tidal velocity (which scales as $L_U{}^{-1}$), (iii) the along-channel rate of change in channel width ($\sim L_w{}^{-1}$), and (iv) the along-channel rate of change in channel depth ($\sim L_h{}^{-1}$). To non-dimensionalize the r.h.s. of equation (3.10), these four terms have all been divided by k. Table 3.1 summarizes the key ratios of temporal and spatial scales (or, in many cases, ratios of spatial rates of change) used to evaluate the importance of the various terms in the momentum and continuity equations.

Table 3.1. *Ratios used to evaluate the importance of terms and likely size of approximation errors in the momentum and continuity equations (in order of appearance in text)*

Ratio	Compact form used in text (if different)	Definition
$\frac{a}{<h>}$	–	$\frac{\text{amplitude of tidal elevation}}{\text{average channel depth}}$
$\frac{r}{\omega}$	–	$\frac{\text{magnitude of friction term in momentum}}{\text{magnitude of acceleration term in momentum}}$
$\frac{U}{c}$	–	$\frac{\text{tidal velocity amplitude}}{\text{tidal phase speed}}$
$\frac{L_U^{-1}}{k}$	$(kL_U)^{-1}$	$\frac{\text{spatial rate of change in } U}{\text{spatial rate of change in tidal phase}}$
$\frac{\Delta b}{<b>}$	–	$\frac{\text{amplitude of tidal variation in estuary width}}{\text{average estuary width}}$
$\frac{L_w^{-1}}{k}$	$(kL_w)^{-1}$	$\frac{\text{spatial rate of change in channel width } w}{\text{spatial rate of change in tidal phase}}$
$\frac{L_h^{-1}}{k}$	$(kL_h)^{-1}$	$\frac{\text{spatial rate of change in } h}{\text{spatial rate of change in tidal phase}}$
$\frac{L}{k^{-1}}$	kL	$\frac{\text{length of estuary}}{\text{length scale over which tidal phase varies}}$
$\frac{L}{L_a}$	–	$\frac{\text{length of estuary}}{\text{length scale for along-channel variation in } a}$
$\frac{k}{L_w^{-1}}$	kL_w	$\frac{\text{spatial rate of change in tidal phase}}{\text{spatial rate of change in } w}$
$\frac{L_h^{-1}}{L_w^{-1}}$	L_w/L_h	$\frac{\text{spatial rate of change in } h}{\text{spatial rate of change in } w}$
$\frac{L_U^{-1}}{L_w^{-1}}$	L_w/L_U	$\frac{\text{spatial rate of change in } U}{\text{spatial rate of change in } w}$
$\frac{\lvert L_w^{-1} - L_b^{-1}\rvert}{L_w^{-1}}$	$\lvert L_w - L_b\rvert/L_b$	$\frac{\text{difference between spatial rates of change in } w \text{ and } b}{\text{spatial rate of change in } w}$
$\frac{\omega}{r}$	–	$\frac{\text{magnitude of acceleration term in momentum}}{\text{magnitude of friction term in momentum}}$
$\frac{L_a^{-1}}{k}$	$(kL_a)^{-1}$	$\frac{\text{spatial rate of change in tidal amplitude}}{\text{spatial rate of change in tidal phase}}$

3.5. Short estuaries

In short tidal estuaries, it is possible to integrate continuity from an arbitrary location x to the head at L in order to solve directly for tidal velocity without considering momentum. Performing a leading-term Taylor expansion on $\partial\eta/\partial t$ and integrating both sides of equation (3.2) in x yields:

$$\frac{\partial\eta_0}{\partial t}\left[1 \pm O\left(kL \pm \frac{L}{L_a}\right)\right]\int_x^L b\,dx = -\left[(uwh)_L - (uwh)_x\right], \tag{3.12}$$

where L is the length of the estuary, and η_0 is tidal elevation at the mouth of the estuary (at $x = 0$).

Assuming no tidal flux at the landward end of the estuary, $(uwh)_L = 0$, and a short tidal estuary ($kL \ll 1$, $L/L_a \ll 1$), equation (3.12) reduces to

$$u(x,t) = \frac{\partial\eta_0(t)}{\partial t}\frac{A_b(x,t)}{A_c(x,t)}, \tag{3.13}$$

where $A_c = wh$ is the wetted channel cross-sectional area at x, and $A_b = \int b\,dx$ is the wetted basin surface area up-estuary of x.

The expected errors in equation (3.13) of size $O(kL, L/L_a)$ are only negligible if the estuary is indeed very short. But equation (3.13) is otherwise fully non-linear, and the solution for $u(x,t)$ given by equation (3.13) can be calculated based just on a knowledge of intertidal topography, local channel bathymetry, and a time series of tidal elevation at a single point. At a given point in x, tidal variations in A_b/A_c can lead to significant distortion in $u(t)$, as illustrated in Fig. 3.3. If h increases dramatically over the tidal cycle, but b does not, the resulting velocity–stage curve will display higher velocities around low water (Fig. 3.3a–b). Conversely, if b changes more strongly than h, higher velocities will be found in the tidal channel around high water (Fig. 3.3c–d). In short estuaries, the momentum equation is inconsequential to the calculation of $u(x,t)$, and tidal phase speed, c, is effectively infinite, since zero phase lag is assumed along the length of the system.

Linearizing equation (3.13) leads to the solution:

$$\eta(t) = a_0 \cos\omega t, \qquad u(x,t) = -U(x)\sin\omega t, \quad U(x) = a_0\omega < A_b(x)/A_c(x) >, \tag{3.14a–c}$$

with additional errors $O(a/<h>, \Delta b/<b>)$. From equation (3.14) it follows that in short estuaries, the linearized component of tidal velocity is 90° out of phase with tidal elevation. Although this is the same phase relation found in a tidal standing wave, it was not necessary to evoke a reflected wave or momentum in any form to derive this result.

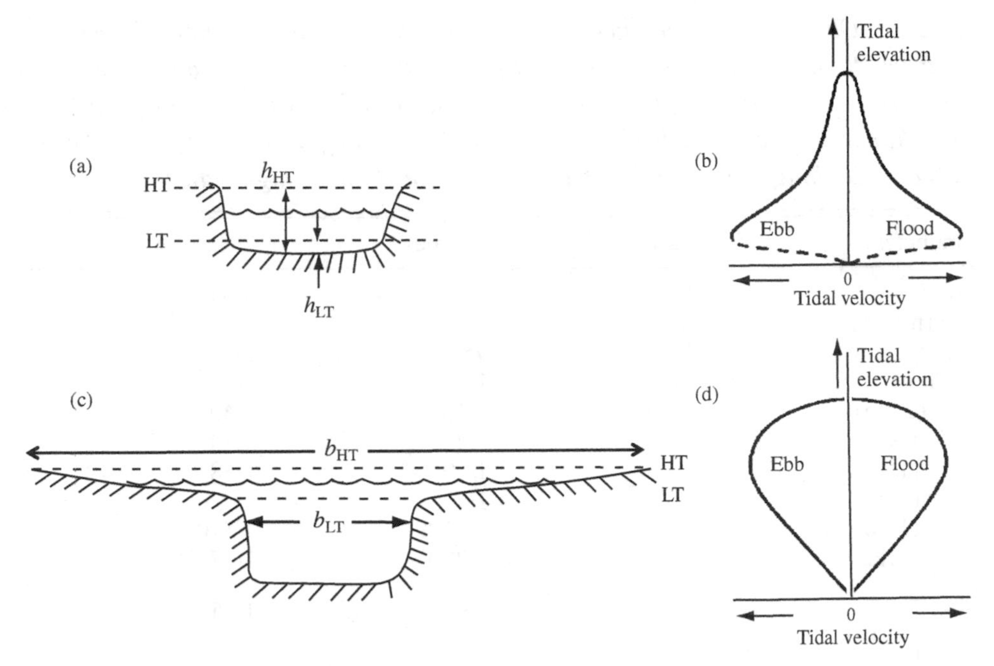

Figure 3.3. (a) Tidal estuary dominated by tidal variations in channel depth. (b) Tidal currents in the channel are strongest closer to low tide when depth is small. (c) Tidal estuary dominated by tidal variations in estuary width. (d) Tidal currents in the channel are strongest closer to high tide when surface area is large. [(b), (d) adapted from Pethick, 1980.]

The short estuary case approximated by equation (3.13) is discussed at length by Pethick (1980). Real-world examples include channelized tidal marsh, mangrove and tidal flat systems, a few kilometers or less in length, found ubiquitously on coastlines around the globe. Specific systems from the literature explicitly examined in the context of the above dynamics include tidal channels near Wachapreague, Virginia, as discussed by Boon (1975), and tidal channels near North Norfolk, England, as discussed by Pethick (1980).

3.6. Long, shallow, funnel-shaped estuaries

A second class of tidal estuaries characterized by a simple solution for tidal velocity that follows directly from continuity are shallow, funnel-shaped estuaries. Figure 3.4 displays along-channel variation in a, kx, $<A_c>$, $<b>$, and U as observed along the Tamar, Delaware, and Thames estuaries (from Friedrichs and Aubrey, 1994). Values for L_a, L_{Ac}, L_b, L_h, L_w, and L_U (based on exponential fits), and k (based on a linear fit) are contained in Table 3.2. From Table 3.2, we see that the largest term

Table 3.2. *Observed tidal and geometric properties for example funnel-shaped estuaries (from Friedrichs and Aubrey, 1994). Values are along-channel averages or otherwise apply to the estuary as a whole. Because low-water elevation in the upper Tamar is kinematically truncated around low water, parameters derived from a (x) and kx along the Tamar are based on the elevation of high water*

Parameter	Tamar	Thames	Delaware
a (m)	2.7	2.0	0.64
$<h>$ (m)	2.9	8.5	5.8
L (km)	21	95	215
k (1/km)	1/64	1/70	1/58
L_w^{-1} (1/km)	1/5.8	1/18	1/40
L_b^{-1} (1/km)	1/4.6	1/25	1/40
L_{Ac}^{-1} (1/km)	1/5.3	1/19	1/38
$\|L_h^{-1}\|$ (1/km)	1/62	1/72	1/720
$\|L_a^{-1}\|$ (1/km)	1/190	1/1500	1/570
$\|L_U^{-1}\|$ (1/km)	1/160	1/280	1/1700
$\frac{L_w^{-1}}{k}$	11	3.9	1.45
$\left\|\frac{L_h^{-1}}{k}\right\|$	1.0	1.0	0.08
$\left\|\frac{L_U^{-1}}{k}\right\|$	0.40	0.25	0.03
$a/<h>$	0.94	0.24	0.11
$\Delta b/<b>$	0.29	0.17	~ 0
$w/<b>$	0.71	0.83	~ 1

on the r.h.s. of equation (3.10) for this class of estuary is $(kL_w)^{-1}$. Assuming $w \sim \exp(-x/L_w)$ and $b \sim \exp(-x/L_b)$, such that $L_w \approx L_b$, the lowest-order, linearized continuity balance becomes

$$\frac{\partial \eta}{\partial t} = \frac{w <h> L_w^{-1}}{<b>} u, \tag{3.15}$$

with associated errors $O(kL_w, L_w/L_h, L_w/L_U, |L_w - L_b|/L_b, a/<h>, \Delta b/<b>)$. Note that the above errors are not necessarily negligible (e.g., $a/<h>$ in the Tamar, kL_w in the Delaware). Nonetheless, the largest linear terms in each case lead to the above simple balance, effectively highlighting fundamental physics. The effects of additional terms on tidal asymmetry and along-channel changes in tidal amplitude will be discussed in detail in later sections.

Plugging equation (3.1) into equation (3.15) immediately gives the solution

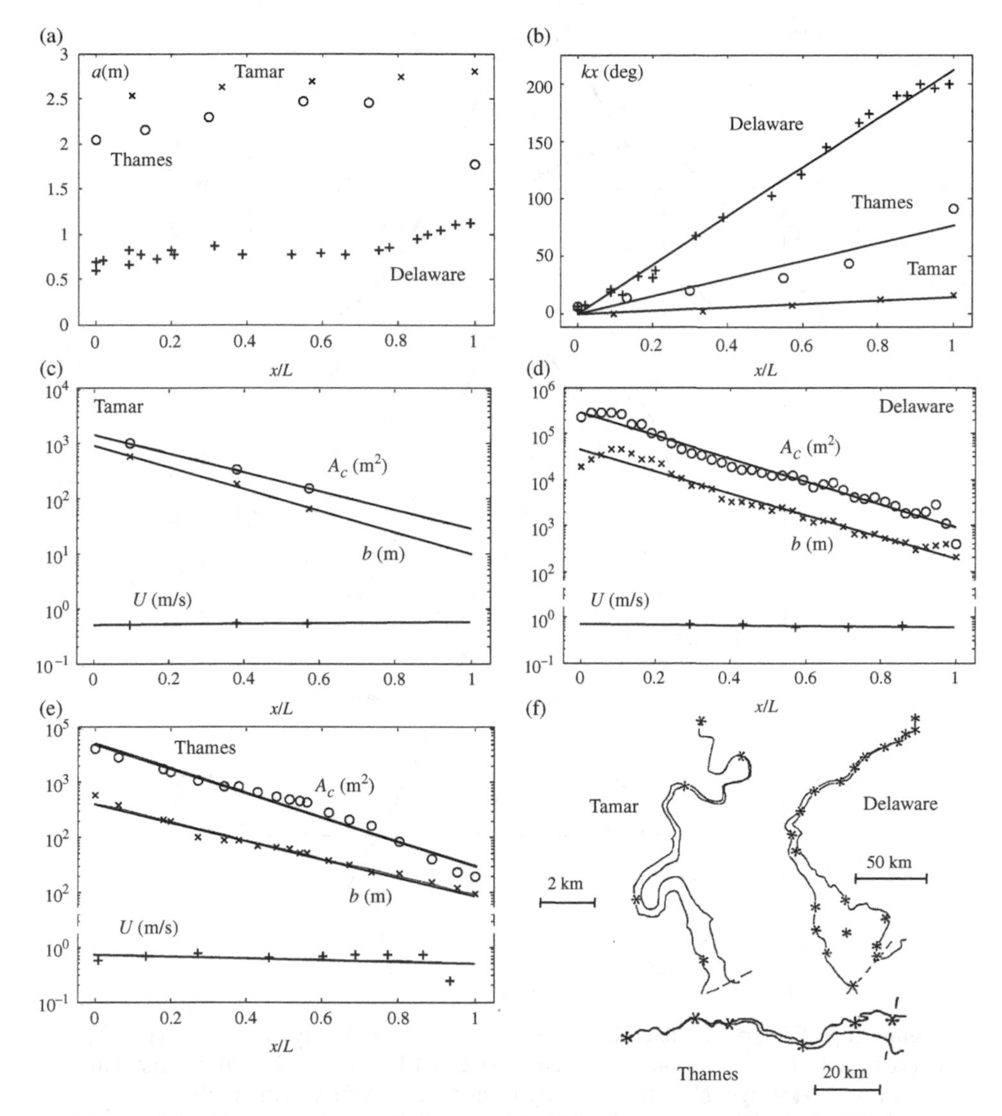

Figure 3.4. Along-estuary variation in: (a) tidal elevation amplitude (× Tamar, o Thames, + Delaware); (b) tidal phase (× Tamar, o Thames, + Delaware); channel cross-sectional area (o), estuary width (×), tidal velocity amplitude (+) for (c) Tamar, (d) Delaware; (e) Thames; (f) tide gauge locations. Adapted from Friedrichs and Aubrey (1994).

$$\eta(t) = a\cos(\omega t - kx), \quad u(x,t) = -U\sin(\omega t - kx),$$
$$U = \frac{a\omega <b>}{<h> L_w{}^{-1} w}. \tag{3.16a–c}$$

As was the case for short channels, we are here likewise able to solve for the magnitude and phase of tidal velocity relative to elevation based on externally

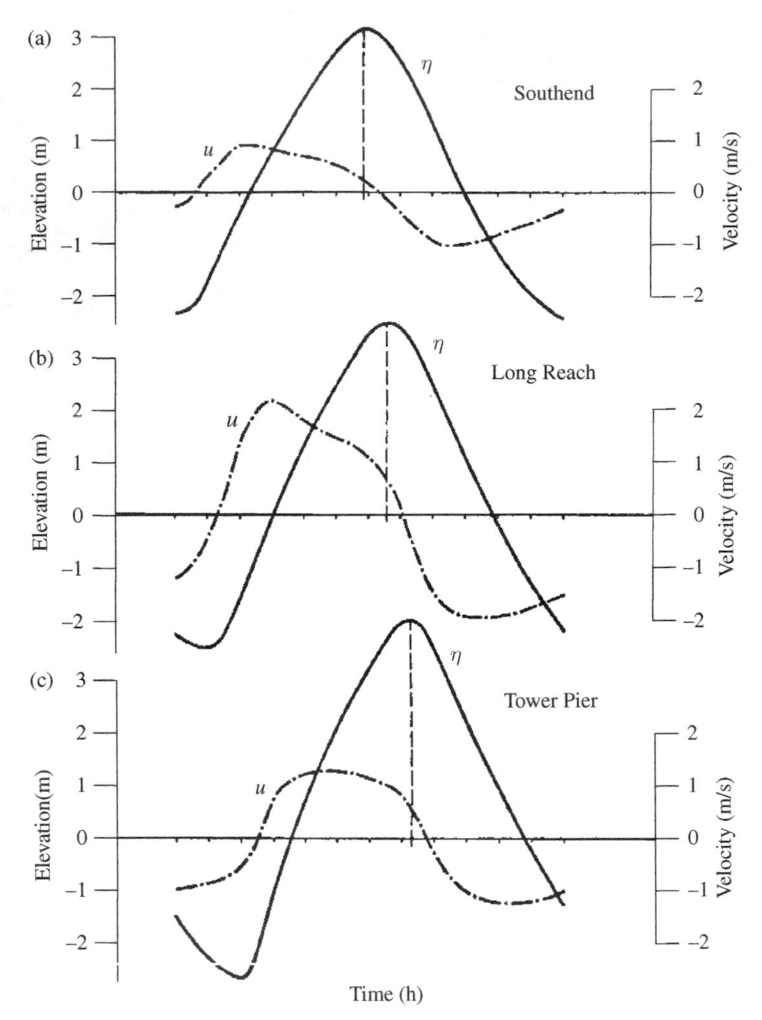

Figure 3.5. Observed tidal elevation and cross-sectionally averaged velocity curves for 11 February 1948 measured (a) 0.2, (b) 13, and (c) 27 km downstream from London Bridge. Note that velocity leads tidal elevation with a relative phase of ~ 90°. Adapted from Hunt (1964).

imposed variables (a, ω, b, w, L_w, h) without evoking momentum. Equation (3.16), like equation (3.14) above, indicates that velocity leads elevation by 90°, as in a classic standing wave. This result is consistent with the relationship between tidal velocity and elevation illustrated by observations from the Thames in Fig. 3.5. Yet kx in Fig. 3.4b clearly indicates that the associated waveform propagates landward. This contradiction reiterates the potential confusion that may result if an estuary as a whole is characterized as having a "standing" or "progressive" wave character based only on the phase relationship between velocity and elevation.

The linearized momentum equation resulting from scaling in Section 3.3 is:

$$\frac{\partial u}{\partial t} + g\frac{\partial \eta}{\partial x} + ru = 0. \tag{3.17}$$

Plugging the known solution given by equation (3.16) into equation (3.17), however, results in no terms available to balance $\partial u/\partial t$. This imbalance can only be neglected at lowest order if ω/r is also a small parameter. Thus we have the interesting result that if width changes dominate along channel changes in depth, velocity amplitude, and tidal phase, it must follow that friction dominates acceleration, even though we do not explicitly know anything about what the bed roughness is.

Equation (3.17) is then approximated as:

$$g\frac{\partial \eta}{\partial x} + ru = 0. \tag{3.18}$$

Combining equations (3.15) and (3.18) to eliminate u then yields:

$$\frac{\partial \eta}{\partial t} + c\frac{\partial \eta}{\partial x} = 0, \tag{3.19}$$

where the tidal phase speed c in equation (3.19) is given by

$$c = \frac{\omega}{k} = \frac{g<h>wL_w^{-1}}{r<b>}. \tag{3.20}$$

The wave speed in equation (3.20) increases with depth, $<h>$, and with increased channel convergence (i.e., larger L_w^{-1}). Conversely, wave speed decreases with greater friction, r, and with greater tidal volume storage in fringing shoals, flats and marsh (i.e., larger $<b>/w$).

Equation (3.19) is an example of a first-order wave equation (termed "first-order" because $\partial/\partial t$ and $\partial/\partial x$ are first derivatives). A first-order wave equation allows propagation of waves in only one direction, in this case up-estuary toward $+x$. Reflected waves (which propagate toward $-x$) are not allowed. Physically, reflected waves do not occur in this type of system because waves propagating toward $-x$ have their energy quickly spread by divergence of width as well as being damped by friction. Waves propagating toward $+x$ are frictionally damped too, but convergence concentrates wave energy enough to maintain a relatively constant amplitude of tidal velocity and elevation.

3.7. Long, deep, non-convergent estuaries

As an example of (an approximately) long, deep and non-convergent estuary channel, we will take the main stem of the 300-km long Chesapeake Bay. From Fig. 3.6, the spatial rate of change in tidal phase in Chesapeake Bay is approximately

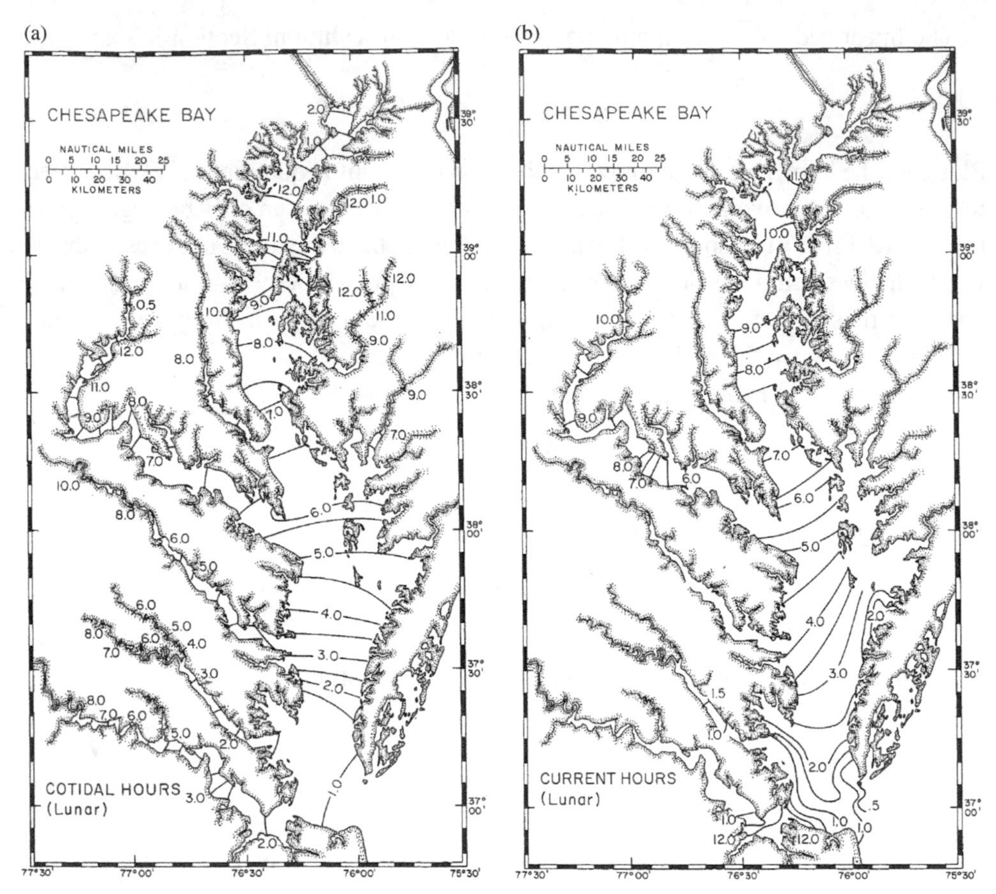

Figure 3.6. Tidal phase in hours for (a) high tide and (b) maximum flood current along the Chesapeake Bay and its major tidal tributaries (from Carter and Pritchard, 1988).

$k \approx 1/(50\text{ km})$ for the semidiurnal tide. From equation (3.10), we then see that as long as spatial rates of change in U, w, and h are all much greater than 1/(50 km), then phase-induced changes in volume flux dominate the r.h.s. of the equation. In other words, if the e-folding lengths of U, w, and h are much greater than 50 km (which they are), then the lowest-order, linearized continuity balance must be between the $\partial\eta/\partial t$ and $\partial(kx)/\partial x$ terms. Plugging equation (3.1) into these two remaining terms immediately yields:

$$\eta(x,t) = a\cos(\omega t - kx), \quad u(x,t) = U\cos(\omega t - kx), \quad U = \frac{a\omega <b>}{<h>kw}. \tag{3.21a--c}$$

The above solution shows that in long, deep, constant-width tidal channels, velocity and elevation are in phase. Figure 3.6 shows that η and u are indeed largely in phase over most of the main stem of the Chesapeake Bay, except within about 100 km of the head, where a reflected wave is indicated by a shift in the velocity–elevation phase difference from ~ 0 to ~ 3 hours at the very top of the bay.

This time, plugging the solution given by equation (3.21) into equation (3.17) results in no terms available to balance the friction term ru. In a manner analogous yet opposite to that presented in Section 3.6, this imbalance can only be neglected if r/ω is a small parameter no larger than $O(kL_w)^{-1}$. Thus, we now have the result that if (i) the spatial rate of change in tidal phase (k) dominates the spatial rates of change in width (L_w^{-1}), depth (L_h^{-1}), and velocity amplitude (L_U^{-1}), then (ii) it must follow that acceleration dominates friction ($r/\omega \ll 1$) in the momentum transporting channel, even though we still do not have (or need) an explicit measure of bottom roughness.

Assuming a momentum balance between the $\partial u/\partial t$ and the $\partial(kx)/\partial x$ component of pressure gradient, a continuity balance between $\partial\eta/\partial t$ and the $\partial(kx)/\partial x$ component of flux convergence, and cross-differentiating to eliminate u then yields:

$$\frac{\partial^2\eta}{\partial t^2} - c^2\frac{\partial^2\eta}{\partial x^2} = 0, \quad c = \frac{\omega}{k} = \left(\frac{g<h>w}{<b>}\right)^{1/2}. \qquad (3.22a,b)$$

Equation (3.22a) is a second-order wave equation, which allows propagation of waves in both $+x$ and $-x$. Physically, reflected waves can occur in this type of system (at least for a finite distance) because a relatively small rate of channel width change and low friction does not cause immediate divergence or dissipation of reflected energy. For the case of $w = <b>$ (i.e., negligible tidal storage outside the central channel), equation (3.22b) reduces to $c = (g<h>)^{1/2}$, the familiar solution for a frictionless, non-convergent, shallow-water wave. However, the presence of inter-tidal areas, lateral shoals, and/or other areas of tidal storage often slows tidal propagation in real estuaries.

We can check the assumption of low channel friction in the main stem of the Chesapeake Bay as follows. With $k \approx 1/(50\text{ km})$ and $a \approx 0.25$ m, equations (3.21c) and (3.22b) can be combined to give $U \approx 35$ cm/s. Based on a high-resolution numerical model, Spitz and Klinck (1998) suggest a median value of $c_d \approx 0.001$ for the Chesapeake. Assuming the channelized portion of the Chesapeake Bay (not including shallow lateral shoals) to have a mean depth of about $<h> \approx O(10\text{ m})$, it then follows from equation (3.5) that $r \approx 3 \times 10^{-5}\ \text{s}^{-1}$. The magnitude of friction relative to local acceleration, r/ω, is then 0.2, which is indeed small at our order of accuracy.

It is interesting to note that assuming $w = b$ in equation (3.22) would result in an unrealistic solution for tidal propagation in the Chesapeake Bay. Based on Fig. 3.6, the phase speed for the tide in the Chesapeake Bay is about 7 m/s. Neglecting lateral tidal storage areas by applying $c = (g<h>)^{1/2}$ would then give

$<h> = 5$ m. However, the mean depth of the Chesapeake Bay is significantly larger at 8.5 m (NOAA, 1985). The simplest explanation is that tidal volume storage in lateral shoals and tidal tributaries slows tidal propagation up the bay as predicted by equation (3.22b).

3.8. Long, shallow, non-convergent estuaries

For the case of long, shallow, non-convergent estuaries, friction dominates acceleration in momentum, and along-channel gradients in tidal velocity dominate along-channel variations in width and depth. The linearized governing equations are then

$$g\frac{\partial \eta}{\partial x} + ru = 0, \quad <b>\frac{\partial \eta}{\partial t} = -w <h>\frac{\partial u}{\partial x}, \tag{3.23a,b}$$

with errors $O(\omega/r, (kL_w)^{-1}, (kL_h)^{-1}, a/<h>, \Delta b/<b>)$. The above balances can be a useful approximation when exploring the dynamics of shallow tidal channels with abundant fringing marsh and/or tidal flats for which the short estuary assumption is not justifiable (Friedrichs and Madsen, 1992).

Differentiating equation (3.23a) in order to eliminate u in equation (3.23b) gives

$$\frac{\partial \eta}{\partial t} = D\frac{\partial^2 \eta}{\partial x^2}, \quad D = \frac{wg <h>}{<b> r}. \tag{3.24a,b}$$

Equation (3.24) has the form of a diffusion equation with the coefficient D describing the upstream evolution of the tidal signal. With the boundary conditions that $\eta = a_0 \cos(\omega t)$ at $x = 0$ and $\eta \to 0$ as $x \to \infty$ (i.e., $kL \gg 1$), the solution to equation (3.24) is then

$$\eta(x,t) = a_0 \exp(-kx)\cos(\omega t - kx), \quad c = \frac{\omega}{k} = \left(\frac{2\omega g<h>w}{r<b>}\right)^{1/2}. \tag{3.25a,b}$$

Plugging equation (3.25a) into equation (3.23a) then yields

$$u(x,t) = U_0 \exp(-kx)\cos(\omega t - kx + \pi/4),$$
$$U_0 = \frac{a_0}{<h>}\left(\frac{\omega g<h><b>}{rw}\right)^{1/2}. \tag{3.26a,b}$$

The relative phase between elevation and velocity in equation (3.25) and equation (3.26) is $\phi = -\pi/4$. In other words, velocity leads elevation by 45°, and the relationship between η and u is intermediate between standing and progressive. However, the partially "standing" behavior has nothing to do with reflection, since there is no reflected wave.

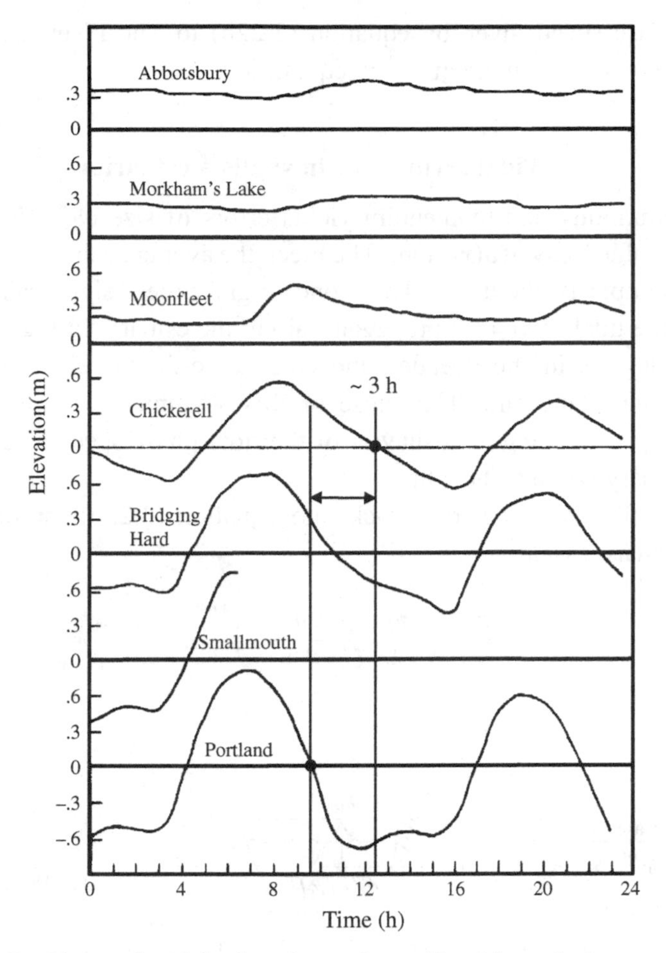

Figure 3.7. Observed tidal elevations along The Fleet during spring tide, 17 December 1967, collected 0, 0.3, 2, 5, 7, 11, and 13 km up-estuary of Portland. Adapted from Robinson *et al.* (1983).

As a real-world example, we take The Fleet, a very shallow, relatively long and straight back-barrier lagoon lined with tidal marsh, located along the south coast of England (Fig. 3.7). From Robinson *et al.* (1983), for The Fleet we have $<h>$ = 0.7 m, $<b>$ = 390 m, w = 130 m, a_0 = 0.7 m, and estuary length L = 13 km. From Fig. 3.7 we see that $c \approx 0.5$ m/s, so from equation (3.25b) we can derive that $\omega/r = 0.05$, confirming that frictional dominance is a good approximation in this case. For the semidiurnal tide, $c \approx 0.5$ m/s gives $k \approx 1/(3$ km), confirming that $kL \ll 1$ and the estuary can be treated as infinitely long. It is interesting to note that Robinson *et al.* (1983) recognized the important role storage in fringing marshes can play in slowing tidal propagation. But they applied the frictionless phase speed given by equation (3.22b) to The Fleet rather than the frictionally dominated solution given by equation (3.25b).

3.9. Tidal asymmetry in shallow estuaries

Linearizing continuity and momentum yields errors of size $O(a/<h>, \Delta b/<b>)$. Accounting for the decay of $a(x)$ along The Fleet, the average value for $a/<h>$ in our real-world example is about 0.7. Thus, one might expect significant non-linear distortion of the tidal wave as it propagates along the estuary. In Fig. 3.8, one can see that the rising tide in The Fleet does indeed become shorter relative to the falling tide with distance landward. The cause of this asymmetry can be understood conceptually by examining the behavior of the tidal phase speed as a function of changes in estuary width and depth.

Substituting the definition for r back into equation (3.25b) for shallow, non-convergent estuaries yields

$$c = \left(\frac{3\pi\omega w g <h>^2}{4c_d U <b>}\right)^{1/2}. \tag{3.27}$$

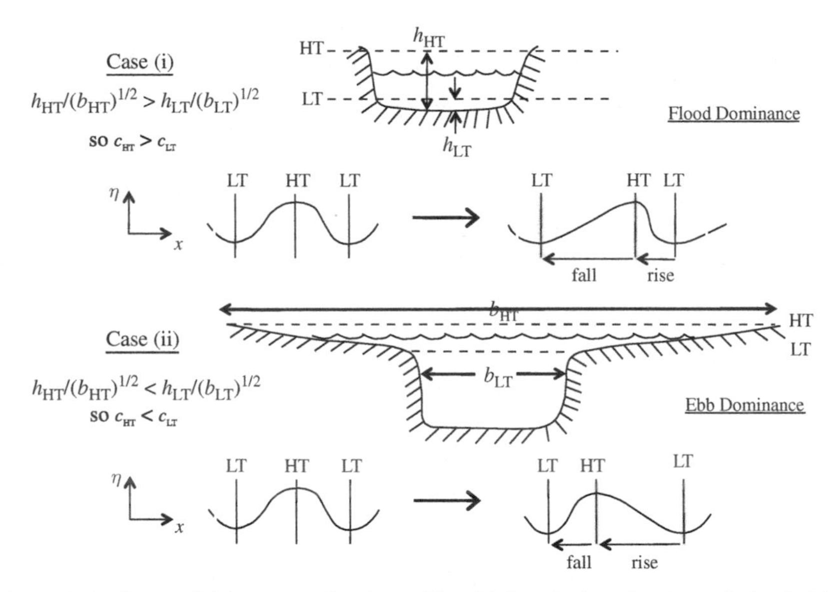

Figure 3.8. (i) In a tidal estuary dominated by tidal variations in channel depth, high tide propagates faster, partially catching up with the previous low tide. The result is a shorter rising tide and flood dominance. (ii) In a tidal estuary dominated by tidal variations in estuary width, low tide propagates faster, partially catching up with the previous high tide. The result is a shorter falling tide and ebb dominance.

The above equation is a linearized solution for wave speed, so c in equation (3.27) is assumed to be constant in time. A non-linear perturbation analysis (following Friedrichs and Madsen, 1992) yields a similar but time-dependent relationship such that

$$c(t) \sim h(t)/[b(t)]^{1/2}, \tag{3.28}$$

allowing c to vary in time at $O(a/<h>, \Delta b/<b>)$. Based on the time-dependent form in equation (3.28), at times when the tidal channel is deeper or the wetted estuary cross-section is narrower (relative to the width of the main channel), the tidal signal will propagate faster; when the channel is shallower or the estuary wider, the tidal signal will propagate more slowly. A qualitatively similar relationship results if one applies this analysis to equation (3.20) for shallow, funnel-shaped estuaries, namely $c \sim h^2/b$.

In other words (and as further represented in Fig. 3.8), if the channel is much deeper around high tide than around low tide ($h_{HT} \gg h_{LT}$), high tide will propagate into the estuary faster. High water will partially "catch up" with the previous low tide, and the duration of the rising tide will be shorter. Conversely, if the estuary is much wider at high tide than at low tide ($b_{HT} \gg b_{LT}$), low tide will propagate into the estuary faster, low tide will partially "catch up" with the previous high tide, and the duration of the falling tide will be shorter.

A leading-term Taylor expansion of $h/b^{1/2}$ over the tidal cycle gives

$$c \sim \frac{h}{b^{1/2}} \approx \frac{<h>[1+(\eta/a)(a/<h>)]}{<b>^{1/2}[1+(\eta/a)(\Delta b/<b>)]^{1/2}} \approx \frac{<h>}{<b>^{1/2}}[1+\gamma(\eta/a)], \tag{3.29}$$

where

$$\gamma = \frac{a}{<h>} - \frac{1}{2}\frac{\Delta b}{<b>} \tag{3.30}$$

is a tidal asymmetry parameter (cf. Friedrichs and Madsen, 1992). If $\gamma > 0$, tidal changes in depth dominate changes in width, and the tide is faster rising. If $\gamma < 0$, changes in width dominate, and the tide is faster falling. Because the phase of tidal velocity relative to elevation is at least partially standing in the frictionally dominated estuaries described by equations (3.29) and (3.30), these systems will tend to be flooding during the rising tide and ebbing during the falling tide. For these cases, $\gamma > 0$ corresponds to a faster rising tide and therefore flood dominance, while $\gamma < 0$ corresponds to a faster falling tide and therefore ebb dominance.

In order to investigate the role of channel depth and intertidal storage in determining flood vs ebb dominance, Friedrichs and Aubrey (1988) solved equations (3.2) and (3.3) numerically, retaining all non-linearities including quadratic friction. They considered 84 combinations of channel depth and total estuarine

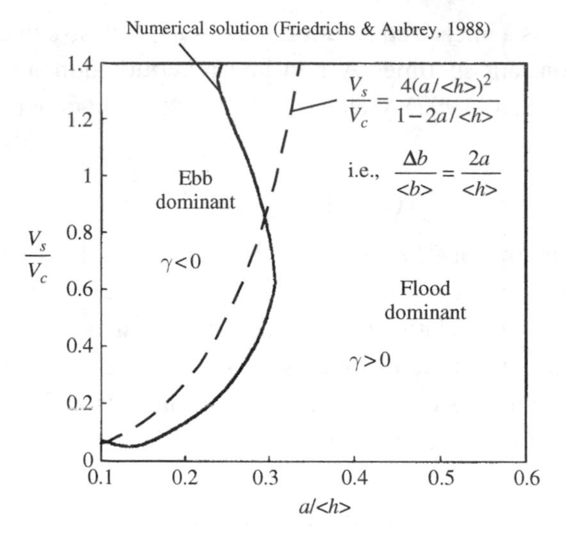

Figure 3.9. Regions of ebb and flood dominance as a function of tidal amplitude to channel depth ratio ($a/<h>$) and volume of storage in intertidal areas relative to volume of channels (V_s/V_c). $\Delta b/<b>$ is the amplitude of estuary width change over the tidal cycle divided by tidally-averaged estuarine width. The two lines separating the ebb- and flood-dominant fields are non-linear numerical results from Friedrichs and Aubrey (1988) (solid line) and equation (3.31) (dashed line).

width, all with $L = 7$ km, $c_d = 0.01$, $w \approx 100 <h>$, $a_0 = 0.75$ m, and linearly sloping intertidal flats. In each case, w, $<h>$, and $<b>$ were uniform along-channel. Based on an intuitive scaling rather than equations (3.28)–(3.30), Friedrichs and Aubrey (1988) suggested the two governing parameters to be $a/<h>$ and V_s/V_c, where V_s is the volume of storage in intertidal areas at high tide, and V_c is the volume in subtidal channels at mean water. Figure 3.9 displays the regions in V_s/V_c and $a/<h>$ "space" that Friedrichs and Aubrey (1988) found to be flood and ebb dominant, respectively.

The theoretical line separating flood- and ebb-dominant systems can alternatively be found by setting $\gamma = 0$ in equation (3.32), i.e., the situation where $\Delta b/<b> = 2a/<h>$. (Note that the very same relationship ends up separating flood- and ebb-dominant cases for shallow, funnel-shaped systems.) Using our present notation, $V_s/V_c \approx (2a\ \Delta b)/[h(<b>-\Delta b)]$ for cases where $w \approx b$ at low tide (like those modeled by Friedrichs and Aubrey, 1988). Setting $\Delta b/<b> = 2a/<h>$ then gives the analytical curve shown in Fig. 3.9, where $\gamma = 0$, namely

$$\frac{V_s}{V_c} = \frac{4(a/<h>)^2}{1 - 2a/<h>} \quad . \tag{3.31}$$

Both equations (3.30) and (3.31) indicate that for $2a/<h> \geq 0.5$, there is no amount of intertidal area or storage volume that can overcome depth-dependence and induce a faster-falling tide. Furthermore, the plot of equation (3.31) superimposed on Fig. 3.9 shows that the simple approximation given by equations (3.29) and (3.30) reasonably reproduces the fully non-linear results of Friedrichs and Aubrey (1988).

3.10. Tides in long, intermediate-depth, equilibrium estuaries

We now generalize the cases from Sections 3.6 and 3.7, i.e., long estuaries that maintain nearly uniform tidal amplitude along the length of the estuary. We term this generalized type of weakly to strongly funnel-shaped estuary an "equilibrium" estuary because uniform tidal velocity amplitude in space is associated with morphodynamic equilibrium (e.g., Friedrichs, 1995). The simplest approach in terms of mathematical analysis is to look for a solution that is the real part of the following (with $i = (-1)^{1/2}$):

$$\eta(x,t) = a \exp\left[i(\omega t - kx)\right], \qquad u(x,t) = U \exp\left[i(\omega t - kx - \phi)\right], \tag{3.32a,b}$$

and assume that at least one of k or L_w^{-1} is $\gg L_h^{-1}$ and L_U^{-1}. We will also assume $a/<h> \ll 1$, $\Delta b/<b> \ll 1$, and $w \sim b \sim \exp(-x/L_w)$. However, we now allow $(kL_w)^{-1} = O(1)$ and $r/\omega = O(1)$.

Plugging equation (3.32) into equations (3.2) and (3.17) with the above assumptions gives the following for continuity and momentum:

$$i\omega ab = ikwhU\mathrm{e}^{-i\phi} + whUL_w^{-1}\mathrm{e}^{-i\phi}, \quad i\omega U\mathrm{e}^{-i\phi} - ikga + rU\mathrm{e}^{-i\phi} = 0, \tag{3.33a,b}$$

where h and b are still tidally averaged, but we have dropped the brackets for brevity. [Note that it follows from equation (3.33a) that $L_a^{-1} \approx L_U^{-1}$.] Equations (3.33a) and (3.33b) can then be rearranged respectively to become

$$\frac{\omega a}{U}\mathrm{e}^{i\phi} = -\frac{kwh}{b} + \frac{iwhL_w^{-1}}{b}, \quad \frac{\omega a}{U}\mathrm{e}^{i\phi} = -\frac{\omega^2}{kg} + \frac{i\omega r}{kg} = 0. \tag{3.34a,b}$$

By separately equating the real and imaginary parts of equations (3.34a) and (3.34b), it immediately follows that

$$c = \frac{\omega}{k} = \left(\frac{ghw}{b}\right)^{1/2} \quad \text{and} \quad c = \frac{\omega}{k} = \frac{ghwL_w^{-1}}{rb}. \tag{3.35a,b}$$

In other words, in equilibrium estuaries, the phase–speed relationships for deep, straight estuaries and shallow, funnel-shaped estuaries hold simultaneously. The fact that $c \approx (gh)^{1/2}$ in such systems (for cases where $b \approx w$) has likely caused confusion in the past. In an equilibrium estuary, the frictionless wave speed holds regardless of the relative importance of friction (cf. Hunt, 1964). But this definitely does not mean the dynamic balance associated with a frictionless wave applies.

Furthermore, by eliminating either ghw/b or ω/k in equation (3.35), we have the added constraints that

$$\frac{r}{\omega} = \frac{L_w^{-1}}{k}, \quad \text{and} \quad L_w = \frac{(ghw/b)^{1/2}}{r}. \tag{3.36a,b}$$

The equality of r/ω and $(kL_w)^{-1}$ in equation (3.36a) emphasizes the balance between the effects of friction and convergence in equilibrium estuaries. As friction becomes more important (as indicated by a greater r/ω), width convergence must similarly strengthen [i.e., $(kL_w)^{-1}$ must increase] in order to maintain tidal amplitude. The near equality of $(kL_w)^{-1}$ and r/ω is apparent for the Delaware, whose geometry and tidal properties are particularly well constrained by data. From equation (3.35b), one can solve for r/ω, plug in observed values from Table 3.1, and get $r/\omega \approx 1.24$. This is quite close to our previously observed value of $(kL_w)^{-1} \approx 1.45$.

Using the identity $\exp(i\phi) = \cos\phi + i\sin\phi$, the phase of velocity relative to elevation is seen from equation (3.34a) to be

$$\phi = -\arctan\left(\frac{L_w^{-1}}{k}\right) = -\arctan\left(\frac{r}{\omega}\right). \tag{3.37}$$

Thus, as convergence and friction increase in equilibrium estuary cases [and $(kL_w)^{-1}$ and r/ω increase], the relative phase goes to 90°, as in Section 3.6. As convergence and friction decrease [and $(kL_w)^{-1}$ and r/ω go to zero], the relative phase goes to 0°, as in Section 3.7. Figure 3.10 displays observed values of the phase of velocity relative to elevation as a function of $(kL_w)^{-1}$ along several real estuaries with varying degrees of along-channel convergence. Figure 3.10 demonstrates that equation (3.37) holds reasonably well, even locally as the convergence rate changes along-channel.

By taking the absolute value of equation (3.34a), it follows from continuity that the amplitude of tidal velocity is

$$U = \frac{a\omega b}{hkw[1 + (kL_w)^{-2}]^{1/2}}, \tag{3.38}$$

which reduces to equations (3.16c) and (3.21c) for the conditions in Section 3.6 $[(kL_w)^{-1}] \gg 1]$ and in Section 3.7 $[(kL_w)^{-1}] \ll 1]$, respectively.

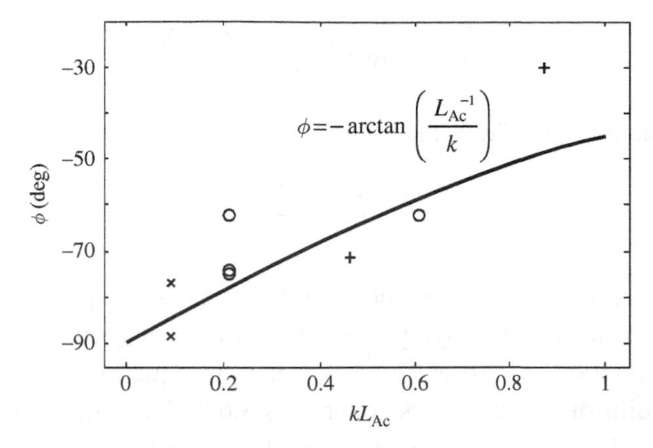

Figure 3.10. Observed phase of tidal velocity relative to tidal elevation as a function of kL_{Ac} along the Tamar (×), Thames (o), and Delaware (+) (modified from Friedrichs and Aubrey, 1994). $L_{Ac} \approx L_w$ is the along-channel e-folding length for the convergence of channel cross-sectional area. For this analysis, Friedrichs and Aubrey (1994) defined L_{Ac} based on an exponential fit to $A_c(x)$ in the local vicinity of the tidal observations. Also shown is the analytical prediction given by equation (3.37).

3.11. Tides in long, non-equilibrium estuaries

We now allow tidal amplitude to change with distance along-channel and assume a solution that is the real part of:

$$\begin{aligned} \eta(x,t) &= a_0 \exp(x/L_a)\ \exp\left[i(\omega t - kx)\right], \\ u(x,t) &= U_0 \exp(x/L_a)\ \exp\left[i(\omega t - kx - \phi)\right] \end{aligned} \tag{3.39a,b}$$

[although we still require nearly constant depth, $a/h \ll 1$, $\Delta b/b \ll 1$, and $b \sim w \sim \exp(-x/L_w)$]. Following a derivation analogous to that used in Section 3.10, equation (3.39) can be substituted into equation (3.2), and one can separately balance the real and imaginary parts to derive:

$$\frac{c^2}{(ghw/b)} = \frac{\omega}{r}\left(\frac{L_w{}^{-1}}{k} - \frac{2L_a{}^{-1}}{k}\right) \quad \text{and} \quad \frac{c^2}{(ghw/b)} = 1 + \frac{L_w{}^{-1}}{k}\frac{L_a{}^{-1}}{k} - \left(\frac{L_a{}^{-1}}{k}\right)^2. \tag{3.40a,b}$$

Eliminating c from equation (3.40) yields:

$$\frac{L_a{}^{-1}}{k} = \frac{\omega}{r} + \frac{1}{2kL_w} - \left(\frac{\omega^2}{r^2} + 1 + \frac{1}{4(kL_w)^2}\right)^{1/2}. \tag{3.41}$$

The real and imaginary parts of continuity can be combined to show:

$$\phi = -\arctan\left(\frac{L_w^{-1}}{k} - \frac{L_a^{-1}}{k}\right). \tag{3.42}$$

And from momentum it follows that:

$$U_0 = gh\frac{a_0}{h}\frac{k}{\omega}\left|\frac{i - (kL_a)^{-1}}{i + r/\omega}\right|. \tag{3.43}$$

Similar solutions for this generalized non-equilibrium case are provided by Savenije *et al.* (2008). Several simpler asymptotes follow from equations (3.40)–(3.42). As $(kL_a)^{-1} \rightarrow 0$, equations (3.40) and (3.42) reduce to equations (3.35) and (3.37), which is the equilibrium solution (as in Sections 3.6, 3.7 and 3.10). As $(kL_w)^{-1} \rightarrow 0$ and $r/\omega \rightarrow \infty$, we have the case of a long, straight channel with strong friction, as in Section 3.8, and equations (3.41) and (3.42) reduce to $(kL_a)^{-1} = -1$ and $\phi = -45°$.

We can also consider arbitrary channel convergence with zero friction ($r/\omega \rightarrow 0$) as long as we require $L_a \rightarrow 2L_w$ [otherwise c becomes undefined in equation (3.40a)]. Then we have

$$\frac{L_a^{-1}}{k} = \frac{L_w^{-1}}{2k}, \quad \frac{c^2}{(ghw/b)} = 1 + \left(\frac{L_w^{-1}}{2k}\right)^2, \quad \text{and} \quad \phi = -\arctan\left(\frac{L_w^{-1}}{2k}\right). \tag{3.44a-c}$$

Keeping in mind the definitions $a \sim \exp(x/L_a)$ and $w \sim \exp(-x/L_w)$, the frictionless asymptote in equation (3.44a) gives $a \sim w^{-1/2}$, which is the classic Green's Law solution for constant depth. The strong increase in tidal range toward the head of the Gulf of Maine, for example, results from strong convergence in combination with relatively low friction. We also see from equation (3.44) that in the absence of friction, strong convergence causes $c \rightarrow \infty$ and $\phi \rightarrow -90°$ (cf. Jay, 1991).

For a long, constant-width channel with unconstrained friction [i.e., $(kL_w)^{-1} \rightarrow 0$ with arbitrary r/ω, cf. Ippen and Harleman, 1966], equation (3.40) and (3.41) reduce to

$$\frac{L_a^{-1}}{k} = \frac{\omega}{r} - \left(\frac{\omega^2}{r^2} + 1\right)^{1/2} \quad \text{and} \quad \frac{c^2}{(ghw/b)} = 1 - \left(\frac{L_a^{-1}}{k}\right)^2. \tag{3.45a,b}$$

In other words, if a channel is non-convergent, long and frictional, equation (3.45a) indicates that L_a^{-1} is always negative (i.e., tidal amplitude always decreases landward), and equation (3.45b) tells us that wave speed is always less than the equilibrium value. For equation (3.45) [and for equations (3.40) and (3.41)], one can explicitly solve for wave speed by solving first for $(kL_a)^{-1}$.

For a long, constant-width channel, the momentum-based velocity amplitude solution given by equation (3.43) remains the same, but the relative phase solution in equation (3.42) reduces to

$$\phi = \arctan\left(\frac{L_a{}^{-1}}{k}\right). \tag{3.46}$$

Thus the phase of velocity relative to elevation in a long, constant-width channel ranges between $\phi = 0°$ for no friction [i.e., $(kL_a)^{-1} \to 0$, equivalent to Section 3.7] and $\phi = -45°$ for high friction [i.e., $(kL_a)^{-1} \to -\infty$, equivalent to Section 3.8].

3.12. Tides in long, near-equilibrium estuaries

For the observationally common case (cf. Prandle, 2004) where along-channel changes in tidal amplitude are small relative to the effects of channel convergence and along-channel tidal phase i.e., $(kL_a)^{-1} < (kL_w)^{-1} = O(1)$, a logical next approximation is to keep terms that are $O(kL_a)^{-1}$, but drop terms of relative size $O(kL_a)^{-2}$. Then the far right-hand term in equations (3.40b) can be neglected, and equation (3.40b) becomes (cf. Friedrichs and Aubrey, 1994):

$$\frac{c^2}{(ghw/b)} = 1 + \frac{L_w{}^{-1}}{k}\frac{L_a{}^{-1}}{k}, \tag{3.47}$$

while the equation for ϕ remains the same as equation (3.42). Combining equation (3.47) and equation (3.40a) then yields

$$\frac{L_a{}^{-1}}{k} = \frac{(kL_w)^{-1} - (r/\omega)}{2 + (r/\omega)(kL_w)^{-1}}. \tag{3.48}$$

By working backward from equation (3.47), it can further be shown that this approximation is equivalent to dropping the $\partial U/\partial x$ term on the r.h.s. of continuity while keeping the $\partial a/\partial x$ term within the pressure gradient in momentum.

Equation (3.48) highlights the competing roles of channel convergence $(kL_w)^{-1}$ and friction (r/ω). If convergence overcomes friction, i.e., $(kL_w)^{-1} > (r/\omega)$, then $(kL_a)^{-1}$ is positive, and tidal amplitude increases as one moves landward. Such an estuary is termed hypersynchronous (Nichols and Biggs, 1985). If friction overcomes convergence, i.e., $(r/\omega) > (kL_w)^{-1}$, then $(kL_a)^{-1}$ is negative, the amplitude decreases landward, and the estuary is hyposynchronous. Equation (3.47) highlights the association between amplitude change and wave speed. If tidal amplitude grows as one moves landward, the tidal phase speed is greater than the classic "frictionless" value of $(ghw/b)^{1/2}$, but if amplitude decays landward, the phase speed is less than $(ghw/b)^{1/2}$. The tipping point for landward growth or decay of tidal amplitude then becomes the following value of the along-channel convergence length scale:

$$L_w = \frac{\omega}{rk} = \frac{(ghw/b)^{1/2}}{r} \tag{3.49}$$

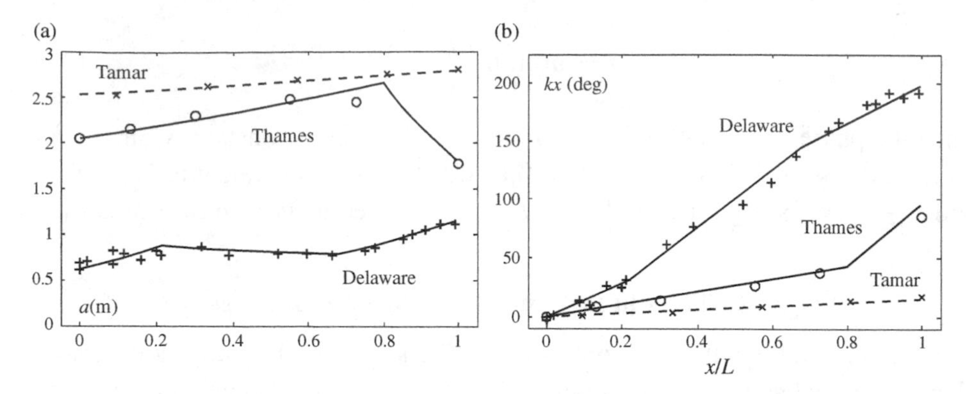

Figure 3.11. Along-estuary variation in (a) tidal elevation amplitude and (b) tidal phase with results of equations (3.39), (3.47), and (3.48) superimposed on segmented, exponentially convergent channels (× Tamar, o Thames, + Delaware). Adapted from Friedrichs and Aubrey (1994).

(cf. Le Floch, 1961). If the e-folding length of channel convergence is less than c/r, then amplitude grows, but if the e-folding length is greater then c/r, amplitude decays.

Equations (3.39), (3.47) and (3.48) are consistent with observed changes in tidal amplitude and phase speed along near-equilibrium estuaries (Fig. 3.11). Based on least-squares exponential fits to along-channel variations in channel geometry, Friedrichs and Aubrey (1994) divided up the Delaware into three segments, divided the Thames into two segments, and represented the Tamar with one segment. As predicted by equation (3.48), observed increases and decreases in tidal amplitude as one moves along-channel are consistent with inferred local values of $(kL_w)^{-1}$ and (r/ω). Furthermore, in the systems fit with more than one segment, changes in the sign of $(kL_a)^{-1}$ from one segment to the next agree with changes in c as predicted by equation (3.47) (Fig. 3.11). Where the next up-estuary segment is marked by a switch from positive to negative $(kL_a)^{-1}$, the change in $(kL_a)^{-1}$ is accompanied by a decrease in c (i.e., an increase in the slope of kx); likewise, a switch from negative to positive $(kL_a)^{-1}$ causes c to increase once more.

3.13. Tides in intermediate-length estuaries

An intermediate-length tidal estuary is defined as one where the inner end of the channel has a major impact on the tide through a notable fraction of the estuary, yet the estuary is too long to apply the short-channel solution (i.e., it is *not* true that both $kL \ll 1$ and $L/L_a \ll 1$). Examples of real tidal embayments that are reasonably represented by this approximation include Long Island Sound, USA (Swanson, 1976) and Gulf St. Vincent, Australia (Bowers and Lennon, 1990). Since intermediate-length, approximately rectangular, "dead-end" channels are not morphodynamically stable, they tend to form within relict features such as tectonic basins or submerged glacial topography rather than within classical river valley estuaries.

The effect of a barrier at the head of a channel can manifest itself away from the inner wall via a reflected wave that can be included in the expression for tidal elevation:

$$\eta(x,t) = a_I(x)\cos(\omega t - k_I x) + a_R(x)\cos(\omega t + k_R x - \phi_R). \qquad (3.50)$$

In equation (3.50), a_I and a_R represent the potentially x-dependent amplitudes of the incident and reflected components, k_I and k_R are the potentially distinct incident and reflected wave numbers, and ϕ_R is the relative phase of the reflected wave. It is important to note that where the reflected wave is significant, the observed wave number, based on the observed phase speed, will be different from k_I and k_R. Up until this point in this chapter, k always represented the directly observed wave number (k_{obs}) that can be observationally derived from the observed variation in tidal phase along the estuary.

The difference between k_{obs} and k_I in the context of an intermediate-length, nearly constant-width channel is demonstrated well by the example of Gulf St. Vincent (Bowers and Lennon, 1990). The observed phase speed, ω/k_{obs}, of ~ 60 m/s over the inner 120 km of the Gulf (Fig. 3.12) is much greater than the theoretical incident

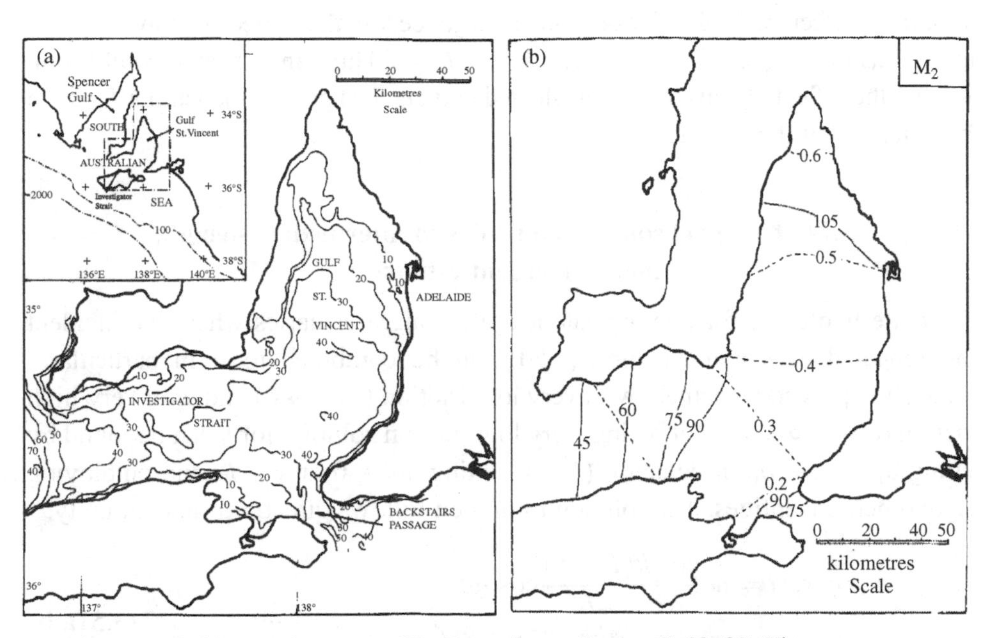

Figure 3.12. (a) Location map of the Investigator Strait – Gulf of St Vincent system, with depth contours in meters. (b) Observed amplitude in meters (dashed line) and phase in degrees (solid line) of M_2 tidal elevation. From Bowers and Lennon (1990).

wave speed, $\omega/k_I \approx (gh)^{1/2}$, of ~ 16 m/s. Also, the doubling of the observed tidal amplitude over this distance indicates that L/L_{a_obs} is not much less than one, so the short-channel solution from Section 3.5 is not applicable.

The region near the channel head where a reflected wave is important is delimited by the point at which a_R/a_I becomes small, or about $a_R/a_I \approx 1/e$ for the leading-order accuracy with which this chapter is concerned. For the case of a non-convergent estuary with an abrupt reflecting wall, we have $k_I = k_R$, and both a_I and a_R will decay equally with propagation distance. It follows that a reflected wave should be included for distances from the head wall less than about $|L_a|/2$, where $|L_a|$ is the absolute value of L_a as predicted by equation (3.45a), not the observed value (L_{a_obs}) discussed in the previous paragraph. Within the region where the reflected wave is important, the total wave solution can be calculated by using equation (3.45) to determine L_a and c for the incident and reflected components separately, adding the two components together, and choosing ϕ_R via continuity to satisfy no volume flux at $x = L$.

For the case of an equilibrium or near-equilibrium estuary, the rate at which friction reduces tidal amplitude $(L_{a\_frict})^{-1}$ can be estimated from equation (3.44a) as equal to the spatial rate of change that would have occurred without friction, i.e., $|(L_{a\_frict})^{-1}| \approx (L_w^{-1})/2$. Once the wave has been reflected, the effect of channel divergence alone also follows from equation (3.44a) as $|(L_{a\_diverg})^{-1}| \approx (L_w^{-1})/2$. The combined effect of both friction and divergence on the reflected wave is then approximately $|(L_{a\_frict})^{-1}| + |(L_{a\_diverg})^{-1}| \approx L_w^{-1}$. Thus, in a nearly equilibrium estuary, the reflected wave can be neglected at distances greater than about L_w from the inner end of the channel.

3.14. Compact solutions for tides in intermediate-length, non-convergent estuaries

There are two cases for intermediate-length, straight estuaries where the incident and reflected waves in equation (3.50) can be combined to form particularly compact expressions, namely with very low friction ($r/\omega \rightarrow 0$) and with very high friction ($r/\omega \rightarrow \infty$). The following, very low friction solution for η can be found by letting $a_I = a_R = a_0$ in equation (3.50), setting $\phi_R = 0$ at $x = L$, and employing trigonometric identities. The solution for u then follows directly from continuity:

$$\eta(x,t) = a_0 \frac{\cos k(L-x)}{\cos kL} \cos \omega t,$$
$$u(x,t) = -\frac{a_0}{h}\left(\frac{ghb}{w}\right)^{1/2} \frac{\sin k(L-x)}{\cos kL} \sin \omega t. \tag{3.51a,b}$$

In equation (3.51), k is not the observed wave number based on observed phase propagation, but rather the equivalent k for an infinite, straight, frictionless channel as in Section 3.7. The solution in equation (3.51) can lead to the classical case of "quarter wave resonance". This name arises from the fact that the denominator of equation (3.51) becomes undefined, and a and $U \rightarrow \infty$ when the estuary length $L = (2\pi/k)/4 = \lambda/4$, where λ is the length of the associated frictionless progressive wavelength. (Of course in practice, tidal amplitude would stop growing because of enhanced frictional damping.) For $L > \lambda/4$, one or more tidal nodes appear within the estuary, where tidal amplitude is zero but velocity is at a maximum. For short systems ($kL \ll 1$), equation (3.51) reduces to the linearized tidal pumping solution given by equation (3.14).

The very high friction case yields the real parts of:

$$\begin{aligned} \eta(x,t) &= a_0 \frac{\cosh\ k'(L-x)}{\cosh\ k'L} \exp(i\omega t) \ , \\ u(x,t) &= U_0 \frac{\sinh\ k'(L-x)}{\cosh\ k'L} \exp[i(\omega t + \pi/4)] \ , \end{aligned} \qquad (3.52a,b)$$

where $k' = (1+i)\ k$ (Friedrichs and Madsen, 1992), and U_0 is given by equation (3.26b). In equation (3.52) k is once more not the observed wavenumber based on observed phase propagation, but this time the equivalent k for a long, constant-width, frictionally dominated channel as in Section 3.8. For long systems ($kL \gg 1$), equation (3.52) reduces to equations (3.25) and (3.26), while for short systems ($kL \ll 1$), equation (3.52) reduces to equation (3.14). Friedrichs and Madsen (1992) showed that equation (3.52a) provides a useful approximation for the propagation of tidal elevation in many shallow estuaries and channelized tidal embayments. However, the quantitative utility of equation (3.52b) for predicting spatial variations in the amplitude of tidal velocity is limited by the constant channel width assumption.

3.15. Discussion and conclusions

This chapter has laid out a methodology by which widely available observations of estuary geometry and along-channel variations in tidal amplitude and phase (e.g., the e-folding length scales of L_w, L_U, L_a, and L_h, and the wave number k) can be used to scale the governing equations of continuity and conservation of mass for barotropic tides in channelized estuaries. The goal for each relevant estuarine scenario was to eliminate as many secondary terms as possible and identify the simplest physically insightful and reasonably realistic governing balances that adequately describe the problem. Table 3.3 summarizes the properties of the solutions discussed in this chapter (other than those that explicitly involve reflected

Table 3.3 *Properties of linearized asymptotic solutions for estuarine tides discussed in this chapter. Except for the short channel solution, all cases assume* $\eta \approx a_0 \exp(x/L_a) \cos(\omega t - kx)$, $u \approx U_0 \exp(x/L_a) \cos(\omega t - kx - \phi)$, $w \sim b \sim \exp(-x/L_w)$, $a/h << 1$, $\Delta b/b \ll 1$, and k and/or $L_w^{-1} \gg L_h^{-1}$. *See text for discussion of cases involving reflected waves*

Case	Additional key assumptions	Remaining r.h.s. terms in continuity	Remaining momentum terms	Spatial rate of change in tidal elevation amplitude relative to change in phase	Phase speed, c	Velocity relative phase, ϕ	Tidal velocity amplitude, U
(i) Short	$\frac{L}{k^{-1}} \ll 1,\ \frac{L}{L_a} \ll 1$	$\frac{\partial w}{\partial x}, \frac{\partial U}{\partial x}$	None	$\frac{L_a^{-1}}{k} = 0$	$c = \infty$	$\phi = -90^\circ$	$U(x) = \frac{a_0 \omega A_b(x)}{A_c(x)}$
(ii) Long, shallow, funnel-shaped	$\frac{L_w^{-1}}{k} \gg 1, \frac{L_U^{-1}}{L_w^{-1}} \ll 1$	$\frac{\partial w}{\partial x}$	$\frac{\partial (kx)}{\partial x}, ru$	$\frac{L_a^{-1}}{k} = 0$	$c = \frac{ghwL_w^{-1}}{rb}$	$\phi = -90^\circ$	$U_0 = \frac{a_0 \omega b}{hL_w^{-1} w}$
(iii) Long, deep, non-convergent	$\frac{L_w^{-1}}{k} \ll 1, \frac{L_U^{-1}}{k} \ll 1$	$\frac{\partial (kx)}{\partial x}$	$\frac{\partial u}{\partial t}, \frac{\partial (kx)}{\partial x}$	$\frac{L_a^{-1}}{k} = -1$	$c = \left(\frac{ghw}{b}\right)^{1/2}$	$\phi = 0^\circ$	$U_0 = \frac{a_0 \omega b}{hkw}$
(iv) Long, shallow, non-convergent	$\frac{L_w^{-1}}{k} \ll 1,\ \frac{r}{\omega} \gg 1$	$\frac{\partial (kx)}{\partial x}, \frac{\partial U}{\partial x}$	$\frac{\partial a}{\partial x}, \frac{\partial (kx)}{\partial x}, ru$	$\frac{L_a^{-1}}{k} = 0$	$c = \left(\frac{2\omega ghw}{rb}\right)^{1/2}$	$\phi = -45^\circ$	$U_0 = \frac{a_0}{h}\left(\frac{\omega ghb}{rw}\right)^{1/2}$
(v) Long, intermediate-depth, equilibrium	$\frac{L_w^{-1}}{k} = O(1), \frac{L_U^{-1}}{k} \ll 1$	$\frac{\partial w}{\partial x}, \frac{\partial (kx)}{\partial x}$	$\frac{\partial u}{\partial t}, \frac{\partial (kx)}{\partial x}, ru$	$\frac{L_a^{-1}}{k} = 0$ $\left(\frac{r}{\omega} = \frac{L_w^{-1}}{k}\right)$	$c = \frac{ghwL_w^{-1}}{rb} = \left(\frac{ghw}{b}\right)^{1/2}$	$\phi = -\arctan\left(\frac{L_w^{-1}}{k}\right)$ ($\phi = -90^\circ$ to 0°)	$U_0 = \frac{a_0 \omega b}{hkw[1 + (kL_w)^{-2}]^{1/2}}$
(vi) Long, intermediate-depth, non-convergent	$\frac{L_w^{-1}}{k} \ll 1,\ \frac{r}{\omega} = O(1)$	$\frac{\partial (kx)}{\partial x}, \frac{\partial U}{\partial x}$	$\frac{\partial u}{\partial t}, \frac{\partial a}{\partial x}, \frac{\partial (kx)}{\partial x}, ru$	$\frac{L_a^{-1}}{k} = \frac{\omega}{r} - \left(\frac{\omega^2}{r^2} + 1\right)^{1/2}$ $(-1 \le (kL_a)^{-1} \le 0)$	$\frac{c^2}{(ghw/b)} = 1 - \left(\frac{L_a^{-1}}{k}\right)^2$	$\phi = \arctan\left(\frac{L_a^{-1}}{k}\right)$ ($\phi = -45^\circ$ to 0°)	$U_O = \frac{a_0 kgh}{h\omega}\left\lvert\frac{i - (kL_a)^{-1}}{i + r/\omega}\right\rvert$

(vii) Long, deep, funnel-shaped	$\frac{L_w^{-1}}{k} = O(1), \frac{r}{\omega} \ll 1$	$\frac{\partial w}{\partial x}, \frac{\partial (kx)}{\partial x}, \frac{\partial U}{\partial x}$	$\frac{\partial u}{\partial t}, \frac{\partial a}{\partial x}, \frac{\partial (kx)}{\partial x}$	$\frac{L_a^{-1}}{k} = \frac{L_w^{-1}}{2k}$	$\frac{c^2}{(ghw/b)} = 1 + \left(\frac{L_w^{-1}}{2k}\right)^2$	$\phi = -\arctan\left(\frac{L_w^{-1}}{2k}\right)$ ($\phi = -90°$ to $0°$)	$U_0 = \frac{a_0 kgh}{h\omega}[1 - (kL_a)^{-2}]^{1/2}$
(viii) Long, intermediate-depth, near equilibrium	$(kL_w)^{-1} = O(1)$ $(kL_U)^{-1} < (kL_w)^{-1}$ Drop terms $O(kL_U)^{-2}$	$\frac{\partial w}{\partial x}, \frac{\partial (kx)}{\partial x}$	$\frac{\partial u}{\partial t}, \frac{\partial a}{\partial x}, \frac{\partial (kx)}{\partial x}, ru$	$\frac{L_a^{-1}}{k} = \frac{(kL_w)^{-1} - (r/\omega)}{2 + (r/\omega)(kL_w)^{-1}}$ $(-O(1) < (kL_a)^{-1} < O(1))$	$\frac{c^2}{(ghw/b)} = 1 + \frac{L_w^{-1}}{k}\frac{L_a^{-1}}{k}$	$\phi = -\arctan\left(\frac{L_w^{-1}}{k} - \frac{L_a^{-1}}{k}\right)$ ($\phi = -90°$ to $0°$)	$U_0 = \frac{a_0 kgh}{h\omega}\left\lvert\frac{i - (kL_a)^{-1}}{i + r/\omega}\right\rvert$
(ix) Long, intermediate-depth, non-equilibrium	$(kL_w)^{-1} = O(1)$ $r/\omega = O(1)$ $L_w/L_u = O(1)$	$\frac{\partial w}{\partial x}, \frac{\partial (kx)}{\partial x}, \frac{\partial U}{\partial x}$	$\frac{\partial u}{\partial t}, \frac{\partial a}{\partial x}, \frac{\partial (kx)}{\partial x}, ru$	$\frac{L_a^{-1}}{k} = \frac{\omega}{r} + \frac{L_w^{-1}}{2k} - \left(\frac{\omega^2}{r^2} + 1 + \left(\frac{L_w^{-1}}{2k}\right)^2\right)^{1/2}$	$\frac{c^2}{(ghw/b)} = 1 + \frac{L_w^{-1}}{k}\frac{L_a^{-1}}{k} - \left(\frac{L_a^{-1}}{k}\right)^2$	$\phi = -\arctan\left(\frac{L_w^{-1}}{k} - \frac{L_a^{-1}}{k}\right)$ ($\phi = -90°$ to $0°$)	$U_0 = \frac{a_0 kgh}{h\omega}\left\lvert\frac{i - (kL_a)^{-1}}{i + r/\omega}\right\rvert$

waves or tidal asymmetries). The scenarios are arranged in Table 3.3 in order of the increasing number of terms retained in each case in the associated continuity and momentum equations. A theme common to all cases is the potential role of the full estuary width, b (including tidal storage in marsh, flats, shoals and tributaries), relative to the width of the momentum transporting channel, w, such that a small value of w/b tends to slow the speed of tidal phase propagation.

The first two cases in Table 3.3 [(i) short estuaries and (ii) shallow, funnel-shaped estuaries] involve scenarios where simple solutions for the amplitude of tidal velocity as a function of external parameters follow directly from the continuity equation without the need to consider momentum. In each of these cases, spatial gradients in volume flux are dominated by the system's abrupt morphology, namely limited channel length ($\sim L^{-1}$) in very short estuaries and rapid width convergence ($\sim L_w^{-1}$) in shallow, funnel-shaped systems. Another property common to cases (i) and (ii) is a "standing wave" relationship between velocity and elevation, such that tidal velocity leads elevation by 90°. Yet in neither case is a reflected wave explicitly present. In case (ii), the tidal waveform is in fact clearly "progressive" in that it propagates unidirectionally up-estuary. This highlights the potential confusion that may result if an estuary as a whole is characterized as having a "standing" or "progressive" wave character, based only on the phase relationship between velocity and elevation.

Cases (ii), (iii), (v), and (viii) in Table 3.3 represent a spectrum of naturally common "equilibrium" to "near-equilibrium" tidal estuaries characterized by small along-channel changes in the amplitude of tidal velocity. In equilibrium to near-equilibrium estuaries, $\partial U/\partial x$ is negligible relative to other sources of flux gradients in the continuity equation. Relatively constant $U(x)$ is favored by negative morphodynamic feedback: if the estuary channel were randomly perturbed to produce highly variable depth, then the locally shallower, constricted, and thus higher-velocity areas would tend to scour, while deeper and lower-velocity areas would tend to accrete, favoring an evolution back toward uniform $U(x)$. The shallow, funnel-shaped estuary [case (ii)] and the deep, constant-width estuary [case (iii)] represent the two extremes where observed channel shape and $\partial U/\partial x \approx 0$ together necessitate the simplest possible dynamic balances: friction (ru) vs the phase-induced pressure gradient [$\partial(kx)/\partial x$] in the first scenario, and $\partial u/\partial t$ vs $\partial(kx)/\partial x$ in the second. In neither case is it necessary to know the size of the friction term a priori in order to infer its relative importance.

Unlike cases (ii) and (iii), cases (v) and (viii) allow along-channel gradients in width and velocity phase to simultaneously contribute to continuity [$(kL_w)^{-1} \approx O(1)$]. Since $\partial U/\partial x$ is still small, it must follow that friction and acceleration can simultaneously contribute to momentum [$r/\omega \approx O(1)$]. In fact, these two cases indicate that for $\partial U/\partial x \approx 0$ in general, it must follow that $(kL_w)^{-1} \approx r/\omega$. This equality leads to the somewhat surprising result that the tidal phase speed, c, in all equilibrium estuaries is equal to the frictionless value for a rectangular channel of the same depth, regardless of the actual size of the friction term. The equilibrium case highlights a balance between the competing roles of channel convergence $[(kL_w)^{-1}]$, which causes amplitude to increase, and friction (r/ω), which causes amplitude to decay. In near-equilibrium estuaries, if friction overcomes convergence, i.e., $(r/\omega) > (kL_w)^{-1}$, then tidal amplitude decreases gradually along the estuary $[(kL_a)^{-1} < 0]$, and c is slightly less than its frictionless value. If convergence overcomes friction, i.e., $(kL_w)^{-1} > (r/\omega)$, then tidal amplitude increases up-estuary $[(kL_a)^{-1} > 0]$, and c is slightly greater than its frictionless value. The rate of channel convergence also determines the phase relationship (ϕ) between velocity and elevation, such that $\phi \approx -\arctan((kL_w)^{-1})$. Strong convergence pushes ϕ toward $-90°$, whereas weak convergence pushes ϕ toward $0°$.

Cases (iii), (iv), and (vi) in Table 3.3 represent classic long, rectangular channel scenarios with weak, strong and intermediate friction, respectively. If friction is present without a reflected wave or channel convergence, then tidal amplitude always decreases with distance, and the tidal phase speed is always less than its frictionless value. As r/ω increases, c decreases, and the rate of amplitude decay with distance (L_a^{-1}) increases. However, with constant width, the spatial rate of amplitude decay is limited by the tidal wave number, i.e., $L_a^{-1} \leq k$. Weak friction pushes ϕ toward $0°$, whereas strong friction pushes ϕ toward $90°$.

Cases (iv), (vi), (vii), and (ix) are all non-equilibrium estuaries characterized by significant changes in tidal velocity amplitude with distance along-channel such that $\partial U/\partial x$ is a leading-order term in the continuity equation. For cases involving long, straight channels with friction [cases (iv) and (vi)], tidal amplitude must decrease with distance up-estuary. Case (vii) is a funnel-shaped estuary without friction, such that channel convergence (or divergence) increases (or decreases) amplitude with distance following Green's Law. Finally, case (ix) is the most general (and most complicated) case, which allows for an arbitrarily large or small increase or decrease in tidal amplitude with distance as a function of convergence, divergence, and/or friction. All the other long-channel scenarios considered in this chapter could alternately be derived by applying simplifying asymptotes to this general case.

There exist some cases (not included in Table 3.3) where a head wall in a real estuary or embayment affects the tides over a notable fraction of the system such that the dynamics are best understood by explicitly including a reflected wave component (see Sections 3.13 and 3.14 of the text). The fraction of the inner estuary affected by a reflected wave can easily be estimated for a constant-width or equilibrium/near-equilibrium estuary. In addition, compact, whole-estuary solutions are available for dead-end, non-covergent channels with very high or very low friction. When a significant reflected wave is present in a real estuary, however, it is important to recognize the differences between the length scales associated with along-channel changes in (a) the observed tidal phase and amplitude, (b) the theoretical incident wave, and (c) the theoretical reflected wave, all three of which may have distinct values for effective k and L_a^{-1}.

Finally, short and shallow channel solutions (see Sections 3.5 and 3.9 of the text) provide insights into how temporal variations in estuary depth and width produce tidal asymmetries in velocity and elevation. Tidal variations in estuary width act to distort the tide in two ways. First, a much greater total estuary width around high water (due to extensive intertidal flats and/or marsh, i.e., large $\Delta b/b$) causes more water to move in and out of the estuary around high tide than around low tide. This leads to a pulse of higher flood and ebb velocities around high water and a shortened period of high-water slack. Second, a much wider estuary around high water also slows the along-estuary propagation speed of high tide (c_{HT}) relative to low tide (c_{LT}). With $c_{HT} < c_{LT}$, up-estuary propagation of low water partially "catches up" with the previous high water, shortening the duration of the falling tide and, via continuity, favoring ebb-dominance. Conversely, a much deeper estuary around high water (due to a large value of a/h) reduces the magnitude of velocity around high tide because the cross-sectional area of the channel is larger then, and the duration of high-water slack is therefore increased. Second, a deeper estuary around high tide increases c_{HT} relative to c_{LT}. With $c_{HT} > c_{LT}$, high tide partially catches up with the previous low tide. The tide then rises more quickly, and flood dominance is favored.

Acknowledgments

Work on this chapter was supported in part by National Science Foundation Grant OCE-0536572. This is contribution no. 3005 from the Virginia Institute of Marine Science, College of William and Mary.

References

Boon, J. D. (1975) Tidal discharge asymmetry in a salt marsh drainage system. *Limnol. Oceanogr.* **20**, 71–80.

Bowers, D. G. and G. W. Lennon (1990) Tidal procession in a near-resonant system – a case study from South Australia. *Est. Coast. Shelf Sci.* **30**, 17–34.

Carter, H. H. and D. W. Pritchard (1988) Oceanography of Chesapeake Bay. In B. Kjerfve (ed.), *Hydrodynamics of Estuaries, 2. Estuarine Case Studies*. CRC Press, Boca Raton, Florida.

Friedrichs, C. T. (1995) Stability shear stress and equilibrium cross-sectional geometry of sheltered tidal channels. *J. Coastal Res.* **11**, 1062–1074.

Friedrichs, C. T. and D. G. Aubrey (1988) Non-linear tidal distortion in shallow well-mixed estuaries: a synthesis. *Est. Coast. Shelf Sci.* **27**, 521–545.

Friedrichs, C. T. and D. G. Aubrey (1994) Tidal propagation in strongly convergent channels. *J. Geophys. Res.* **99**, 3321–3336.
Friedrichs, C. T. and O. S. Madsen (1992) Nonlinear diffusion of the tidal signal in frictionally-dominated embayments. *J. Geophys. Res.* **97**, 5637–5650.
Friedrichs, C. T., B. A. Armbrust and H. E. de Swart (1998) Hydrodynamics and equilibrium sediment dynamics of shallow, funnel-shaped tidal estuaries. In J. Dronkers and M. Scheffers (eds), *Physics of Estuaries and Coastal Seas*. Balkema Press, Rotterdam, pp. 315–328.
Hunt, J. N. (1964) Tidal oscillations in estuaries. *Geophys. J. Roy. Astron. Soc.* **8**, 440–455.
Ippen, A. T. and D. R. F. Harleman (1966) Tidal dynamics in estuaries. In A. P. Ippen (ed.), *Estuary and Coastline Hydrodynamics*. McGraw-Hill, New York, pp. 493–545.
Jay, D. A. (1991) Green's Law revisited: tidal long-wave propagation in channels with strong topography. *J. Geophys. Res.* **96**, 20,585–20,598.
Lanzoni, S. and G. Seminara (1998) On tide propagation in convergent estuaries. *J. Geophys. Res.* **103**, 30,793–30,812.
Le Floch, J. (1961) *Propagation de la marée dans l'estuaire de la Seine et en Seine-Maritime*. ScD thesis, University of Paris, 507 pp.
Nichols, M. M. and R. B. Biggs (1985) Estuaries. In R. A. Davis (ed.), *Sedimentary Environments*, 2nd edn. Springer-Verlag, New York, pp. 77–186.
NOAA (1985) *National Estuarine Inventory: Data atlas. Volume 1. Physical and Hydrologic Characteristics*. NOAA/NOS Strategic Assessment Branch, Rockville, MD, 103 pp.
Parker, B. B. (1991) The relative importance of the various nonlinear mechanisms in a wide range of tidal interactions. In B. B. Parker (ed.), *Tidal Hydrodynamics*. John Wiley & Sons, New York, pp. 237–268.
Pethick, J. S. (1980) Velocity surges and asymmetry in tidal channels. *Est. Coast. Mar. Sci.* **11**, 331–345.
Prandle, D. (2003) Relationships between tidal dynamics and bathymetry in strongly convergent channels. *J. Phys. Oceanogr.* **33**, 2738–2750.
Prandle, D. (2004) How tides and river flows determine estuarine bathymetries. *Progr. Oceanogr.* **61**, 1–26.
Robinson, I. S., L. Warren and J. F. Longbottom (1983) Sea-level fluctuations in the Fleet, an English tidal lagoon. *Est. Coast. Shelf Sci.* **16**, 651–668.
Savenije, H. H. G., M. Toffolon, J. Haas and J. M. Veling (2008) Analytical description of tidal dynamics in convergent estuaries. *J. Geophys. Res.* **113**, C10025, doi:10.1029/2007JC004408.
Savenije, H. H. G. and E. J. M. Veling (2005) Relation between tidal damping and wave celerity in estuaries. *J. Geophys. Res.* **110**, C04007, doi:10.1029/2004JC002278.
Spitz, Y. H. and J. M. Klinck (1998) Estimate of bottom and surface stress during a spring–neap tide cycle by dynamical assimilation of tide gauge observations in the Chesapeake Bay. *J. Geophys. Res.* **103**, 12,761–12,782.
Swanson, R. L. (1976) *Tides*. Marine EcoSystems Analysis Program, MESA New York Bight Atlas Monograph 4, New York, Sea Grant Institute, Albany, NY, 44 pp.

4

Estuarine variability

DAVID A. JAY
Portland State University

4.1. Introduction

The variability between estuaries, and within a single system over time, is staggering. In North America alone, there are thousands, with the number depending on how one counts. The classification of these many estuaries is discussed in Chapter 1. Here we provide some examples of time and space variability and describe it more formally for typical shallow estuaries in a simple analytical framework. The discussion considers only positive estuaries, where river flow plus precipitation exceeds evaporation, so that salinity decreases toward the head of the estuary. Estuaries in arid climates are considered in Chapter 9.

4.2. Examples of estuarine variability

Estuarine variability can be described as being either:

- Intratidal variability, that occurs at tidal frequencies of 12–25 hours or on even shorter time scales. The diurnal (daily) and semidiurnal (twice-daily) astronomical tides and their "overtides" (circulation driven by non-linear processes and occurring at sums or multiples of the basic astronomical frequencies) are the most obvious examples. Also in this category are variations in scalar properties (e.g., salinity, temperature and density) driven directly by tidal currents, the effects of a daily sea breeze, harbor seiches with periods of minutes to hours, internal waves, inertial motion at the local pendulum frequency (periods of 12–20 hours at mid-latitudes) caused by impulsive wind forcing (large estuaries only), and variations in currents and scalar properties driven by tidal variations in vertical mixing and the along-channel density gradient. This last category includes the related topics of tidal straining, internal asymmetry, and strain-induced periodic stratification.
- Subtidal variability, that occurs at frequencies of < 1 cycle/day. In this category are low-frequency tidal motions with periods of 13–15 days and at 27–31 days, variability related to weather systems (typical periods 3–10 days), the response to rapid changes in river flow, e.g., floods related to severe storms and rapid snow-melt events, and

changes in the mean circulation related to the annual river flow cycle and the balance of evaporation vs precipitation and inflow. The term "mean circulation" (or "residual circulation") covers a wide variety of processes that are often difficult to distinguish. These include gravitational circulation (Chapter 2), wind-driven circulation (Chapter 6), internal asymmetry, non-linear tidal motions, the Stokes drift, and the Stokes compensation flow. Internal asymmetry and non-linear tidal circulation are non-linear; thus, they drive both subtidal and overtide motions. They are discussed later in the chapter.

Variations in the likelihood and strength of these diverse motions contribute to the diversity of estuaries, and the importance of each type of event is related to the estuarine setting. Tides, for example, are universal, though much stronger in some systems than others. Other processes are much more regionally variable. In estuaries of the Gulf of Mexico, hurricanes (and the very high flows caused by the heavy rains they bring) occur with some frequency, and snow-melt flood events are uncommon. Precipitation intensity generally decreases to the north, but other weather processes still cause high-flow events. Estuaries along the northwest coast of the USA are rarely affected by hurricanes, but floods caused by rain-on-snow events occur with some frequency.

The following sections consider tidal and subtidal variability, and then internal asymmetry and non-linear tidal motion. Internal asymmetry and non-linear tidal motion are considered separately because they contribute to both intratidal and subtidal variability. In order to present a relatively coherent picture of estuarine variability, most of the descriptive material presented below is taken from observations in one system, the Columbia River estuary.

4.2.1 Intratidal variability

The dominant circulation of most temperate, shallow to moderate-depth estuaries is driven by the tides acting at periods of 12–13 and 23–25 hours. Current patterns are complex, with strong vertical and lateral variability. Among the factors that contribute to this variability are topography and density stratification, and variations in vertical mixing related to both. Typical examples of lateral (Fig. 4.1a) and vertical (Fig. 4.1b) current variability are shown for the Columbia River estuary. These patterns are "unique" in the sense that they would not be duplicated precisely in any other system, but "typical" in that the phenomena shown are seen in many estuaries. As in many narrow estuaries and tidal rivers, the tidal currents are nearly rectilinear or reversing, with the major axis oriented nearly along the channel. Lateral currents are much weaker but not unimportant. They influence, for example, lateral density variations and the lateral variability of along-channel flow.

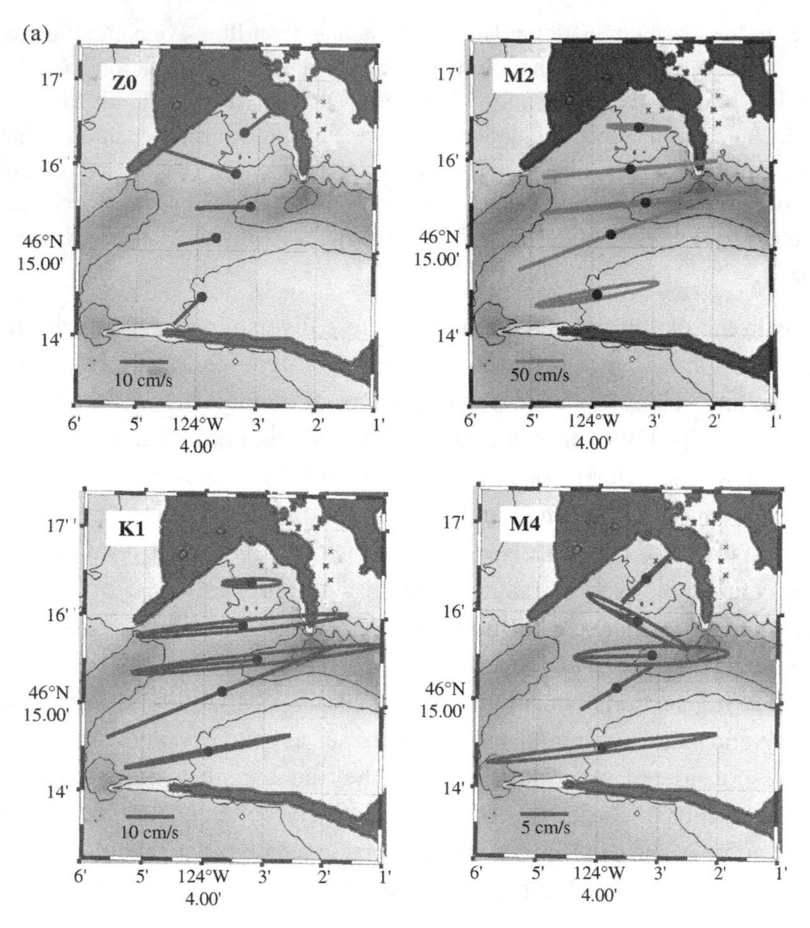

Figure 4.1(a) Lateral variability of estuarine currents: vertically averaged currents at the mouth of the Columbia River estuary for the mean flow (Z_0), the largest diurnal tidal constituent K_1, the largest semidiurnal tidal constituent M_2, and M_4, the first non-linear overtide of M_2. The ocean is to the left, the river to the right. The major axis of the ellipse of each tidal current indicates the predominant direction of each component, while the minor axis indicates the strength of the currents normal to the major axis. The data were collected with acoustic Doppler current profilers in place for the month of August 2005. The stations are numbered 1 to 5 from north to south. Bed depth is indicated by contours and shading, with light shading indicating deeper water; figure courtesy of Ed Zaron.

Figure 4.1a shows vertically averaged flow conditions near the mouth of the Columbia River estuary during a summer upwelling period of weak winds and low river flow. The flow over the month of August 2005 has been divided into components by frequency (mean, K_1, M_2, and M_4), using tidal harmonic analysis (Pawlowicz *et al.*, 2002). In Fig. 4.1a, the lateral distribution of flow is strongly controlled by the channel configuration (water depth is indicated in Fig. 4.1a) and

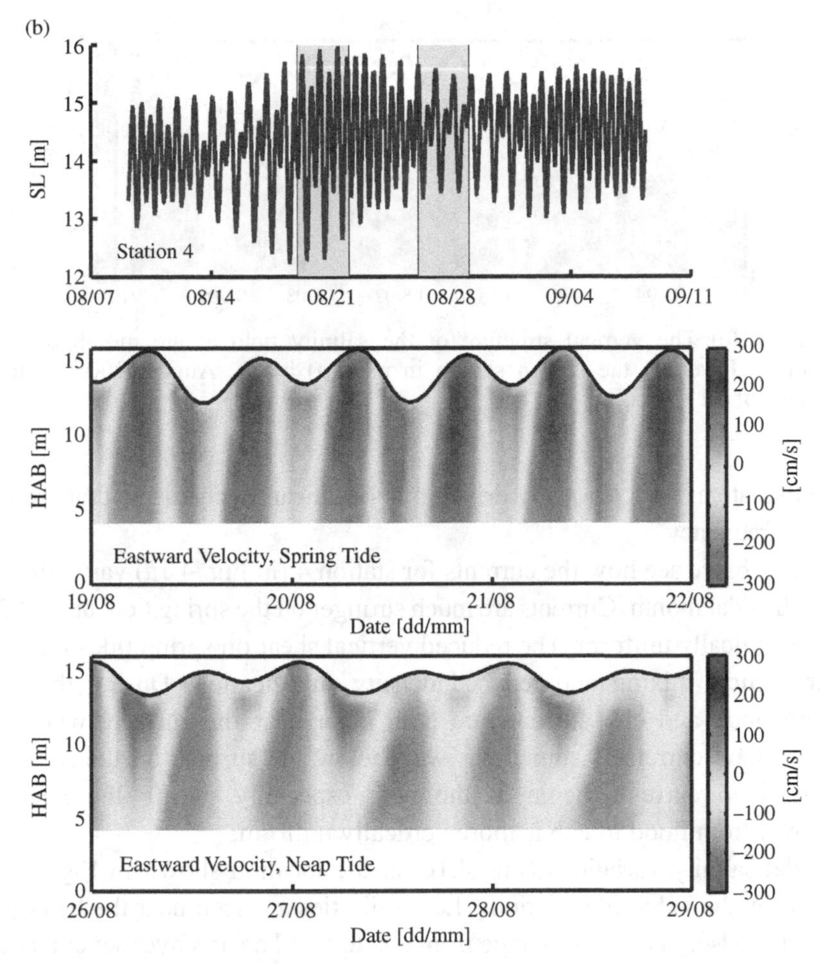

Figure 4.1(b) The vertical structure of currents at the mouth of the Columbia River, at station 4 [shown in part (a)], August 2005. Top: surface elevation, with the spring tide period and the neap tide period. Middle: currents for a three-day period near a spring tide. Bottom: currents for a three-day period near a neap tide; figure courtesy of Ed Zaron.

three stone jetties, one each to the north and south of a deep navigation channel, and one extending south from the north side of the channel at ~124° 2.5′ W (Jetty A). It is also influenced by the density stratification (Fig. 4.1c) and consequent variations in vertical mixing. The westward tidal flow on ebb diverges laterally after it passes a constriction at Jetty A, while the eastward flood flow must converge strongly to pass through this constriction. The river causes the mean flow to be seaward at all locations, except at the northernmost station in an eddy close to shore – this area is known as Deadman's Cove for its ability to trap floatables. The quarterdiurnal (M_4) flow is more convergent than that for K_1, M_2

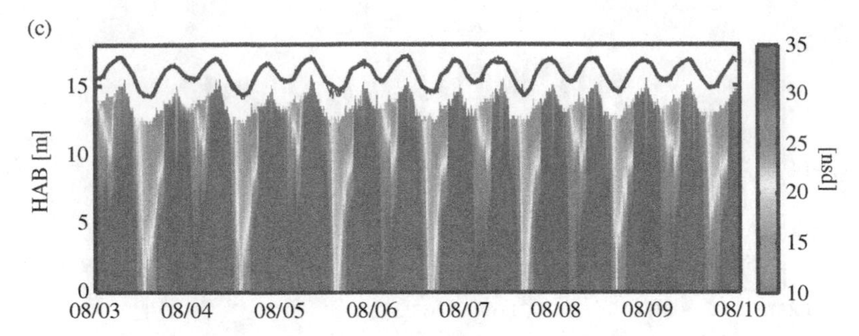

Figure 4.1(c) The vertical structure of the salinity field at the mouth of the Columbia River, at the station shown in part (a) by ⊗, August 2005; figure courtesy of Ed Zaron.

because the lateral flow is mostly driven by non-linear mechanisms that are strong at the M_4 frequency.

In Fig. 4.1b, we see how the currents for station 4 (in Fig. 4.1a) vary with depth and over the tidal month. Currents are much stronger on the spring tide, up to 2.5 m/s, and more vertically uniform. The reduced vertical shear on spring tides reflects the stronger vertical mixing and reduced density stratification. During the spring, maximum flood and ebb currents are both at or near the surface, whereas the maximum flood current is sometimes well below the surface on the neap. Note also that flood currents begin at the bed, especially during the neap. The change-over from flood to ebb is more vertically uniform.

The tidal salinity variations (Fig. 4.1c) at the location shown in Fig. 4.1b are influenced by the velocity variations. Low salinities are seen near the bed only on greater ebbs. Also, just as the change from ebb to flood occurs over several hours in Fig. 4.1b, we see in Fig. 4.1c that the flood onset near the bed after strong ebbs causes a period of high stratification, because seawater is moving landward near the bed while estuarine water is still moving seaward near the surface.

4.2.2 Subtidal variability

The estuarine mean flow varies with the river flow on tidal monthly and seasonal time scales. In order to understand this variability, some averaging is needed. Instead of viewing the actual time evolution of the flow, we examine its frequency structure, considering separately the mean flow and the diurnal (K_1) and semidiurnal (M_2) tides, based on a series of short tidal analyses, each of about 1 week's duration.

In Fig. 4.2a, we see the mean flow in a tidal channel of the Columbia River estuary over a 7.5-month period during a very high-flow year (1997). Between d125

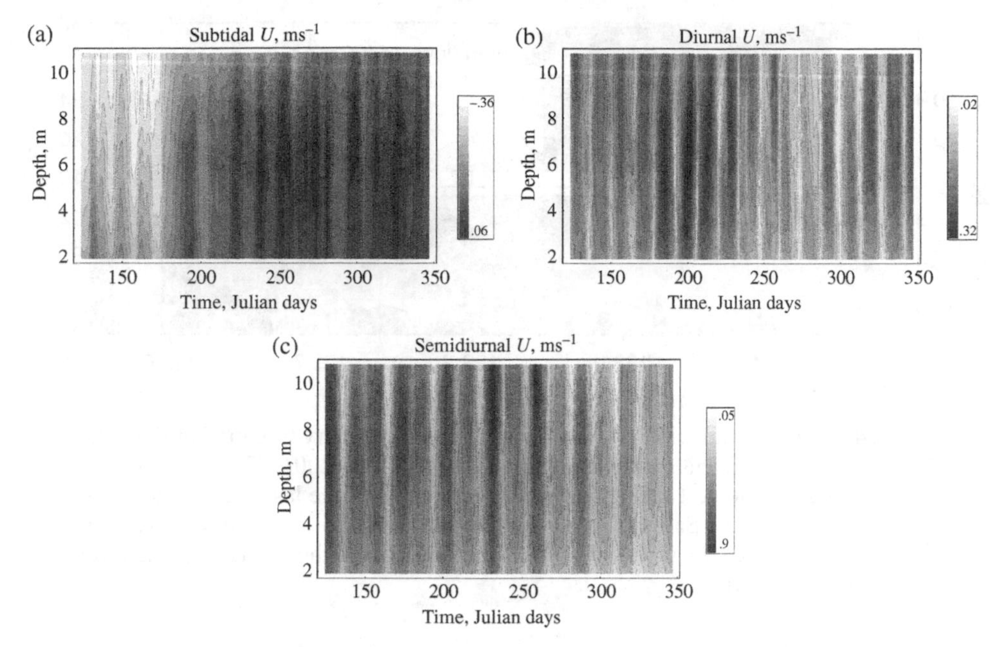

Figure 4.2. Top to bottom: mean flow, K_1, and M_2 along-channel currents as a function of depth (vertical axis) and time (Julian days, 1997) at a location in the navigation channel of the Columbia River about 15 km from the ocean. The year 1997 had very high spring flows that peaked before day 180.

and 175, the outflow is very strong, and very little two-layer flow is seen – upstream bottom flow is present only during the neaps, e.g., around d135 and d150. By d200, river inflow has decreased considerably, surface outflow is weak, and upstream bottom flow extends within 2–4 m of the surface on neap tides. Semidiurnal (M_2) currents are strong on spring tides throughout the record, but are strongest before d275. The seasonal variability of the M_2 currents is sufficiently large that we suspect an interaction with the mean flow – this ADCP was located at a station on the outside of a major bend in the channel. It appears that the strong river outflow early in the record shifted the lateral location of maximum outflow toward the ADCP, resulting in generally stronger currents during the high-flow (freshet) season. On the other hand, diurnal (K_1) currents are strongest below the surface at the end of the freshet season. The reason for this is related to the mixed tides that prevail on the west coast of the USA. As the flow decreased from its seasonal maximum, the salinity distribution adjusted landward. During intermediate flows between d180 and d230, all the salt was removed on most larger ebbs but not on lesser ebbs, resulting in one strong and one weak ebb throughout most of the water column, and strong diurnal currents. This phenomenon, related both to tidal properties and the river flow level, recurred in a weaker form after d320.

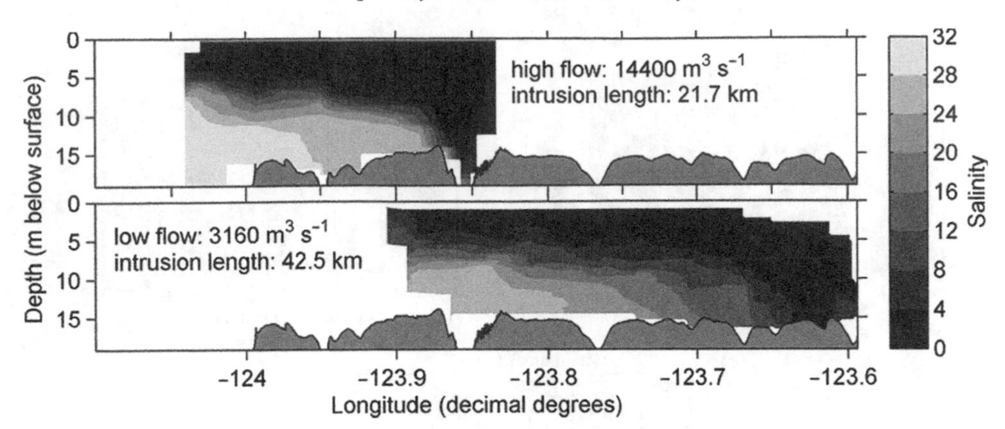

Figure 4.3. Salinity intrusion into the Columbia River estuary (see map in Fig. 4.2) during two very different spring freshet periods, 1997 and 2001. These correspond to the highest and lowest spring freshets respectively, of the last decade; tides are weak in both cases. Salinity intrusion doubles with a fourfold decrease in river flow; distance is measured from the mouth of the estuary (based on Chawla *et al.*, 2008).

Another major change related to river flow variability is the length of the salinity intrusion into the estuary. In Fig. 4.3, we can see that an ~fourfold decrease in flow resulted in an approximate doubling of salinity intrusion length into the Columbia estuary. (Note that the estuary, defined to extend to the head of the tide at Bonneville Dam, is 245 km long. Salinity rarely extends >50 km into the system.) This suggests that salinity intrusion varies, in this system, with the square root of river flow Q_R as Q_R^{-n}, with $n = ½$. For reasons that remain incompletely understood, the power n of the salinity dependence on flow varies substantially between systems from n =~1/7 in San Francisco Bay (Monismith *et al.*, 2002) to n =~1 in other systems (see also Chapter 2).

Tidal range also strongly influences salinity intrusion length. Intuitively, one would expect that salinity intrusion length would increase on spring tides, on the grounds that stronger tidal currents should cause greater advection of the salinity field. If the stratification is either very strong or very weak, and there are no large neap–spring changes in stratification, this expectation is fulfilled. Many estuaries are, however, in an intermediate condition much of the year – the stronger vertical mixing during spring tides reduces density stratification, allowing higher dissipation throughout the water column. The higher mixing levels actually inhibit salinity advection, and salinity intrusion is stronger on neap throughout much of the annual flow cycle, as seen in Fig. 4.4 for the Columbia River estuary (from Jay and Smith, 1990a, for low-flow conditions).

The importance of atmospheric forcing is highly variable between estuaries. In general, wide estuaries with low buoyancy input rates are more strongly affected

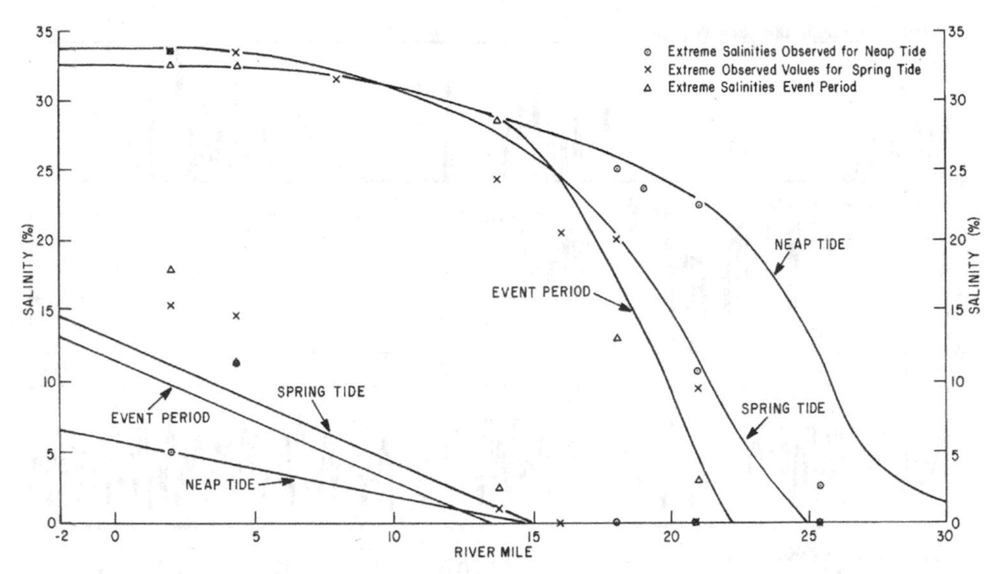

Figure 4.4. Example of neap vs spring salinity intrusion into the main navigation of the Columbia River estuary, for low-flow conditions (3500–4000 m^3/s). Stratification and maximum salinity intrusion both increase on the neap, despite a smaller tidal excursion. Minimum intrusion length changes rather little for this flow level. Also shown is the result of an adjustment to an increase in flow of brief duration (the "event period"). Because the residence time of the system is only 3–4 days for this flow level, a rapid adjustment to the flow change is seen, and the response to the neap–spring cycle is almost in phase with the tidal forcing.

than narrow estuaries and river estuaries with high buoyancy inputs. We do not expect atmospheric forcing, therefore, to be of first-order importance in the Columbia River estuary, with its strong river flow. Still, the impacts of atmospheric forcing can be discerned, and they are surprisingly complex, because the main role of atmospheric forcing is to change the water masses that enter the estuary from the ocean, and continental shelf processes are complicated and highly three-dimensional.

The impact of winds on salinity intrusion can be seen in a 3-year, near-bed record of daily maximum salinity taken along the north side of the estuary (Fig. 4.5, based on Chawla *et al.*, 2008). This station is ~40 km from the estuary mouth. Salinity intrusion occurs primarily during low-flow neap-tide periods (as a result of enhanced upstream near-bed flow), but wind forcing modifies the pattern. There is also, however, a subtle decrease in salinity intrusion during the winter stormy periods. This occurs during the winters of 2001–2002 and 2003–2004. We see that salinity drops during downwelling periods (winds to the north) as winds to the north decrease the offshore salinity at the depth of the entrance sill, ~18–20 m. Prolonged strong downwelling winds have a contrary effect. The recovery in salinity

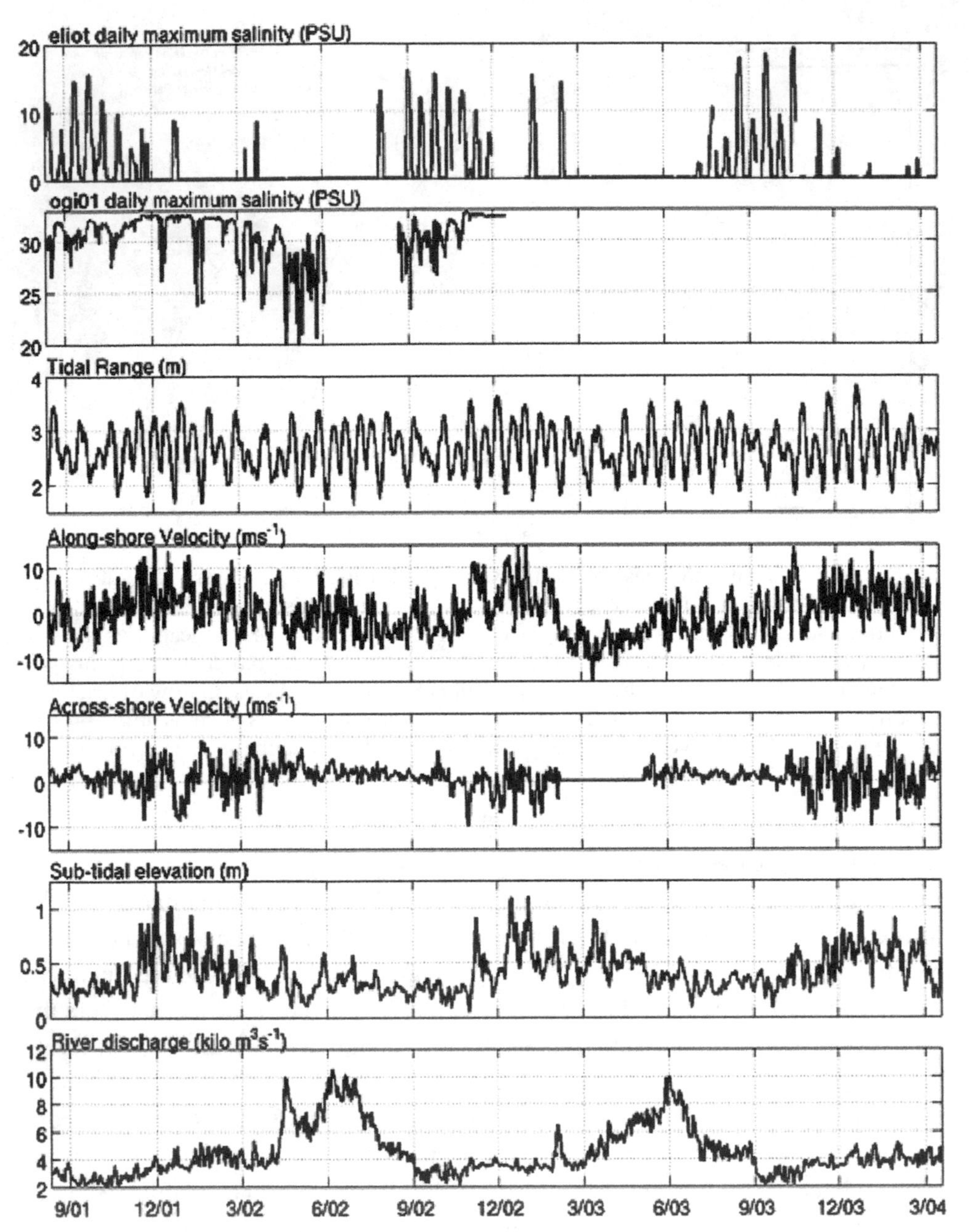

Figure 4.5. Time series of (top to bottom) daily maximum salinity at 13.6 m at a station about 25 km from the ocean in the Columbia River estuary, daily minimum surface salinity offshore of the estuary mouth, along-shore (positive to the south) and onshore (positive to the west) wind components, subtidal elevation and river discharge. The time series extend from September 2001 through March 2004.

during two neap tides in late December 2002 and early January 2003, even under strong winds, is probably due to persistent strong downwelling. Under these conditions, the freshwater that emerges from the estuary to form a buoyant plume is pushed along-shore and is unable to move back toward the mouth of the estuary. Some of this is seen in the time series of offshore daily minimum salinity. Unfortunately, the offshore mooring was lost during a winter storm in December 2002.

4.2.3 Non-linear tidal variability and internal asymmetry

Below, we will formally describe the residual flow as a sum of linearly separable modes, each related to a particular type of forcing. Here, we describe the various types of residual flow that derive from quadratic non-linearities in the tidal mass and momentum conservation equations. Just as in the vertically integrated tidal equations (Chapter 3) and depth-dependent tidal equations (Chapter 6), a quadratic non-linearity will transfer tidal energy to both tidal and overtide frequencies. Thus, each type of residual flow described here has an overtide counterpart, the distribution of which may be useful in separating the various residual modes. Let us begin with internal tidal asymmetry.

Internal tidal asymmetry is driven by correlations between tidal shear and tidal vertical mixing. The velocity (and its vertical shear) and eddy viscosity have time-averaged, tidal, and overtide components:

$$
\begin{aligned}
U(\vec{x},t) &= \bar{U}(\vec{x}) + U_T(\vec{x},t) + U_{OT}(\vec{x},t),\\
A_z(\vec{x},t) &= \overline{A}_z(\vec{x}) + A_{zT}(\vec{x},t) + A_{zOT}(\vec{x},t),
\end{aligned}
\tag{4.1a}
$$

where an overbar indicates a time average, and the subscripts "*T*" and "*OT*" indicate tidal and overtide-frequency terms, respectively. Thus, the stress in the along-channel momentum equation may be written:

$$
\begin{aligned}
\textit{Turbulent stress} &= A_z\frac{\partial U}{\partial z} = \left(\overline{A}_z + A_{zT} + A_{zOT}\right)\left(\frac{\partial \bar{U}}{\partial z} + \frac{\partial U_T}{\partial z} + \frac{\partial U_{OT}}{\partial z}\right)\\
&\approx \overline{A}_z\frac{\partial \bar{U}}{\partial z} + \left(A_{zT}\frac{\partial \bar{U}}{\partial z} + \overline{A}_z\frac{\partial \bar{U}_T}{\partial z}\right) + A_{zT}\frac{\partial U_T}{\partial z} + \overline{A}_z\frac{\partial U_{OT}}{\partial z} + \ldots
\end{aligned}
\tag{4.1b}
$$

(For simplicity, terms involving the overtide eddy diffusivity A_{zOT} are neglected until we have a means to systematically reject small terms.) The first term on the last line of equation (4.1b) will contribute only to the mean circulation, the two terms in parentheses only to the tidal circulation, the third term to both the residual and overtide circulations, and the final term only to the overtide circulation. The mean stress is then given by:

$$Mean\ turbulent\ stress = \overline{A}_z \frac{\partial \bar{U}}{\partial z} + \left\langle A_{zT} \frac{\partial U_T}{\partial z} \right\rangle, \tag{4.2}$$

where the angle brackets indicate the tidal average of a quadratic term. Thus, the residual stress has two contributions, the first stemming from the product of the mean eddy viscosity and mean shear, and the second being the time average of the product of the tidally varying shear and eddy viscosity. Complicating the interpretation of equation (4.2) is the fact that the mean shear in the first term on the r.h.s. of equation (4.2) contains contributions from all the types of mean flow, linear and non-linear. Assuming, however, that all of the mean-flow modes are small, we can neglect their interactions. Thus, only the internal asymmetry velocity appears in the stress term in the equation for the internal asymmetry velocity. Like gravitational circulation, internal asymmetry is an internal mode (integrating to zero over the cross-section) that has only a small effect on the surface elevation.

Why does internal asymmetry occur? The same essential physics have been described in several ways. The essence of the matter is that, given a stratified, positive estuary, the tidal currents reverse tidally (from flood to ebb and back again), but the density gradient never changes sign – the density is always higher on the ocean side. Consider the conceptual sketch in Fig. 4.6. Flood currents advect surface waters more strongly than near-bed waters, making the density contours more nearly vertical and reducing the stratification $\partial\rho/\partial z$. Also, tidal currents are usually stronger at the ocean entrance of an estuary and decrease landward. The more seaward salinity contours are advected further than the more landward ones, compressing the density field and increasing $\partial\rho/\partial x$. Ebb currents advect the surface waters and the more oceanic density contours more strongly, increasing $\partial\rho/\partial z$ and decreasing $\partial\rho/\partial x$. This tidal alteration of the density field and stratification has been called "tidal straining" (Simpson *et al.*, 1991). Recognizing that the periods of high stratification are periodic, and that vertical mixing is reduced during slack water (allowing some restratification), this process has also been called "strain-induced periodic stratification" or SIPS (Stacey *et al.*, 1999).

Now, let us consider what happens to the velocity field in a tidal flow with tidal straining or SIPS (Fig. 4.7, from Jay, 1991). The eddy viscosity is higher on flood than on ebb, because the stratification inhibits vertical mixing more on ebb than on flood. Given a relatively constant horizontal density gradient over the tidal cycle, the flood and ebb distributions of the tidal flows are different. On flood, the barotropic ($-g\ \partial\eta/\partial x$) and baroclinic $(-g/\rho_0)\int_z^{\eta}\frac{\partial\rho}{\partial x}dz'$ components of the pressure gradient $-1/\rho_0\partial P/\partial x$ are both >0 and reinforce one another at depth. On ebb, $-g\ \partial\eta/\partial x < 0$, while the baroclinic pressure gradient remains positive, so the two pressure gradient components oppose one another at depth. The nearly constant (in time), positive baroclinic pressure gradient means that the near-bed flow will be relatively more

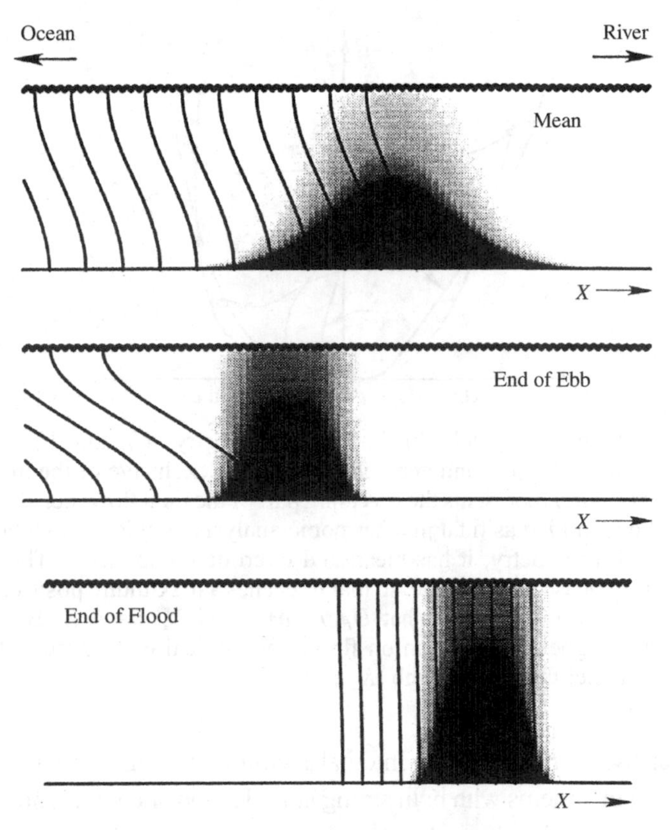

Figure 4.6. Conceptual sketch of tidal straining, and its influence on an estuarine turbidity maximum. Salinity is shown as contours, suspended particulate matter as shading. The density contours are much more nearly vertical at the end of flood than at the end of ebb, allowing considerable resuspension at and behind the head of the advancing salinity intrusion. Within the salinity intruded part of the estuary, vertical mixing is considerably inhibited on ebb. However, resuspension occurs on ebb landward of salinity intrusion, where bed stress is high. The mean salinity field is intermediate between the flood and ebb extremes, and the mean sediment concentration is smeared out, relative to flood and ebb.

landward than the surface flow on both flood and ebb. This results in a nearly uniform flow on flood (or perhaps even a stronger flow near the bed) and a highly sheared ebb flow, as shown in Fig. 4.7.

It is important to realize that the vertically averaged tidal flow integrates to zero over a tidal cycle – as a wave motion must. However, the vertical distributions of the flood and ebb flows are quite different. The net result is that, at any given depth, there are residual and overtide flows. For an M_2 tide, a mean flow, M_4 and M_8 are generated. For a mixed tide, a wide variety of overtides appear. This process is known as "internal tidal asymmetry" (Jay, 1991; Jay and Musiak, 1996). Its

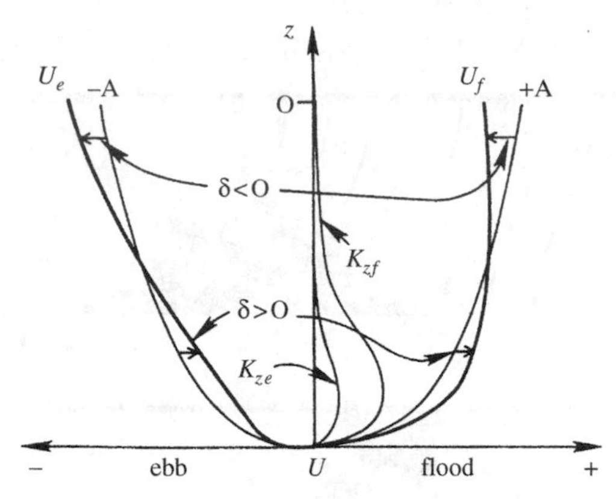

Figure 4.7. Definition sketch for internal asymmetry. U_f and U_e represent, respectively, the peak flood and ebb velocity profiles exclusive of the mean flow; from Jay (1991). $A(z)$ represents the reversing part of the tidal flow; i.e., that portion that would be identified as tidal in a harmonic analysis. $\delta(z)$ is the two-layer flow due to internal asymmetry; it has mean and overtide components. The overtide component is in phase with the tide, so that it reaches a maximum, positive value at peak flood and peak ebb. Note that $U_f(z) = \delta + A$ and $U_e(z) = \delta - A$. Because the stratification is higher on ebb than on flood, the vertical eddy diffusivity K_z on flood (K_{zf}) is higher than that on ebb (K_{ze}).

dynamics are analyzed in Section 4.6. Internal asymmetry occurs in many estuaries, but is most important in systems with both strong stratification and strong tides, especially, therefore, in river estuaries. In such systems, there is strong advection of the density contours over the tidal cycle and, therefore, strongly variable stratification.

Like internal asymmetry, non-linear tidal circulation comes out of a quadratic non-linear term in the momentum equation (Ianniello, 1977; see also Chapter 6). The analogue to equation (4.2) is:

$$Mean\ tidal\ stress = \left\langle U_T \frac{\partial U_T}{\partial x} \right\rangle + \left\langle W_T \frac{\partial U_T}{\partial z} \right\rangle, \tag{4.3a}$$

where W_T is the vertical tidal velocity. As with the turbulent stress, there is also a corresponding overtide term. Like the mean turbulent stress in equation (4.2), the mean tidal stress in equation (4.3a) will drive a mean flow mode.

Another tidal non-linearity, the Stokes drift (and the mean flow that compensates for it) can be seen in the vertically integrated continuity conservation equation, averaged over a tidal cycle:

$$\int_{-h}^{0} \bar{U}_{ST}\, dz + \left\langle U_T|_{\zeta}\, \eta_T \right\rangle = 0, \tag{4.3b}$$

where the angle brackets $\langle \; \rangle$ indicate a tidal average, $\bar{U}_{ST}$ is the Eulerian Stokes compensation flow, and the second term on the l.h.s. is the Stokes drift. Whenever the ratio $\varepsilon_T = \eta / H$ of tidal amplitude η to depth H is finite, and there is a correlation between tidal elevation and tidal current, there will be a Lagrangian net landward transport due to the tides. That is, even if the current is perfectly reversing at each depth, the flood flow is deeper than the ebb flow and moves mass landward. This Lagrangian transport of water is compensated by a return flow set up by the surface slope that the Stokes drift sets up. See also Chapter 6.

In shallow systems with strong tides and weak stratification, these tidally induced circulations are known to be important. There is little evidence that they are important in deeper stratified systems over a level bed. Intuitively, however, they seem likely to be very important in a stratified system in the vicinity of topographic features, e.g., in the lee of a sill or a large sand ridge.

4.3. A formal analysis of estuarine circulation

Here we provide a formal description of the subtidal, tidal, and overtide variability in estuarine circulation and develop the appropriate along-channel momentum and mass conservation equations to describe it. Because estuarine circulation is complex and influenced by many factors, we need an analytical theory that is flexible and incorporates the following factors:

- Variations in external forcing by ocean tides, river flow, winds, and atmospheric pressure.
- A realistic representation of both the tidal and mean velocity and salinity fields.
- The response of the estuary in terms of velocity structure, salinity intrusion length, stratification, and vertical mixing.
- Simple variations in topography, like width convergence or landward shoaling.

The theory presented here includes the following factors:

- External forcing is specified in terms of variations in river flow, winds, and tidal forcing at the ocean end of the system.
- Vertical and temporal variations in vertical mixing are specified parametrically through an eddy viscosity.
- Tidal, residual, and overtide circulation modes are described within a formal perturbation structure.

The following assumptions are necessary:

- The mean density gradient $\partial \bar{\rho} / \partial x$ is spatially uniform and $\partial \bar{\rho} / \partial x < 0$ (x is positive landward), i.e., the estuary is positive. Physically, this is the "central regime" of Hansen and Rattray (1965), a situation typical of partially mixed and weakly stratified estuaries.
- The estuary is shallow (<~20 m deep or so), "strongly forced" such that the bottom boundary layer extends throughout most of the water column for most of the tidal cycle.

- The estuary is narrow such that the external Kelvin number (width B over external Rossby radius $\frac{\sqrt{gH}}{f}$) is $\ll 1$.
- The tidal and overtide solutions are "non-accelerated", meaning that the acceleration term is either neglected or included as a small correction. This restricts analysis to strongly forced systems of moderate depth.
- The channel is narrow and the various flow mechanisms are laterally uniform. All flow properties are functions of x, z, and t only.
- The channel is of uniform width and depth. (This is the simplest of the restrictions to relax.)
- The tide consists only of a landward propagating wave with no reflected wave. This is consistent with convergent topography as well as the uniform channel considered here, as long as the channel is dissipative and there are no sharp constrictions.
- The model is harmonic, so tidal monthly evolution of the flow is described in terms of harmonic tidal constituents.

The formal structure of the analysis is called a "perturbation expansion". In this approach to approximating complex physical processes, one process (in this case, the tidal circulation) is considered to be the dominant [denoted "order one" or "$O(1)$"] factor controlling or influencing the other smaller processes (in this case overtide currents, gravitational circulation, internal asymmetry, and the other residual flow modes). Each smaller process [denoted "order epsilon" or "$O(\varepsilon_{sub})$", with a subscript "*sub*" to identify the driving process] is affected by the tidal circulation through vertical mixing and (in some cases) non-linear driving terms. The smaller circulation modes are assumed to be independent of one another and do not affect the larger tidal circulation (see Chapter 6 for another application of a perturbation analysis).

The following circulation modes are considered, though not all are developed in this chapter:

- A lowest-order tidal velocity $U_T(x,z,t)$ forced by the ocean tide.
- Internal circulation modes that integrate to zero:
  - An $O(\varepsilon_I)$ overtide velocity $U_{IA}(x,z,t)$ driven by internal asymmetry.
  - An $O(\varepsilon_I)$ residual circulation $U_{IA}(x,z)$ due to internal asymmetry.
  - An $O(\varepsilon_i)$ residual gravitational circulation $U_G(z)$ driven by the mean horizontal density gradient.
  - An $O(\varepsilon_T)$ residual tidal non-linear circulation $U_{NL}(x,z)$ driven by the convective accelerations.
  - An $O(\varepsilon_w)$ residual wind-driven circulation $U_W(z)$; see also Chapter 6.
- External circulation modes that do not integrate to zero:
  - An $O(\varepsilon_R)$ mean river flow $U_R(z)$.
  - An $O(\varepsilon_T)$ residual Stokes compensation flow $U_{ST}(x,z)$.

We neglect external overtides as small (relative to internal overtides) in the lower reaches of river estuaries, though this mode could easily be added.

The resulting velocity field is written as:

$$U(x,z,t) = U_T(x,z,t) + \varepsilon_R \bar{U}_R(z) + \varepsilon_i \bar{U}_G(z) + \varepsilon_I(\bar{U}_{IA}(x,z) + U_{IA}(x,z,t)) + \varepsilon_T(\bar{U}_{NL}(x,z) + \bar{U}_{ST}(x,z)) + \varepsilon_W \bar{U}_W(z). \quad (4.4)$$

The following salinity modes are considered:

- A lowest-order background (central regime) salinity field $\bar{S}(x) = S_0(1 - x/L_S)$ that is a function of x only and decreases linearly from the ocean landward.
- A lowest-order tidally advected and vertically uniform regime that is exactly out of phase with the velocity field, $S_T(x,t)$. While $S_T(x,t)$ is the dominant part of the tidal salinity field, it does not contribute to the mean stratification, because it is uncorrelated with velocity.
- An $O(\varepsilon_I)$ tidal mode $S_{TZ}(x,z,t)$ due to shear advection and vertical mixing; this mode contributes to the mean stratification because it is not in quadrature with the velocity field.
- The $O(\varepsilon_i)$ $s_G(z)$ mean stratification or density defect due to the combined effects of tidal and residual advection of the salinity field. This is the component of the salinity field commonly associated with gravitational circulation, but all parts of the residual flow contribute to its maintenance.

The salinity and density fields are represented as:

$$S(x,z,t) = S_0\left(1 - \frac{x}{L_S}\right) + S_T(x,t) + \varepsilon_I S_{TZ}(x,z,t) + \varepsilon_i s_G(z), \quad (4.5a)$$

$$\rho(x,z,t) = \rho_0 + \Delta\rho_H\left(1 - \frac{x}{L_S}\right) + \rho_T(x,t) + \varepsilon_I \rho_{TZ}(x,z,t) + \varepsilon_i \rho_G(z). \quad (4.5b)$$

Here we only analyze the lowest-order part of the salinity and density fields, i.e., the components that are vertically uniform.

Crucial to modeling the velocity field is the representation of the eddy viscosity A_z, because it is the vertical mixing that structures estuarine circulation. The actual time-space distribution is complex and poorly known, because it is strongly influenced by topography, stratification, and shear. In general we think of estuarine mixing as driven by: (a) shear at bed, (b) interfacial shear and instabilities, and (c) winds at the surface. Here, a simple form of A_z is used that yields all the modes of estuarine circulation defined above. A_z is also simplified by assuming that it varies primarily with z and t. It seems logical to assume that A_z varies temporally about an average $\kappa_m(z)$, and that there are both tidal and overtide variations. The overtide variations A_{zOT} at twice the dominant tidal frequency are caused by alternation of periods of strong and weak currents – mixing is strong at peak flood and ebb but is weak at both slack waters. This causes A_{zOT} to be in phase with U_T. The tidal variations A_{zT} are caused by tidal straining at the dominant tidal frequency – but because shear is also important in setting A_{zT}, the phase difference φ between U_T and A_{zT} must be assumed or specified from data. Thus:

$$A_z(z,t) = \kappa_m(z) + \varepsilon_I A_{zT}(z,t) + (1-\varepsilon_I) A_{z0}(z,t)$$
$$\kappa_m(z)\left(1 + \varepsilon_I f_1(z)\, e^{i(t+\phi)} + (1-\varepsilon_I) f_2(x,z)\, e^{2it}\right). \tag{4.6a}$$

The factor of $(1-\varepsilon_I)$ in equation (4.6a) arises because there is not a 100% variation in A_z between periods of peak currents and slack water – there is always some bed stress (proportional to ε_I) produced by the residual flow. For simplicity, we represent the time-averaged eddy viscosity $\kappa_m(z)$ as:

$$\kappa_m(z) = ku_* z\,(H + a - z), \tag{4.6b}$$

where u_* is the shear velocity, $k = 0.41$ is van Karman's constant, and $a > 0$ is a small constant that prevents κ_m from vanishing at the free surface ($z = H$). While equation (4.6a) does not explicitly represent the impact on vertical mixing of density stratification, this is implicitly present in the functions f_1 and f_2.

4.4. The tidal circulation

We solve here the tidal frequency equations for conservation of mass and momentum to determine the tidal flow in two dimensions (x and z). The analysis builds on Ianniello (1977) and Jay and Smith (1990b). The motion is confined to a narrow channel of uniform width and depth, and the ratio ($\varepsilon = \zeta/H$) of tidal amplitude ζ to depth H is small, allowing the convective accelerations to be neglected. Without the convective accelerations, the lowest-order tidal frequency momentum equation is:

$$\frac{\partial U_T}{\partial t} = -g\frac{\partial \zeta_T}{\partial x} - \frac{g}{\rho_0}\int_z^{\zeta} \frac{\partial \rho_T}{\partial x} dz + \frac{\partial}{\partial z}\left(\kappa_m \frac{\partial U_T}{\partial z}\right), \tag{4.7}$$

where $\zeta_T(x,t)$ is tidal variation of surface elevation. Note that only the lowest order, vertically uniform tidal density variation ρ_T has been included. Other components of the total density $\rho(x,z,t)$ are either not at tidal frequency (and cannot contribute to a tidal frequency motion) or have been neglected because they are smaller than $O(1)$. The corresponding local and vertically integrated continuity equations for U_T are:

$$\frac{\partial U_T}{\partial x} + \frac{\partial W_T}{\partial z} = 0, \quad \frac{\partial}{\partial x}\int_{z_0}^{h} U_T dz + \frac{\partial \zeta_T}{\partial t} = 0. \tag{4.8a, b}$$

Note that there are three equations – (4.7) and (4.8a,b) – that must be solved to specify the three unknowns: ζ_T, U_T, and W_T.

We want to solve these equations in non-dimensional form so that the parameter dependence of the solution is clear. We first non-dimensionalize all the variables (those with hats):

$$U_T = U_0\hat{U}_T = \frac{u_*}{C_D{}^{1/122}}\hat{U}_T, \quad W_T = U_0\tfrac{H}{L_x}\hat{U}_T, \quad \zeta_T = \zeta_0\hat{\zeta}_T$$
$$x = L_x\hat{x} = \frac{\varepsilon_T g H^2}{k u_* U_0}\hat{x}, \quad z = H\hat{z}, \quad t = \tfrac{\hat{t}}{\omega}$$
$$\varepsilon_T = \frac{\zeta_0}{H}, \quad \kappa_m = k u_* H \hat{\mathrm{K}}_m, \quad \frac{\partial \rho_T}{\partial x} = \frac{\Delta\rho_H}{L_S}\frac{\partial\hat{\rho}}{\partial\hat{x}}. \tag{4.9}$$

The value of the horizontal length scale L_x has been chosen so that the amplitudes of the surface slope and stress divergence term balance at times of peak current. At this time, acceleration is very small, and the baroclinic integral also vanishes, because ρ_T is out of phase with U_T – this causes the baroclinic integral to vanish at peak flood and ebb. Thus, at times of peak current, the surface slope and stress divergence are the only two significant terms. The tidal velocity scale U_0 has been related to the bed stress amplitude.

Dropping the hats on the non-dimensional variables, the scaled, non-dimensional equations are:

$$m^2\tfrac{\partial U_T}{\partial t} = -\tfrac{\partial \zeta_T}{\partial x} - \pi_1 \int_z^1 \tfrac{\partial \rho}{\partial x}dz + \tfrac{\partial}{\partial z}\left(\kappa_m \tfrac{\partial U_T}{\partial z}\right),$$
$$\tfrac{\partial U_T}{\partial x} + \tfrac{\partial W_T}{\partial z} = 0, \quad \tfrac{\partial}{\partial x}\int_{z_0}^1 U_T dz + m^2\tfrac{U_\lambda}{U_0}\tfrac{\partial \zeta_T}{\partial t} = 0, \tag{4.10a–c}$$

where $U_\lambda = \varepsilon_T(gH)^{1/2}$ is a velocity scale based on the tidal wave speed, and the non-dimensional mean depth is 1. In estuaries with strong friction, $U_\lambda > U_0$ and over-estimates the tidal velocity (which is controlled by bed friction).

There are two scaling parameters in equation (4.10):

$$m^2 = \frac{\omega H}{k u_*}, \quad \pi_1 = \frac{g H^2 \Delta\rho_H}{k u_* U_0 \rho_0 L_S} \tag{4.11}$$

(where ρ_0 is the constant, background density). The first parameter, m^2, is a ratio of acceleration to the stress divergence (loosely speaking, friction); it may also be thought of as a depth non-dimensionalized by the eddy diffusivity. Thus, the effective depth of a system depends on the strength of the vertical mixing. The second parameter, π_1, is the ratio of the baroclinic density forcing to the stress divergence, which varies with the horizontal density difference $\Delta\rho_H$ across the central regime and the square of depth H.

Solutions can be found for equation (4.10) using a separation of variables, assuming that all properties vary harmonically in time. Thus:

$$U_T = \mathrm{Re}[i\,M'[x]\;P[z]\;e^{it}], \quad \zeta_T = \mathrm{Re}[M[x]\;e^{it}],$$
$$W_T = -\tfrac{H}{L_x}\mathrm{Re}\left[i\,M''[x]\int_{z0}^{z} P[z]\;dz'\;e^{it}\right]. \tag{4.11a–c}$$

These functional forms are motivated by the form of the continuity equations [(4.10a) and (4.10b)]. Physically, $P(z)$ is a velocity structure function, and $M(x)$ describes the propagating tidal wave.

Consider the momentum equation (4.10a), substituting the solution form (4.11a), with the real part assumed. This results in an ordinary differential equation with two (inhomogeneous) forcing terms on the r.h.s.:

$$(\mathrm{K}_m P')' - im^2 P = -i - \pi_1(1 - z). \tag{4.12}$$

The terms in equation (4.12) represent (from left to right) the stress divergence, the acceleration, the surface slope, and the baroclinic pressure gradient. Physically, equation (4.12) says that the tidal motion is forced by the surface slope and baroclinic pressure gradient, and that the motion accelerates if the stress divergence is not in balance with the pressure gradient.

The boundary conditions for solution of equation (4.12) are: (a) no slip at the bed (actually, at $z=z_0$, the scale roughness height) and (b) no stress at the mean free surface at $z=1$. Thus:

$$P|_{z_0} = 0, \qquad\qquad P'|_1 = 0. \tag{4.13a,b}$$

It is possible to solve equation (4.12) with equation (4.13) using the parabolic eddy diffusivity (4.6b). However, the solution, in terms of Legendre polynomials, is lengthy. We use physical reasoning to define a simpler problem. Specifically, we note that m^2 is typically much less than 1 and neglect the acceleration term. Considering a 10–15 m deep tidal channel with u_*= 0.03 m/s and M_2 tidal forcing, then m^2 = ~0.12–0.18. Neglecting the acceleration term in equation (4.12), the resulting ODE is:

$$(\mathrm{K}_m P')' = -i - \pi_1(1 - z). \tag{4.14}$$

.

Because equation (4.14) neglects acceleration, we refer to it (and to the solutions for other components of the tidal flow that use the same approximation) as "non-accelerated".

Equation (4.14) can be integrated twice, with boundary conditions equation (4.13); the resulting solution is:

$$P_{full}(z) = \frac{i}{(1+a)} \left(\log \left[\frac{\left(\frac{z}{z_0}\right)^{1+\frac{\pi_1}{2}}}{\left(\frac{1+a-z_0}{1+a-z}\right)^{\left(a-a^2\frac{\pi_1}{2}\right)}} \right] - \frac{\pi_1}{2}(1+a)\,(z - z_0) \right), \tag{4.15a}$$

where log indicates the natural log function. Note that for $a = \pi_1 = 0$, we recover a log profile $P_{log} = i \log(z/z_0)$, the expected solution for a boundary layer, but multiplied by $i = -1^{1/2}$. Referring to equation (4.11a), we see that there is another factor of i in the solution for U_T. This solution to equation (4.14) differs slightly from the full solution to equation (4.12), but is much simpler! For simplicity in calculating non-linear flow modes, we will set $\pi_1 = 0$, yielding:

$$P(z) = \frac{i}{(1+a)} \left(\log \left[\frac{\left(\frac{z}{z_0}\right)}{\left(\frac{1+a-z_0}{1+a-z}\right)^a} \right] \right). \tag{4.15b}$$

Figure 4.8 shows both the full solution and the solution for $\pi_1 = 0$. Note that the presence of a baroclinic pressure gradient causes more shear close to the bed, so that

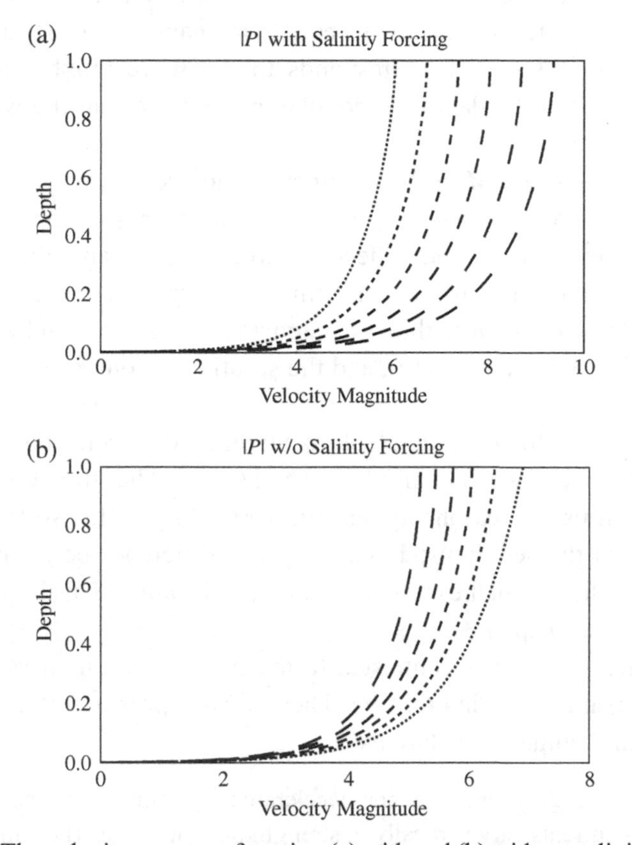

Figure 4.8. The velocity structure function (a) with and (b) without salinity forcing. In (a), $a = 0.1$ and π_1 varies from 0 to 1 (left to right). In (b), $\pi_1 = 0$ and a varies from 0 to 0.25 (left to right). The absolute value of P is not important, because this is normalized through the horizontal solution (below), but the vertical distribution is critical.

the solution in the outer part of the flow is more vertically uniform. For any given value of a, $\pi_1 > 0$ causes P to be larger near the bed than it is without density forcing. Completing the solution for U_T requires solving (equation 4.10c). Substituting the functional forms for U_T and ζ_T specified by equations (4.11a) and (4.11b) yields a wave equation and solution for $M(x)$:

$$M''(x) + q^2 M(x) = 0, \quad M(x) = A_I e^{-iqx} + A_R e^{iqx} = A_I e^{-iqx}, \qquad (4.16a,b)$$

$$q = \frac{n}{\left(\int_{z_0}^{1} P(z)\, dz\right)^{\frac{1}{2}}}, \quad n = m\sqrt{\frac{U_\lambda}{U_0}}. \qquad (4.16c)$$

Here q is the complex wave number (set so that the profile and wave solutions are consistent) and A_I is the amplitude of the tide at $x = 0$, determined by imposing a boundary condition on ζ_T at $x = 0$. Because $P(z)$ contains a factor of i, $q = k + ip$ is a complex number with equal (in magnitude) real and imaginary parts. The real part k (which corresponds to the wave number in an inviscid wave) is positive, but the damping modulus p is < 0, so that the wave damps as it travels.

The function ζ_T = Re[$M(x)e^{it}$] describes an incident wave propagating in a channel. Because equation (4.16a) is a second-order solution, there are two solutions, for the incident and reflected waves. Given an infinite channel of constant width and depth (or a uniformly converging channel), the reflected wave amplitude, $A_R = 0$. Now that $M(x)$ and $P(z)$ are known and A_I is specified, W_T is specified by equation (4.11c), and the solution is complete – ζ_T, U_T, and W_T are all known.

What does the resulting solution look like over a tidal cycle? Figure 4.9a shows the solution described by equations (4.15b) and (4.16). The strongest feature of this solution is that it uses a reasonably realistic vertical A_z profile so that the shear is concentrated near the bed. It also has a 45° phase difference between the transport and surface elevation, with the current leading the elevation. This is quite typical for tidal channels (see Chapter 3).

This analytical solution is contrasted to the result of the numerical solution of equations (4.12) and (4.16) in Fig. 4.9b. There are two primary differences between the analytical and numerical solutions:

- Vertical phase progression: The non-accelerated solution shows no vertical phase progression – currents vary vertically in strength but not phase. The numerical solution suggests that the vertical phase progression in the current is ~9° or ~20 min for the M_2 tide, but there is no difference between flood and ebb in terms of the phase progression. Both of these results show far less phase progression than does Fig. 4.4, and the differences between flood and ebb seen in Fig. 4.4 are absent. The most important are

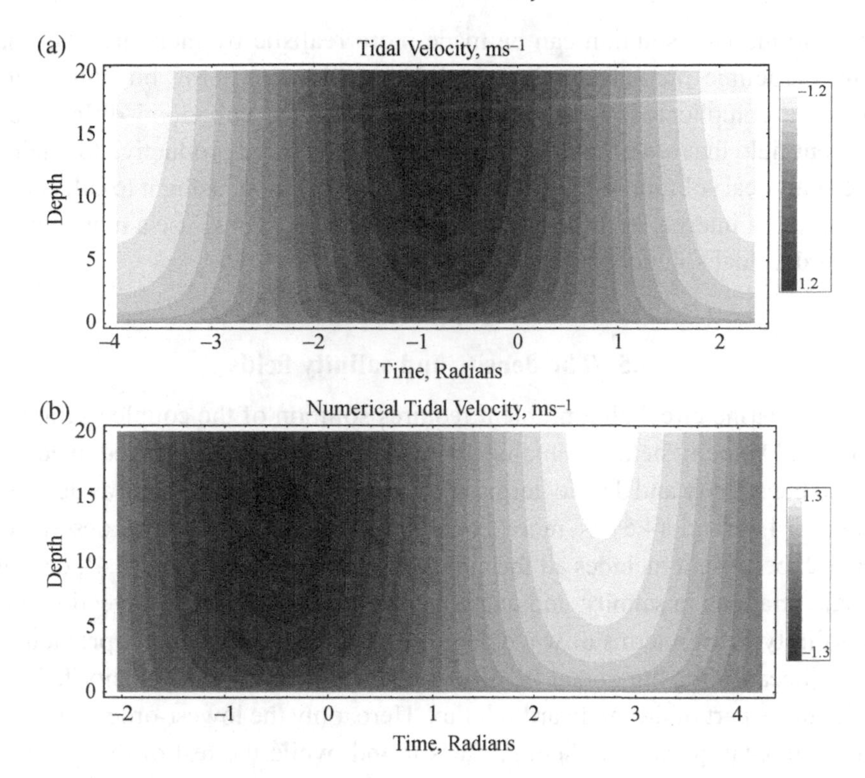

Figure 4.9. Dimensional velocity solutions. In (a), tidal velocity solution over a tidal cycle as specified by equations (4.15) and (4.16). In (b), tidal velocity solution as determined by numerical solution of equation (4.12) with equation (4.16). Parameters used in both solutions were: $U_0 = 1$m/s, $\pi_1 = 0$, $z_0 = 0.15$ m, $u_* = 0.03$ m/s, $H = 15$ m, and $a = 0.05$. The tidal amplitude at $x = 0$ was 1 m.

the subsurface velocity maximum on flood and the very different vertical phase progression on flood vs ebb, especially for the neap tide. We will see that the correct phase progression is only generated when internal tidal asymmetry is included in the solution.

- Phase difference between height and flow: The non-accelerated solution exhibits a transport vs height phase difference of about 45°. This is actually closer to reality than the result for the numerical solution (~37° at $z = 1$), but for the wrong reasons. In real channels the 45° phase difference comes from a combination of friction and channel convergence, the latter of which is not considered here. Because it does not consider convergence, the numerical solution does not yield a realistic phase difference. Neglect of acceleration in the analytical solutions compensates for the lack of convergent geometry to give better results.

In summary, a relatively simple analytical ("non-accelerated") solution to the tidal velocity profile has been developed; it is closely related to the log boundary layer solution. This solution can be made more realistic by including the tidally varying baroclinic pressure gradient and the acceleration term, but this yields a much more complicated result without introducing the ebb–flood differences in the current field that are observed in reality. It will be more productive to retain the simple analytical solution for the lowest-order tidal solution, using it to calculate the contribution of internal asymmetry to the velocity field than to use a more complex lowest-order tidal solution.

4.5. The density and salinity fields

The full estuarine circulation problem requires solution of the coupled salinity (or density) and velocity fields, including all the circulation modes defined in equation (4.4) and the salinity and density terms in equation (4.5). The problem as defined by equations (4.4) and (4.5) is more complicated than the problem described in Chapter 2, because it includes all the circulation modes in equation (4.4), describes the tidal variations in salinity, and quantifies the influence of tidal advection on the mean salinity field in terms of wave fluxes. The salinity and velocity problems are coupled, because the horizontal density gradient influences the velocity field, and the tides drive part of the landward salt flux. Here, only the lowest-order component of the horizontal density gradient $\partial\rho/\partial x$ is found, while the rest of the problem is described qualitatively.

From equation (4.5) the lowest-order density field consists of two terms: (a) the background, linear, "central regime" component $\Delta\rho_H(1-x/L_S)$ and (b) a vertically uniform, tidally advected component ρ_T that is related to the vertically averaged tidal flow. To determine $\partial\rho/\partial x$, we first need to determine $\rho_T(x,t)$. The dimensional mass conservation equation for ρ_T is:

$$\frac{\partial\rho\,(x,t)}{\partial t} = -\{U_T\}\frac{\partial\bar{\rho}}{\partial x} = -\{U_T\}\,\frac{\partial\left(\Delta\rho_H\left(1-\frac{x}{L_S}\right)\right)}{\partial x} = \frac{\rho_0}{L_S}\{U_T\}, \qquad (4.17a)$$

where the curly bracket { } indicates a vertical average. Equation (4.17a) says that the dominant part of the tidal salinity variation is due to the vertically uniform tidal advection of a vertically uniform mean salinity field. In non-dimensional form, equation (4.17a) can be written (dropping the hats over the non-dimensional variables) as:

$$\frac{\partial\rho_T(x,t)}{\partial t} = \frac{\rho_0}{\omega\,L_S}\{U_T\}. \qquad (4.17b)$$

We then assume a harmonic salinity variation in time and use equation (4.11a) to define $\{U_T\}$. This yields the following solution:

$$\rho_T(x,t) = \frac{\rho_0}{\omega L_S} \mathrm{Re}\left[M'(x)\{P\}\, e^{it}\right] = \frac{\rho_0}{\omega L_S} \mathrm{Re}\left[M'(x) \int_{z_0}^{1} P(z)\, dz\, e^{it}\right]. \quad (4.18)$$

An important consequence of equation (4.18) (and the corresponding relationship for salinity) is that the lowest-order tidal salinity and density fields are 90° (a quarter of a tidal cycle) out of phase with the tidal current. This means that the salinity is maximum at the end of flood and minimum at the end of ebb. Because ρ_T and U_T are in quadrature, there is no correlation between the tidal flow and this component of the salinity field. Thus, there is no contribution to the mean stratification from the tidal advection of this salinity component. The contribution of tidal advection to the mean stratification is related to tidal variations in vertical mixing and, therefore, in stratification.

We can now define the lowest-order part of $\partial\rho/\partial x$ as:

$$\frac{\partial\rho}{\partial x} = -\frac{\rho_0}{L_S} - iq\,\frac{\rho_0}{\omega L_S} \mathrm{Re}\left[M'(x) \int_{z_0}^{1} P(z)dz\, e^{it}\right]. \quad (4.19)$$

We will use equation (4.19) in the analysis of internal tidal asymmetry below.

The $O(\varepsilon)$ salinity field is too complex to analyze here completely, but we can set up the relevant differential equations to understand where these variations come from. The tidal variations in stratification are described by:

$$\frac{\partial\rho_{TZ}(x,z,t)}{\partial t} = -(U_T - \{U_T\})\frac{\partial\bar{\rho}}{\partial x} - \bar{U}\frac{\partial\rho_T(x,t)}{\partial x} + \frac{\partial}{\partial z}\left(K_\rho \frac{\partial\rho_{TZ}}{\partial z}\right), \quad (4.20a)$$

where K_ρ is the turbulent eddy diffusivity for scalar properties. Equation (4.20a) says that tidal variations in stratification are a balance between [in order on the right-hand side of equation (4.20a)]: (a) tidal shear advection of the mean density field, (b) mean shear advection of tidal density variations, and (c) vertical mixing. The first two terms typically increase stratification while the last reduces it. Unfortunately, the mean shear advection term couples this mode to all of the residual circulation modes, leading to a difficult calculation.

The mean stratification is governed by:

$$\frac{\partial\bar{\rho}_G(z)}{\partial t} = 0 = -\frac{\partial(\bar{U}\bar{\rho}_0)}{\partial x} - \frac{\partial\langle U_T\rho_{TZ}\rangle}{\partial x} + \frac{\partial}{\partial z}\left(K_\rho \frac{\partial\bar{\rho}_G}{\partial z}\right). \quad (4.20b)$$

Equation (4.20b) says that the mean stratification is a balance between: (a) the gradient in horizontal shear advection by the mean flow, (b) the gradient in horizontal tidal wave fluxes, and (c) vertical mixing (which again tends to reduce the stratification). A vertical integral of equation (4.20b) is used to determine the salinity intrusion length.

4.6. The circulation caused by internal tidal asymmetry

Here, we develop the equations necessary to describe internal tidal asymmetry, a mode of circulation related to interactions of time-varying mixing and density stratification. Again neglecting the convective accelerations, we can write the part of the dimensional along-channel momentum conservation equation that describes internal asymmetry as:

$$\varepsilon_I \frac{\partial U_{IA}}{\partial t} + \frac{\partial}{\partial z}\left(A_z\left[\frac{\partial U_T}{\partial z} + \varepsilon_I \frac{\partial U_{IA}}{\partial z}\right]\right)$$
$$= \varepsilon_I \frac{\partial U_{IA}}{\partial t} + \frac{\partial}{\partial z}\left(\kappa_m(z)\left(1 + \varepsilon_I f_1[z]\, e^{i(t+\phi)} + (1-\varepsilon_I) f_2[x,z]\, e^{2it}\right)\right.$$
$$\left.\left[\frac{\partial U_T}{\partial z} + \varepsilon_I \frac{\partial U_{IA}}{\partial z}\right]\right)$$
$$= g\varepsilon_I \frac{\partial \zeta_{IA}}{\partial x} + \frac{g}{\rho_0}\int_z^1 \frac{\partial \rho}{\partial x} dz,$$
$$\varepsilon_I = \frac{\Delta\rho_V}{\Delta\rho_H}. \qquad (4.21\text{a,b})$$

Equation (4.21) describes $O(\varepsilon_I)$ motions at overtide and zero frequencies, as will become clear as the analysis develops. There are three tidally varying properties in equation (4.21a) – A_z, U_T, and ρ – and it is their interactions that drive the motion due to internal asymmetry.

The strength of the currents in equation (4.21a) due to tidal asymmetry is scaled by ε_I, the ratio of the vertical and horizontal density differences in the estuary [equation (4.21b)]. In very strongly stratified systems, ε_I can approach unity, so the motion is not so small relative to the dominant tidal motion. The validity of the formal perturbation approach is saved only by the quadratic nature of the non-linear terms driving the motion – the shear and the vertical mixing (for example) are not perfectly correlated, so the actual currents due to internal asymmetry remain considerably smaller than the tidal current.

Some algebra is needed to see how the intratidal variations in A_z, U_T, and ρ can drive internal asymmetry at overtide and zero frequencies. To begin with, we will simplify the problem by neglecting the acceleration term $\partial U_{IA}/\partial t$ in equation (4.21a). This is strictly correct at zero frequency, and even if it is an approximation, it greatly simplifies the analysis of the overtide mode. The analysis will then yield "non-accelerated" solutions, exactly as in Section 4.2.1. If $\partial U_{IA}/\partial t$ is neglected, then equation (4.21a) may be integrated once from the free surface down to an arbitrary depth z. We apply the boundary condition that the free surface (at $z = 1$) is unstressed. This yields:

$$\kappa_m(z)\left(1+\varepsilon_I f_1\, e^{i(t+\phi)}+(1-\varepsilon_I) f_2\, e^{2it}\right)\left(\frac{\partial U_T}{\partial z}+\varepsilon_1 \frac{\partial U_{IA}}{\partial z}\right)$$
$$=-g\varepsilon_I \frac{\partial \zeta_{IA}}{\partial x}(1-z)-\frac{g}{\rho_0}\int_z^1\int_z^1 \frac{\partial \rho}{\partial x} dz'\, dz. \tag{4.22}$$

We now divide by $A_z(z)$ and consider only the $O(1)$ part of $\partial \rho/\partial x = \partial \rho_0/\partial x + \partial \rho_T/\partial x$:

$$\frac{\kappa_m\, \varepsilon_1 f_1(z)\, e^{i(t+\phi)} \frac{\partial U_T}{\partial z}+\varepsilon_1 \kappa_m \frac{\partial U_{IA}}{\partial z}}{\kappa_m\left(1+\varepsilon_1 f_1(z)\, e^{i(t+\phi)}+(1-\varepsilon_I) f_2(x, z)\, e^{2it}\right)} =$$
$$-\frac{g\varepsilon_I}{\kappa_m\left(1+\varepsilon_1 f_1(z)\, e^{i(t+\phi)}+(1-\varepsilon_I) f_2(x, z)\, e^{2it}\right)} \frac{\partial \zeta_{IA}}{\partial x}(1-z)$$
$$-\frac{g/\rho_0}{\kappa_m\left(1+\varepsilon_1 f_1(z)\, e^{i(t+\phi)}+(1-\varepsilon_I) f_2(x, z)\, e^{2it}\right)}\int_z^1\int_z^1 \frac{\partial\left(\rho_0+\rho_T\right)}{\partial x} dz'\, dz. \tag{4.23}$$

We then separate equation (4.23) into residual and overtide parts, using the binomial theorem to bring $(1+\varepsilon_I f_1\, e^{i(t+\varphi)}+(1-\varepsilon_I) f_2\, e^{i2t})$ into the numerator, and discard terms of the wrong frequency. The resulting zero-frequency part is:

$$\frac{\partial \overline{U}_{IA}}{\partial z}=-\frac{g}{\kappa_m} \frac{\partial \overline{\zeta}_{IA}}{\partial x}(1-z)-\left\langle f_1(z)\, e^{i(t+\phi)} \frac{\partial U_T}{\partial z}\right\rangle+$$
$$\left\langle\frac{g}{\rho_0}\int_z^1\int_z^1 \frac{\partial \rho_T}{\partial x} dz'\, dz\, \frac{f_1(z)\, e^{i(t+\phi)}}{\kappa_m}\right\rangle. \tag{4.24}$$

Equation (4.24) says that the residual flow due to internal asymmetry is driven by two terms: (a) the correlation of tidal shear with tidal variations in vertical mixing (second term on the r.h.s.) and (b) the correlation of tidal salinity variations due to tidal straining with tidal variations in vertical mixing (third term on the r.h.s.). The surface slope (the first term on the r.h.s.) is passive – it guarantees that, as an internal mode, overtide internal asymmetry has no net transport.

The overtide part of internal asymmetry (indicated in compound terms by the subscript "*OT*") is:

$$\frac{\partial U_{IA}}{\partial z} = -\frac{g}{\kappa_m}\frac{\partial \zeta_{IA}}{\partial x}(1-z) - \left(f_1(z)\,e^{i(t+\phi)}\frac{\partial U_T}{\partial z}\right)_{OT} + \left(\frac{g}{\varepsilon_I \rho_0}\int_z^1\int_z^1 \frac{\partial \rho}{\partial x}dz'\,dz\;\frac{\left(-\varepsilon_I f_1(z)\,e^{i(t+\phi)} - (1-\varepsilon_I)\,f_2(x,z)\,e^{2it} - \ldots\right)}{\kappa_m}\right)_{OT}. \tag{4.25}$$

Equation (4.25) says that the overtide flow due to internal asymmetry is driven by three terms: (a) the correlation of tidal shear with tidal variations in vertical mixing (second term on the r.h.s.), (b) the correlation of tidal salinity variations due to tidal straining with tidal variations in vertical mixing (third term on the r.h.s.), and (c) the zero-frequency pressure gradient interacting with overtide variations in vertical mixing (last term on the r.h.s.)

The corresponding overtide and residual local and integral mass conservation equations are:

$$\frac{\partial U_{IA}}{\partial x} + \frac{\partial W_{IA}}{\partial z} = 0, \quad \frac{\partial}{\partial x}\int_{z_0}^{1} U_{IA}dz = \int_{z_0}^{1} U_{IA}dz = 0, \tag{4.26a,b}$$

$$\frac{\partial \overline{U}_{IA}}{\partial x} + \frac{\partial \overline{W}_{IA}}{\partial z} = 0, \quad \frac{\partial}{\partial x}\int_{z_0}^{1} \overline{U}_{IA}dz = \int_{z_0}^{1} \overline{U}_{IA}dz = 0, \tag{4.27a,b}$$

where W_{IA} and $\overline{W}_{IA}$ are, respectively, the overtide and residual vertical velocities associated with internal asymmetry. Thus, the residual and overtide currents due to internal asymmetry are: (a) internal modes (integrate to zero) and (b) non-divergent (invariant in x). The latter condition means that they are local disturbances, not waves propagating through the estuary.

We need now to non-dimensionalize equations (4.24) and (4.25) and define functional forms for U_{IA} and $\bar{U}_{IA}$; the scaling is the same as in equation (4.9), with the addition of:

$$U_{IA} = \varepsilon_I U_0 \hat{U}_{IA} = \frac{\varepsilon_I u_*}{C_D^{1/2}}\hat{U}_{IA}, \qquad \zeta_{IA} = \varepsilon_I \zeta_0 \hat{\zeta}_{IA},$$
$$\frac{\partial \rho}{\partial x} = \frac{\Delta\rho_H}{L_S}\frac{\partial \hat{\rho}}{\partial x}. \tag{4.28}$$

The scaled versions of equations (4.24) and (4.25) are:

$$\frac{\partial \overline{U}_{IA}}{\partial z} = -\frac{(1-z)}{\kappa_m}\frac{\partial \overline{\zeta}_{IA}}{\partial x} - \left\langle f_1(z)\, e^{i(t+\phi)}\frac{\partial U_T}{\partial z}\right\rangle +$$

$$\left\langle \pi_6 \int_z^1 \int_z^1 \frac{\partial \rho_T}{\partial x} dz'\, dz \frac{f_1(z)\, e^{i(t+\phi)}}{\kappa_m}\right\rangle, \tag{4.29a}$$

$$\frac{\partial U_{IA}}{\partial z} = -\frac{(1-z)}{\kappa_m(z)}\frac{\partial \zeta_{IA}}{\partial x} - \left(f_1(z)\, e^{i(t+\phi)}\frac{\partial U_T}{\partial z}\right)_{OT} +$$

$$\left(\frac{(1-\varepsilon_I)\,\pi_8}{\varepsilon_I}\int_z^1 \int_z^1 \frac{\partial \overline{\rho}}{\partial x} dz'\, dz \frac{f_2(x,z)\, e^{2it}}{\kappa_m}\right)_{OT} +$$

$$\left(\pi_6 \int_z^1 \int_z^1 \frac{\partial \rho_T}{\partial x} dz'\, dz \frac{f_1(z)\, e^{i(t+\phi)}}{\kappa_m}\right)_{OT}, \tag{4.29b}$$

$$\pi_8 = \frac{gH}{\varepsilon_I k u_*\, U_0}\frac{\Delta\rho_H}{\Delta\rho_{00}}\frac{H}{L_S}, \quad \pi_6 = \pi_8 \frac{U_0}{\omega L_S}. \tag{4.29c}$$

The parameters π_6 and π_8 scale the strength of, respectively, the tidal and overtide variability in vertical mixing with, respectively, the tidal and mean density gradient variability. Also, we need a way to determine the mean and overtide parts of the product of two harmonically varying (in time) functions. For any complex a and b:

$$\mathrm{Re}[ae^{it}] \equiv \frac{1}{2}\left(ae^{it} + a^* e^{-it}\right), \quad \mathrm{Re}[be^{it}] \equiv \frac{1}{2}\left(be^{it} + b^* e^{-it}\right)$$

$$\mathrm{Re}[ae^{it}]\mathrm{Re}[be^{it}] = \frac{1}{4}\left(abe^{2it} + a^* b^* e^{-2it} + ab^* + a^* b\right)$$

$$= \frac{1}{2}\mathrm{Re}[abe^{2it} + a^*\, b]. \tag{4.30}$$

Thus, the product of two harmonics at tidal frequency contains residual and overtide components; note that the complex conjugate of a appears only in the residual part.

Finally, we need functional definitions for the internal asymmetry variables, and for f_1 and f_2:

$$U_{0T} = \mathrm{Re}\,[i M'[x]\, T_4[z]\, e^{2it}], \quad \zeta_{0T} = \mathrm{Re}\,[M[x]\, SL_4\, e^{2it}],$$

$$f_1(z) = \mathrm{K}_{MT0}\, z\,(1-z), \quad f_2(x,z) = M'[x]\, \mathrm{K}_{MT0T},$$

$$\overline{U}_{IA} = \mathrm{Re}\,[i M'[x]\, T_0[z]], \quad \overline{\zeta}_{IA} = \mathrm{Re}\,[M[x]\, SL_0]. \tag{4.31a–c}$$

Note that the $M'(x)$ dependence of f_2 is necessary so that all terms of the differential equation have the same horizontal structure. Since this term is directly related to the strength of the tidal currents, this structure is plausible.

Substituting equations (4.11) and (4.31) into equation (4.29), we find the ordinary differential equations that describe, respectively, the residual and overtide components of internal asymmetry:

$$T_0' = +i\frac{(1-z)}{z\,(1+a-z)}\;SL_0 - \left(\frac{\kappa_{MT0}\,\pi_6\,e^{-i\phi}}{4}\;qz\,(1-z)^3\,\langle P(z)\rangle\right)$$
$$-\frac{\kappa_{MT0}}{2}e^{-i\phi}(z\,(1-z)\,P'), \tag{4.32a}$$

$$T_4' = +i\frac{(1-z)}{z\,(1+a-z)}\;SL_4 - \left(\frac{\kappa_{MT0}\,\pi_6\,e^{i\phi}}{4}\;q\;\frac{(1-z)^3}{(1+a-z)}\;\langle P(z)\rangle\right)$$
$$-\frac{i(1-\varepsilon_I)\;\kappa_{MT0T}\,\pi_8}{2}\left(\frac{(1-z)^2}{z\,(1+a-z)}\right)-\frac{\kappa_{MT0}}{2}e^{i\phi}(z\,(1-z)\,P'). \tag{4.32b}$$

As mentioned above, there are three forcing terms for the overtide internal asymmetry in equation (4.32a) and two for the residual asymmetry in equation (4.32b). These appear on the r.h.s. of the respective equations, along with a surface slope that is required to guarantee that both are internal modes. The surface slopes are determined such that the solutions satisfy the respective integral continuity constraints (4.26) and (4.27); i.e., have zero integral transport. The vertical velocities (not given) are determined from equations (4.26a) and (4.27a).

Equations (4.23a) and (4.23b) are first-order linear differential equations that can be integrated with a no-slip boundary condition at z_0; the results are lengthy and are not presented here. To obtain some insight into the structure of the currents described by the solutions to equations (4.31) and (4.32), we plot the solutions for typical conditions – in this case, the estuary is 20 m deep, the tidal range is 1 m, and $\varepsilon_I = 0.6$. Figure 4.10 shows that the amplitudes of the internal overtide and residual are about 0.1 and 0.05 m s^{-1} just above the bed, respectively, relative to a tidal current of ~0.75 m s^{-1}. These amplitudes are large enough to considerably alter the total flow.

Since U_{OT} is oscillatory at twice the dominant tidal frequency, it has an amplitude and a phase, and the phase relative to the tidal current (shown in Fig. 4.9) matters. The time-varying structure of the internal overtide plus mean flow is shown in Fig. 4.11a. Comparing this to the tidal current alone in Fig. 4.9, it is evident that the internal overtide has landward near-bed and seaward near-surface currents at the times of peak flood and ebb, respectively. This can cause the total flood flow (tidal plus internal asymmetry) to be almost vertically

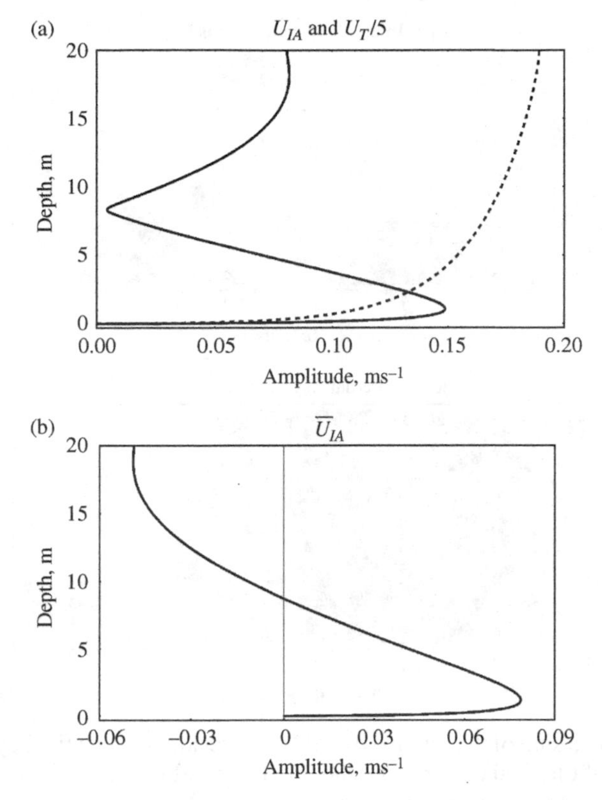

Figure 4.10. Amplitudes of the tidal current/5 and internal overtide current (a) and residual current (b) in a 20 m deep estuary with strong density forcing. The internal overtide reverses phase by 180° at ~8 m above the bed.

uniform and the ebb flow to be strongly sheared, as can be seen in Fig. 4.11b. There is even some vertical progression of the peak flood current in Fig. 4.11b, though this remains weak. Comparing Fig. 4.11b to Fig. 4.1b, we see that the total flood flow (tidal plus internal asymmetry, Fig. 4.11b) more closely resembles the actual flow (Fig. 4.1b) than does the tidal flow (Fig. 4.9) alone. However, there are still substantial differences. The following sections develop several more residual flow modes, to see if we can develop a more fully realistic total estuarine flow model.

4.7. Residual circulation modes in general

It was noted in connection with equation (4.4) that residual flow modes either: (a) integrate to zero (so that they are internal modes without a net flow) or (b) integrate to a specific, known discharge. Each residual flow mode must have a driving mechanism that can arise in one of three ways: (a) from one of the

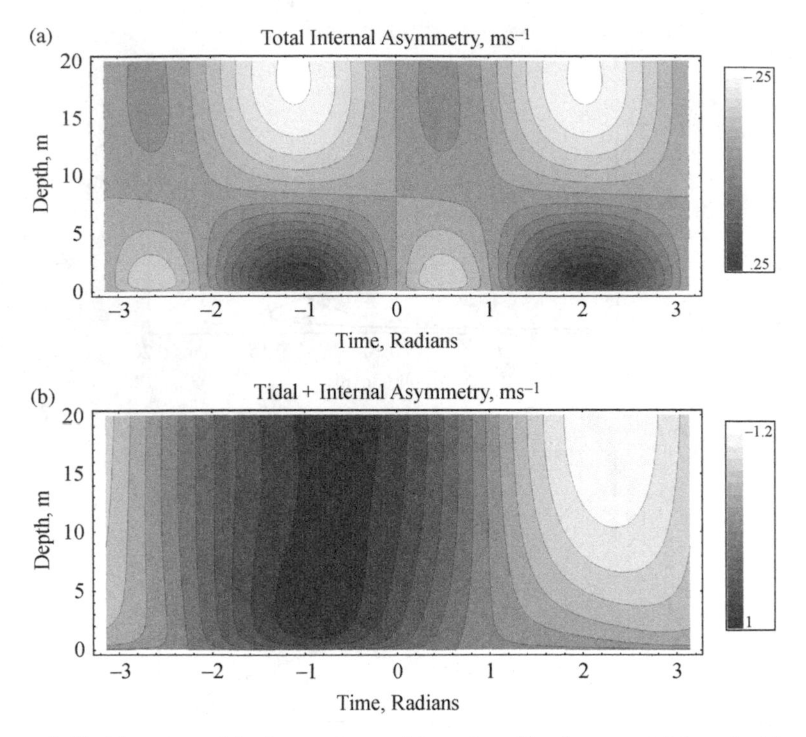

Figure 4.11. The sum of the internal overtide and residual over a tidal cycle (above) and the sum of the tidal current, and the internal overtide and residual over a tidal cycle (below).

terms in the momentum equation (e.g., the correlation between tidal shear and tidal vertical mixing that in part drives internal asymmetry), (b) from a non-zero net discharge (like the river flow), or (c) from a boundary condition (wind-driven circulation). Having now solved the complicated internal asymmetry problem (overtide and residual), we need to summarize the solution technique used there, as it is common to all the residual flow problems. In each case, there are three unknowns: the horizontal and vertical velocity profiles, and the surface slope that provides the discharge appropriate to the mode. For a flat bed, the vertical velocity vanishes at all depths for some modes. The steps are:

- Identify the driving term or terms.
- Scale and non-dimensionalize the relevant mass conservation (local and integral) and along-channel momentum conservation equations.
- Define a functional form for the circulation mode; e.g., equation (4.31), separate variables, and set out the second-order ordinary differential equation that results from substitution of these functional forms into the along-channel momentum conservation equation.

- Integrate from the surface to an arbitrary depth z, using the surface boundary condition that the surface stress vanishes (or has a specific, known value). This defines the shear.
- Integrate a second time to define the vertical structure of the velocity field.
- Integrate a third time to define the integral transport. Set this transport equal to zero (for an internal mode) or the known transport (for an external mode), and use this constraint to determine the surface slope that provides the correct integral transport.
- Determine the vertical velocity from local continuity.

These steps will be followed in analysis of the remaining residual flow modes. Two points require clarification:

- It is obvious that a positive surface slope (surface elevation increasing landward) is needed to drive the river flow seaward – "water runs downhill". It is less intuitive that a surface slope is also needed for an internal mode. Taking the gravitational circulation as an example, the reason becomes clear – there is a landward-directed baroclinic pressure gradient force (pressure is higher on the seaward side) that pushes seawater landward. A steady state can only be achieved with zero net discharge, and this requires a positive surface slope (surface elevation increasing landward) to balance the baroclinic pressure gradient.
- We did not present the solution for the vertical velocity associated with internal asymmetry. The integral conditions (4.26b) and (4.27b) that specify zero net discharge are non-divergent (x-invariant), but this does not mean that flow at each level is x-invariant. The appearance of $M[x]$ and $M'[x]$ in equation (4.31a) guarantees that the internal overtide will be x-dependent, even over a flat bed. Typical estuarine beds are not flat, and substantial x-variations in internal asymmetry are common, e.g., in association with local reversals of $\partial\rho/\partial x$ in the wake of a topographic high. Some of the residual modes described in the following sections also have x-variations.

4.8. The Hansen and Rattray problem

Hansen and Rattray (1965) defined what we will treat as three circulation modes: river outflow, gravitational circulation, and steady wind-driven circulation. They lumped, however, river outflow and gravitational circulation into one broader term, density-driven flow. These two terms are also treated together in Chapter 2. In fact, however, the two phenomena exist separately in many systems. Most river estuaries have a tidal river segment landward of salinity intrusion, where there is a net river discharge but no gravitational circulation. Many low-inflow estuaries (Chapter 9) will have little or no river inflow, but will have a longitudinal density gradient and a density-driven discharge due to evaporation or prior river discharge. It is also useful to separate the two modes, because one (gravitational circulation) is an internal mode, while the other is not.

We begin with the river discharge mode. The relevant dimensional along-channel momentum and mass conservation equations for river flow velocity U_R are:

$$0 = -g\,\varepsilon_R \frac{\partial \varsigma_R}{\partial x} + \varepsilon_R \frac{\partial}{\partial z}\left(\kappa_m \frac{\partial U_R}{\partial z}\right),$$
$$0 > Q_R = \int_{z_0}^{h} U_R\,dz, \quad \varepsilon_R = \frac{U_{R0}}{U_0}. \tag{4.33a,b}$$

where U_{R0} is the river flow velocity magnitude. In the absence of tributaries, precipitation and evaporation, the river flow Q_R is independent of x, allowing U_R and surface slope $\partial\zeta_R/\partial x$ also to be independent of x; this horizontal non-divergence leads to $W_R \equiv 0$. This similarity in x occurs, however, only in a channel of constant width and depth. Otherwise, $\partial U_R/\partial x$ and $W_R \neq 0$.

Following the procedure defined in Section 4.3, we scale and non-dimensionalize, and integrate once using the boundary condition that the free surface is unstressed. We then divide by κ_m and integrate a second time to define the river flow velocity profile:

$$\overline{U}_R = -\,\varepsilon_R \frac{\pi_2}{1+a}\frac{\partial \zeta_R}{\partial x}\left(\log\left[\frac{z}{z_0}\right] + \left[\frac{\log(1+a-z)}{\log(1+a-z_0)}\right]^a\right),$$
$$\pi_2 = \frac{g\,\zeta_0}{k\,u_*\,U_0}\frac{H}{L_S}, \quad \frac{\partial \zeta_R}{\partial x} = \frac{1}{\pi_2}\frac{(1+a)}{(1+a)(1-z_0) + \log\left(z_0\left[\frac{a}{1+a-z_0}\right]^{a^2}\right)}. \tag{4.34}$$

If $a = 0$, a log velocity profile is recovered. The shape of this profile is the same as for the non-accelerated tidal flow. Solution of equation (4.33) with a vertically constant κ_m would yield a quadratic velocity profile, but with the shear farther from the bed than in equation (4.34).

The relevant dimensional along-channel momentum and mass conservation equations for the gravitational circulation U_G are:

$$0 = -g\,\varepsilon_i \frac{\partial \varsigma_G}{\partial x} - \frac{g}{\rho_0}\int_z^h \frac{\partial \rho_0}{\partial x}dz + \varepsilon_i \frac{\partial}{\partial z}\left(\kappa_m \frac{\partial U_G}{\partial z}\right),$$
$$0 = \int_{z_0}^{h} U_G\,dz, \quad \varepsilon_i = \left(\frac{\Delta\rho_H}{\rho_0}\right)^{1/2}. \tag{4.35}$$

The solution for equation (4.35) is similar to equation (4.33), except that: (a) there is no integral transport and (b) the integrations are complicated by the presence of the horizontal density gradient term:

$$\bar{U}_G = -\frac{\varepsilon_i U_i \pi_3}{2(1+a)}\left((1+a)(z-z_0) + \left(1 - 2\pi_2\frac{\partial\zeta_G}{\partial x}\right)\log\left[\frac{z}{z_0}\right] + a\left(a + 2\pi_2\frac{\partial\zeta_G}{\partial x}\right)\log\left[\frac{1+a-z}{1+a-z_0}\right]\right),$$

$$\pi_3 = \frac{gH}{\varepsilon_i k u_* U_i}\frac{\Delta\rho_H}{\rho_0}\frac{H}{L_s},$$

$$\frac{\partial\zeta_G}{\partial x} = \frac{1}{\pi_2}\frac{(1+a)(1-z_0)(3-2a-z_0) - 2\left(a^3\log\left[\frac{a}{1+a-z_0}\right] + \log[z_0]\right)}{4\left((1+a)(1-z_0) + a^2\log\left[\frac{a}{1+a-z_0}\right] + \log[z_0]\right)}. \tag{4.36}$$

In equation (4.36), the gravitational circulation scale velocity is $U_i = (gH\,\Delta\rho_V/\rho_0)^{1/2}$. This solution is quite similar in form to that given in Chapter 2, but the shear is concentrated closer to the bed, because of the vertically variable eddy diffusivity. Because of the central regime assumption equation (4.36) is, like equation (4.34), independent of x if the bed is flat.

We can examine the relative importance of the river outflow and gravitational circulation by comparison of equations (4.34) and (4.36):

$$\frac{\overline{U}_R}{\overline{U}_G} = \varepsilon_R\varepsilon_T\frac{\Delta\rho_H}{\rho_0} = \frac{U_R}{U_0}\frac{\varsigma}{H}\frac{\Delta\rho_H}{\rho_0}. \tag{4.37}$$

Increasing river flow and tidal amplitude (relative to depth) increases the importance of river outflow relative to gravitational circulation, while increasing horizontal density forcing increases the importance of the gravitational circulation. Figure 4.12 shows the profiles of the river outflow and gravitational circulation for three depths of estuaries (8, 20 and 50 m), with $U_{R0} = -0.1*8/H$ m s^{-1} (so that the river flow transport $Q_R = U_{R0}\,H$ is the same for all three) and $\zeta = 1$ m. For a flow 8 m deep, gravitational circulation is quite small, and its primary role is to increase the shear in the total outflow; there is no upstream bottom flow in the total residual current (sum of river outflow and gravitational circulation). At 20 m deep, the two components are similar in strength, and the total current shows a small upstream bottom flow. For a 50 m deep flow, river outflow slightly modifies the gravitational circulation, and the total current shows strong upstream bottom flow.

The steady wind-driven circulation is an internal mode. The solution resembles equation (4.36), but includes an additional term related to the surface wind stress in the velocity, which also modifies the surface slope. We do not discuss it here because: (a) it is not very realistic – winds are rarely steady long enough for a steady two-layer flow to be established, and (b) wind-driven circulation is not a major feature of river estuaries. While a two-layer, wind-driven exchange mode is sometimes observed, atmospheric forcing more often causes alternating, almost barotropic set-up and set-down of the estuarine surface, driven by changes in water level on the continental shelf, as winds alternate between upwelling and downwelling favorable. These oscillations are dynamically similar to low-frequency

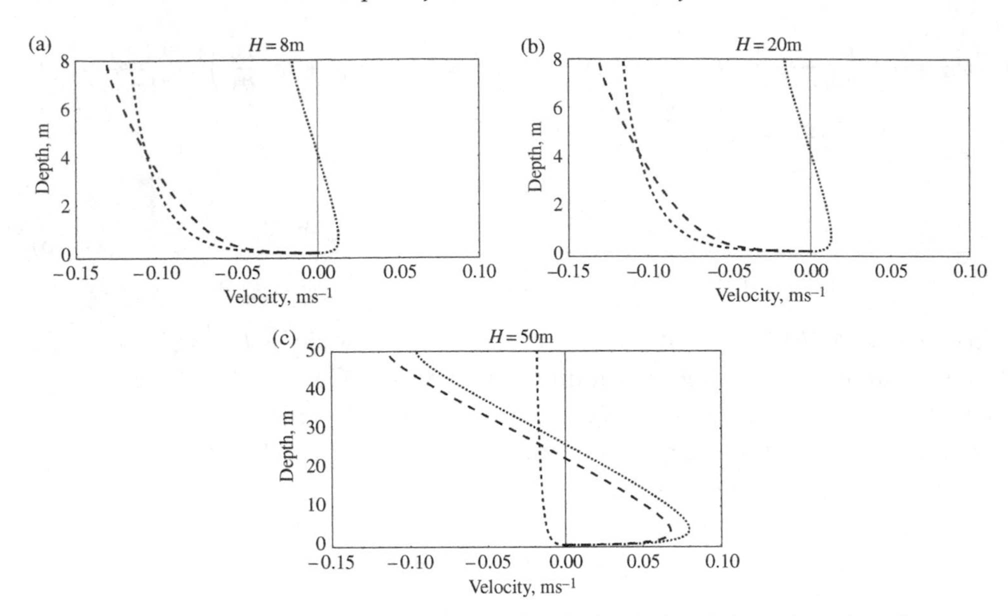

Figure 4.12. River outflow (---), gravitational circulation (...), and total outflow (–––) for estuaries that are (a) 8 m, (b) 20 m, and (c) 50 m deep.

tides, except that they are broadband in frequency, and lack the precisely harmonic structure of tides. See Chapter 6 for a treatment of wind-driven circulation in broad, shallow estuaries, where it is an important factor.

4.9. The circulation due to tidal non-linearities

Here, we develop the equations necessary to the flows forced by: (a) the Stokes drift (which drives the Stokes compensation flow) and (b) the time average of the convective accelerations (which drives the tidal non-linear circulation). These two residual flow modes are important in shallow systems with values of $\varepsilon_T = \zeta / H$ (the ratio of tidal amplitude to mean depth) that approach unity (Ianniello, 1977). This type of circulation is independent of the density field.

The Stokes compensation flow is an external mode and transports mass seaward. Thus, like the river flow that it closely resembles, it is forced from the continuity equation, but by the Stokes drift. The dimensional equations of motion are:

$$0 = -g\varepsilon_T \frac{\partial \zeta_T}{\partial x} + \varepsilon_T \frac{\partial}{\partial z}\left(\kappa_m \frac{\partial U_{ST}}{\partial z}\right),$$

$$0 > Q_{ST} = -\langle U_T|_h \zeta_T \rangle = \int_{z_0}^{h} U_{ST}\, dz. \tag{4.38}$$

The Stokes drift $\langle U_T|_h \zeta_T\rangle$ results from the correlation of tidal height and transport, averaged over a tidal cycle. Like equation (4.34) for the river outflow, the Stokes drift return flow exhibits a modified log-layer structure. Unlike the river outflow, the Stokes drift return flow is divergent (a function of x), because the Stokes drift is. The solution is:

$$U_{ST} = \varepsilon_T U_0 \,\mathrm{Re}[M'[x]\, M^*[x]\, T_{ST}[z]]. \tag{4.39}$$

For a flat bottom, U_{ST} has exactly the same profile as the river outflow with the same discharge (i.e., for $Q_R = Q_{ST}$). $T_{ST}(z)$ is not given here.

The tidal non-linear circulation is an internal mode driven by the time average of the convective accelerations; the relevant dimensional momentum and mass conservation equations are:

$$\varepsilon_T \left\langle U_T \frac{\partial U_T}{\partial x} + W_T \frac{\partial U_T}{\partial z} \right\rangle = -g\,\varepsilon_T \frac{\partial \zeta_{NL}}{\partial x} + \varepsilon_T \frac{\partial}{\partial z}\left(\kappa_m \frac{\partial U_{NL}}{\partial z}\right),$$

$$0 = \int_{z_0}^{h} U_{NL}\, dz. \tag{4.40a,b}$$

The solution is:

$$U_{NL} = \varepsilon_T U_0 \,\mathrm{Re}[M'[x]\, M^*[x]\, T_{NL}[z]]. \tag{4.41}$$

$T_{NL}[z]$ is complicated and not specified here. U_{ST} and U_{NL} are shown in Chapter 6 with varying bathymetry. For the 10–20 m deep systems considered here with $\varepsilon_T \sim 0.5$–0.2, U_{ST} will modestly augment the river outflow. For these systems, U_{NL} is typically quite small, <1 cm s^{-1}.

4.10. The total estuarine circulation

As remarked at the beginning of this chapter, the physical variability between estuaries is daunting, and even the temporal variability within a single estuary may cause it to appear to be a very different system over time. Clearly, it is impractical to model the variability of even a small subset of systems. Still, it is desirable to see how well the model developed in this chapter actually describes the system that has been used for examples, the Columbia River estuary.

Figure 4.13 shows an attempt to reproduce the spring and neap along-channel currents shown in Figure 4.2b,c. The model uses all the circulation modes described above (tidal, residual, and internal asymmetry), except the steady wind-driven flow. The model does show a substantial difference in vertical structure between the flood

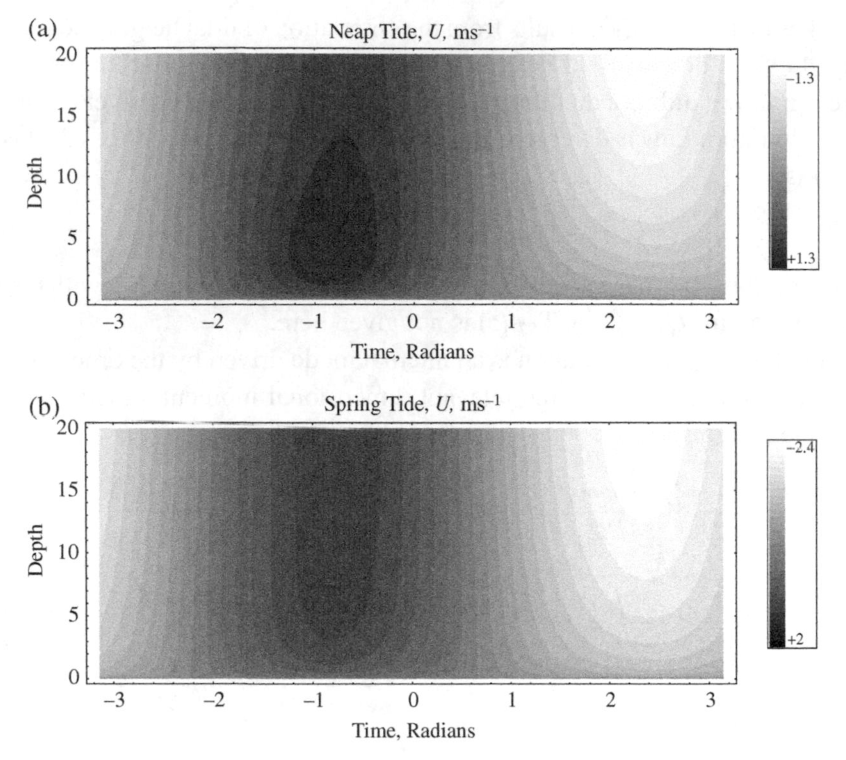

Figure 4.13. The modeled vertical structure of currents over a tidal cycle at the mouth of the Columbia River for conditions similar to Fig. 4.2: (a) currents for a spring tide, and (b) currents for a neap tide. The model used here includes all the velocity modes included in equation (4.4), except the wind-driven circulation.

and ebb, and there are neap–spring differences that resemble reality – a major part of the ability of the model to produce such differences stems from the inclusion of internal tidal asymmetry. However, the ebb–flood and neap–spring differences are too small, the ebb vertical shear is too weak (so surface currents are much too low), there is too little contrast between slack after flood and slack after ebb, and the velocity progression during neap floods is not realistic. Such limitations are expected from the linearized model described here. While this model gives a general idea of the processes at work, it cannot fully reproduce reality. Among the reasons are that: (a) the time and depth variability of vertical mixing is too simple and not properly linked to the stratification, (b) the variability of the baroclinic pressure gradient has not been fully described, and (c) lateral circulation and channel curvature, not included here, influence the along-channel flow. Circulation in a highly stratified estuary with strong tides is, however, quite a difficult modeling problem, and even 3D numerical models of such systems are not altogether successful.

4.11. Summary

This chapter has developed a systematic approach to describing and modeling estuarine variability. The model approach consists of a series of linked linear models of tidal, overtide, and residual flow variability in which the more important processes drive the smaller features. Among the most important model features is inclusion of overtide and residual internal tidal asymmetry, a very important feature of estuarine circulation that is not usually included in analytical modeling approaches. This allows the model to produce fairly realistic ebb–flood and neap–spring differences in the vertical structure of the current field. As expected with any linearized model, however, the results fall considerably short of success in providing a complete description of the complexities of estuarine circulation.

References

Chawla, A., D. A. Jay, A. M. Baptista and M. Wilkin (2008) Seasonal variability and estuary–shelf interactions in circulation dynamics of a river-dominated estuary. *Est. Coasts* **31**, doi:10.1007/s12237-007-9022-7.

Hansen, D. V. and M. Rattray, Jr. (1965) Gravitational circulation in straits and estuaries. *J. Mar. Res.* **23**, 104–122.

Ianniello, J. (1977) Tidally induced residual currents in estuaries of constant breadth and depth. *J. Mar. Res.* **35**, 755–786.

Jay, D. A. (1991) Internal asymmetry and anharmonicity in estuarine flows. In B. B. Parker (ed.), *Progress in Tidal Hydrodynamics*. John Wiley & Sons, New York, pp. 521–543.

Jay, D. A. and J. D. Musiak (1996) Internal tidal asymmetry in channel flows: origins and consequences. In C. Pattiaratchi (ed.), *Mixing Processes in Estuaries and Coastal Seas, An American Geophysical Union Coastal and Estuarine Sciences Monograph*, pp. 219–258.

Jay, D. A. and J. D. Smith (1990a) Circulation, density distribution and neap–spring transitions in the Columbia River Estuary. *Progr. Oceanogr.* **25**, 81–112.

Jay, D. A. and J. D. Smith (1990b) Residual circulation in shallow, stratified estuaries. II. Weakly-stratified and partially-mixed systems. *J. Geophys. Res.* **95**(C1), 733–748.

Monismith, S. G., W. Kimmerer, J. Burau and M. Stacey (2002) Structure and flow-induced variability of the subtidal salinity field in northern San Francisco Bay. *J. Phys. Oceanogr.* **32**, 3003–3019.

Pawlowicz, R., B. Beardsley and S. Lentz (2002) Classical tidal harmonic analysis including error estimates in MATLAB using T_TIDE. *Comput. Geosci.* **28**, 929–937.

Simpson, J. H., J. Sharples and T. P. Rippeth (1991) A prescriptive model of stratification induced by freshwater run-off. *Est. Coast. Shelf Sci.* **33**, 23–35.

Stacey, M., S. G. Monismith and J. Burau (1999) Observations of turbulence in a partially stratified estuary. *J. Phys. Oceanogr.* **29**, 1950–1970.

5

Estuarine secondary circulation

ROBERT J. CHANT
Rutgers University

5.1. Introduction

While the majority of theories developed to describe the dynamics of estuarine circulation are devoted to the study of along-channel flows at both tidal (Friedrichs and Aubrey, 1988) and subtidal frequencies (Pritchard, 1956; Hansen and Rattray, 1966; Geyer *et al.*, 2000; MacCready, 2004), in recent years numerous studies have concentrated on secondary flows and their importance in the along-channel dynamics (Lerczak and Geyer, 2004) and along-channel dispersion (Smith, 1977, 1978; Geyer *et al.*, 2008). While detailed theoretical work by Smith (1977) preceded these more recent studies by several decades, the recent modeling and observational studies discussed here have more clearly elucidated the complex interplay between lateral mixing, along-channel dynamics and dispersion. These studies have also emphasized the importance of the pioneering work by Ronald Smith (1977), who was awarded the BH Ketchum award in 1996 for this seminal work. Thus, in addition to the recent work described in detail in this chapter, the reader is encouraged to study the work of Smith (1977, 1978).

Secondary flows are defined by flow that is normal to the main along-channel flow. For natural flows the directionality of this may be defined by flows normal to channel orientation. This is often somewhat ambiguous, so in practice the "cross-channel direction" is usually defined by either principal component analysis or tidal ellipse analysis of current meter data. Typically, the strength of secondary flows is $<10\%$ of the strength of along-channel flows. However, because cross-channel gradients in velocity, salt, turbidity and other tracers are often larger than their respective along-channel gradients, the magnitude of cross-channel advective terms in the momentum and tracer equations is often as large or larger than the respective along-channel counterparts.

Moreover, it is through secondary flows that material is mixed across a channel, and thus accurately modeling secondary flows is imperative to make detailed

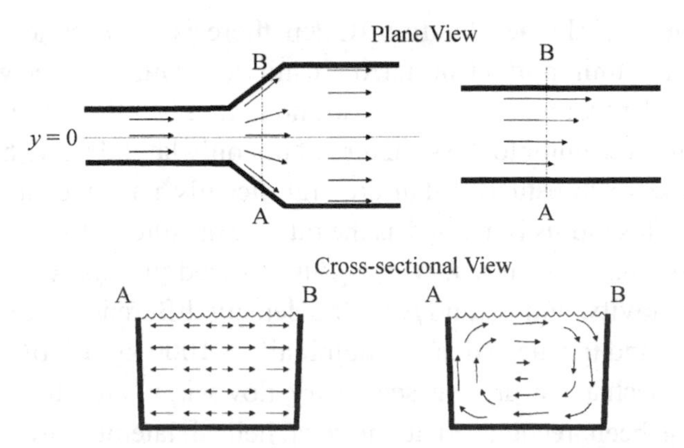

Figure 5.1. Left panels show channel configuration that can produce depth-averaged cross-channel flows. Right panels depict the type of secondary flows that are discussed in this chapter. These flows have zero depth-averaged component and are characterized by closed cells of cross-channel recirculation.

estimates of the dispersive nature of an estuary and to determine the fate and transport of material discharged along an estuarine shore. An early and insightful example of the role of lateral processes driving along-channel dispersion is contained in the pioneering work of Okubo (1973), who demonstrated that the advection of material into and out of shoreline coves can drive strong along-channel dispersion. In addition, the interaction between lateral shear and lateral mixing associated with secondary flows drives an along-channel dispersion that can dominate the dispersive nature of many estuaries (Fischer, 1972; Smith, 1977; Geyer *et al.*, 2008). These processes are discussed in more detail later in this chapter.

Finally, lateral circulation shapes channel morphology and often produces channels that are laterally asymmetric around the channel axis. Moreover, cross-channel variations in channel depth play a central role in driving secondary flows through differential advection (Nunes and Simpson, 1985; Lerczak and Geyer, 2004) and thus provide a positive feedback between secondary flow and channel morphology (Huijts *et al.*, 2006; Fugate *et al.*, 2007). It is noteworthy that secondary flows are typically characterized as baroclinic features with a zero depth-averaged component. While depth-averaged cross-channel flows will develop when along-channel gradients in channel morphology exist (Fig. 5.1a), in this chapter only the case when cross-channel flows are characterized by closed circulation cells with zero cross-section averaged flows (Fig. 5.1b,c) is considered.

5.2. Driving mechanisms

In this section the major mechanisms that drive secondary flows are discussed. One mechanism is Ekman forcing, which represents a dynamical balance between friction and the Coriolis acceleration. Often there is a misconception that the Earth's rotation is unimportant in narrow estuaries. This chapter will hopefully convince the reader that this is not the case and that in fact the Earth's rotation can be an important contributor to the structure of not only lateral flows but also along-channel flows, even in estuaries that are significantly narrower than an internal Rossby radius. This radius is defined as the ratio of the internal wave speed $(g'h)^{1/2}$ to the local Coriolis frequency f, where g' is reduced gravity, $g\ \Delta\rho/\rho$, g is acceleration due to gravity, and $\Delta\rho$ and ρ are the density difference between the upper and lower layers and the mean density, nominally $\sim$1000 kg/m^3, of the two layers.

The second mechanism driving secondary flows is related to flow curvature, which has long been recognized to drive a helical lateral flow normal to the streamwise flow (Rozovski, 1957). The third mechanism is linked to cross-channel baroclinic pressure gradients that arise from differential advection of the longitudinal density gradient. For completeness, the forcing of secondary flows by diffusive boundary layers is also discussed briefly. While this mechanism appears to be important on continental shelves and slopes (Garrett *et al.*, 1993), it does not appear to be a major contributor to secondary flows in estuaries (Lerczak and Geyer, 2004).

The mechanisms driving secondary flows are clearly identified by analysis of a cross-stream momentum equation written in a curvilinear coordinate system:

$$\frac{\partial u_n}{\partial t} + u_s\frac{\partial u_n}{\partial s} - \frac{u_s^2}{R} + fu_s + \frac{1}{\rho}\frac{\partial P}{\partial n} - \frac{\partial \tau}{\partial z} = 0, \tag{5.1}$$

where u_n is the cross-stream (secondary) flow, u_s the streamwise flow, R the radius of curvature in the streamwise flow, f the local Coriolis parameter, n the cross-stream direction, s the streamwise direction, z the vertical direction, P is pressure and τ stress. Equation (5.1) assumes the cross-stream flows u_n are much weaker than the streamwise flows. The first term represents the acceleration of the secondary flows, the second term represents the streamwise advection of streamwise gradients in the cross-stream flow. The third term is centrifugal acceleration, the fourth term is the Coriolis acceleration, and the fifth and sixth terms represent the pressure gradient and a vertical stress divergence, respectively. The pressure gradient has contributions both from cross-stream sea-level slopes and from cross-stream density gradients:

$$\frac{1}{\rho}\frac{\partial P}{\partial n} = g\frac{\partial \eta}{\partial n} + \frac{g}{\rho}\frac{\partial \rho}{\partial n}z. \tag{5.2}$$

Taking the vertical gradient of equation (5.1) and using equation (5.2) to describe the pressure gradient, we find

$$\underbrace{\frac{\partial}{\partial t}\left(\frac{\partial u_n}{\partial z}\right)}_{A}+\underbrace{\frac{\partial u_s}{\partial z}\frac{\partial u_n}{\partial s}}_{B}+\underbrace{u_s\frac{\partial}{\partial s}\left(\frac{\partial u_n}{\partial z}\right)}_{C}-\underbrace{\left(\frac{2u_s}{R}+f\right)\frac{\partial u_s}{\partial z}}_{D}+\underbrace{\frac{g}{\rho}\frac{\partial \rho}{\partial n}}_{E}+\underbrace{\frac{\partial^2 \tau}{\partial z^2}}_{F}=0. \quad (5.3)$$

Equation (5.3) concisely describes the three major forcing mechanisms. Term A is the local acceleration of the secondary flows; term B represents the straining of secondary flows by vertical shear in the streamwise flow; and term C is the streamwise advection of streamwise gradients in secondary flows. The next three terms represent the main forcing mechanisms that generate estuarine secondary flows. Term D represents forcing by flow curvature and the Earth's rotation. The forcing represented by term D increases with vertical shear, and the relative importance of flow curvature to the Earth's rotation is represented by a Rossby number ($2u_s/f R$), which expresses the relative importance of inertia to rotation. The centrifugal acceleration term is an advective term that takes this form in the curvilinear coordinate system (in a Cartesian coordinate system the centrifugal acceleration is contained in the horizontal advective terms). Note that because velocity is squared, the centrifugal acceleration has the same sign on flood and ebb, while the Coriolis acceleration changes sign between flood and ebb. Thus, for purely harmonic tidal motion the tidal average of the Coriolis term will be zero. In contrast, the effects of flow curvature acting on an oscillatory flow result in a strong rectified motion due to the non-linear centrifugal acceleration. Furthermore, the sum of the Coriolis and centrifugal accelerations will augment during one phase of the tide and compete on the opposite phase, thus producing tidal asymmetries in secondary flows in regions of flow curvature.

Term E represents a forcing associated with cross-stream density gradients and is often a dominant term driving secondary circulations. Cross-stream density gradients, however, can also shut down secondary flows, similar to the shelf dynamics associated with the arrested Ekman layer (MacCready and Rhines, 1991). Finally, term F in equation (5.3) represents friction, which typically balances one of the forcing terms that drive secondary flow.

5.3. Development of lateral density gradients

It is instructive to discuss mechanisms that produce cross-channel baroclinic forcing before discussing the details of the dynamics of secondary flows because lateral buoyancy forcing plays a critical role in both the generation and the arresting of lateral flows. This discussion begins with the salt balance equation:

$$\frac{\partial s}{\partial t}+u\frac{\partial s}{\partial x}+v\frac{\partial s}{\partial y}+w\frac{\partial s}{\partial z}=\frac{\partial}{\partial z}\left(K_v\frac{\partial s}{\partial z}\right), \quad (5.4)$$

where K_v is the vertical diffusivity of salt, u is the along-channel (x) flow, v is the cross-channel (y) flow, and w is the flow in the vertical direction z. Taking the cross-channel gradient of equation (5.4) yields:

$$\underbrace{\frac{\partial}{\partial y}\frac{\partial s}{\partial t}}_{A} = \underbrace{\frac{\partial}{\partial t}\frac{\partial s}{\partial y}}_{A} = -\underbrace{\frac{\partial u}{\partial y}\frac{\partial s}{\partial x}}_{B} - \underbrace{u\frac{\partial}{\partial y}\frac{\partial s}{\partial x}}_{B'} - \underbrace{\frac{\partial v}{\partial y}\frac{\partial s}{\partial y}}_{C} - \underbrace{v\frac{\partial^2 s}{\partial y^2}}_{C'} - \underbrace{\frac{\partial w}{\partial y}\frac{\partial s}{\partial z}}_{D}$$
$$- \underbrace{w\frac{\partial}{\partial y}\frac{\partial s}{\partial z}}_{D'} + \underbrace{\frac{\partial}{\partial y}\frac{\partial}{\partial z}\left(K_v\frac{\partial s}{\partial z}\right)}_{E}. \tag{5.5}$$

Term A in equation (5.5) is the "tendency" term and represents the time rate of change of cross-channel salinity gradients. Note that the primed terms cannot generate cross-channel density gradients but rather can only produce a local change in the cross-channel gradient by the advection of existing cross-channel gradients. Term C modifies existing cross-channel salinity gradients by the compression of isohalines (Fig. 5.2a). In contrast, terms B and D actually generate

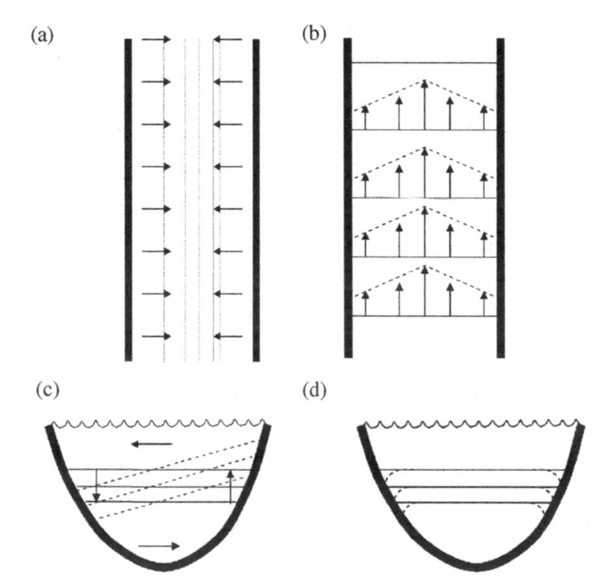

Figure 5.2. Tendency equation for secondary flows. In all figures the dashed lines depict the tendency for the flow field to modify the salt field, depicted by solid lines. (2a) Plan view depicting compaction/compression of isohalines by term C in equation (5.5). (2b) Generation of cross-channel density gradients by differential advection as described by term B in equation (5.5). (2c) Generation of cross-channel density gradient by term D in equation (5.5). (2d) Generation of cross-channel density gradients associated with diffusive boundary layer [term E in equation (5.5)].

cross-channel density (salinity) gradients. Term B generates cross-channel density gradients due to lateral shears in the along-channel flow $\left(\frac{\partial u}{\partial y}\right)$ acting on the along-channel salinity gradient $\left(\frac{\partial s}{\partial x}\right)$ (Fig. 5.2b), also known as differential advection. Later it will be shown that this is often a dominant mechanism driving secondary flows, as described in Nunes and Simpson (1985). Term D acts to tilt the vertical stratification by cross-channel variability in the vertical motion (Fig. 5.2c) and requires the existence of secondary flow. This term has been shown in a number of studies to arrest secondary flows (Chant and Wilson, 1997; Seim and Gregg, 1997; Chant, 2002; Lerczak and Geyer, 2004). Finally, term E represents cross-channel variations in mixing that would generate cross-channel salinity gradients and thus drive secondary flows. However, while this mechanism has been suggested to be important in shelf/slope dynamics (Phillips *et al.*, 1986; Garrett *et al.*, 1993), it appears to only play a minor role in generating estuarine secondary flows (Lerczak and Geyer, 2004). Even though there may be some estuarine environments where cross-channel gradients in mixing are important contributors to secondary flows, the importance of this term in driving estuarine secondary flows has yet to be demonstrated.

In summary, there are three major mechanisms that drive estuarine secondary flows: differential advection, Coriolis acceleration, and flow curvature. In addition, cross-channel salinity gradients can only be generated by three mechanisms (although only two are believed to be dominant): (a) differential advection, which is the development of cross-channel gradients set up by lateral shears in the along-channel flow acting on the along-channel salinity gradient; (b) the tilting of vertical stratification by cross-channel variability in vertical motion associated with the secondary flows themselves; and (c) cross-channel variations in mixing. Thus, from the point of view of the mass field, it is only differential advection that can initiate secondary flows because the second mechanism requires their existence and typically acts to shut down secondary flows. Considering only these two terms, the tendency equation (5.5) can be simplified to:

$$\frac{\partial}{\partial t}\frac{\partial s}{\partial y} = -\frac{\partial u}{\partial y}\frac{\partial s}{\partial x} - \frac{\partial w}{\partial y}\frac{\partial s}{\partial z}. \tag{5.6}$$

The first term on the right-hand side is typically associated with the generation of secondary flows due to differential advection, while the second term on the right-hand side tends to shut down existing secondary flows.

5.4. Differential advection

Depth-averaged along-channel tidal currents tend to be strongest over deeper parts of the channel. Consequently, an isohaline in the middle of the channel will advance further upstream (downstream) on the flood (ebb) tide than on the flanks and produce tidal period variations in cross-channel density gradients (Fig. 5.3). A classic example of differential advection was identified by Nunes and Simpson (1985), who showed that during the flood tide a pair of counter-rotating secondary flow cells develop in the vertical plane. The flow cells result in surface flows that converge over the deep channel, becoming visible as a line of flotsam oriented in the along-channel direction over the deep channel (Fig. 5.3a). This secondary flow is driven by a cross-channel baroclinic pressure gradient that is characterized by heavy, more saline waters in the deep channel relative to the flanks. The deep water in mid-channel then sinks under the influence of gravity and is replaced by fresher fluid from the flanks. During ebb tide the opposite may occur, with differential advection producing fresher water in the main channel relative to the flanks. Here, the saline water on the flanks sinks toward mid-channel while the freshwater in mid-channel rises and spreads to the flanks, producing an opposite lateral flow pattern to the flow that occurs during flood (Fig. 5.3b).

Smith (1977) and Nunes and Simpson (1985) assume that the two dominant terms in the cross-stream momentum balance for secondary flows driven by differential advection are pressure gradient and friction. This is analogous to the along-channel momentum balance assumed in Hansen and Rattray (1966), MacCready (2004), and summarized in Lerczak and Geyer (2004). Moreover, similar to the Hansen and Rattray (1966) theory, a constant vertical eddy viscosity is assumed and a steady-state assumption is made. With these simplifying assumptions, the cross-channel momentum balance can be written as:

$$\frac{g}{\rho}\frac{\partial \rho}{\partial y}z = \frac{\partial}{\partial z}\left[\frac{\tau}{\rho}\right] = K\frac{\partial^2 u}{\partial z^2}. \tag{5.7}$$

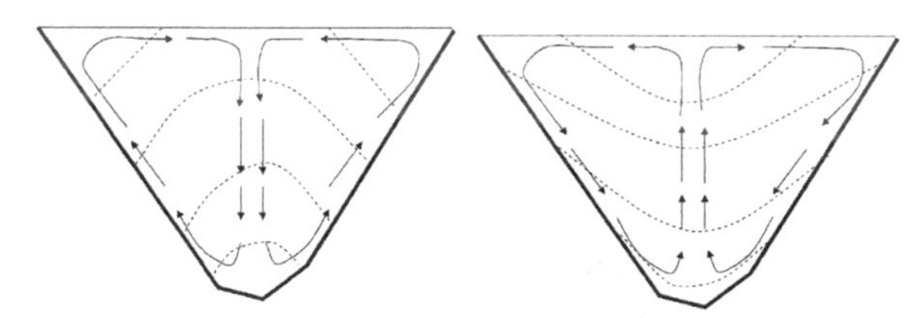

Figure 5.3. Salinity (dashed line) and sense of lateral flows (arrows) associated with differential advection during flood tide (right panel) and ebb tide (left panel).

The cross-channel density gradient, which produces the lateral baroclinic forcing, is

$$\frac{\partial}{\partial t}\frac{\partial \rho}{\partial y} = \frac{\partial u}{\partial y}\frac{\partial \rho}{\partial x}. \tag{5.8}$$

The lateral shear ($\partial u/\partial y$) associated with the oscillatory tide is defined as $\frac{2\Delta u}{B}\sin \sigma t$, where σ is the tidal frequency, Δu is the amplitude of the velocity difference between mid-channel ($y = B/2$) and the channel edges ($y = 0, B$), and B is the channel width. Thus, the cross-channel density gradient scales as $\frac{2\Delta u}{\sigma B}\frac{\Delta \rho}{\Delta x}$, which corresponds to a density difference between the main channel and the channel edges of $\Delta \rho = 2\frac{\Delta \rho}{\Delta x}\frac{\Delta u}{\sigma}$. Integrating equation (5.6) twice vertically, one obtains a scale for the cross-channel flow V_{DA} in an expression identical to the along-channel scale (Hansen and Rattray, 1966; Lerczak and Geyer, 2004):

$$V_{DA} \sim \frac{1}{48}\frac{gH^3}{A_v \rho_o}\frac{\partial \rho}{\partial y} = \frac{1}{48}\frac{gH^3}{A_v B}\frac{\Delta \rho(y)}{\rho_o} = \frac{1}{24}\frac{gH^3}{\rho_o A_v}\frac{\Delta \rho}{\Delta x}\frac{\Delta u(y)}{\sigma B}. \tag{5.9}$$

Equation (5.9) indicates that when lateral friction balances the cross-channel pressure gradient (set up by differential advection), the strength of the secondary flows is linearly proportional to the lateral shear, inversely proportional to the vertical eddy viscosity A_v, and varies with the cube of the channel depth H. Equation (5.9) also suggests that if the lateral shear shows no tidal period variability, then the secondary flows should be of the same magnitude on flood as on ebb. The flood tide causes convergence in the middle of the channel and divergence at depth, while the ebb tide causes divergence at the surface and convergence at depth. As will be shown later, however, lateral circulation in fact varies in strength with the tidal period because of tidal asymmetries in vertical stratification of density (Lerczak and Geyer, 2004).

The tendency for stratification to suppress secondary flows and thus the physics missing in equation (5.7) becomes apparent by comparing the strength of secondary flows predicted by equation (5.9) with field observations. For example, during neap tides in the Hudson River estuary, estimates of vertical eddy viscosity A_v in the halocline are between 10^{-5} and 10^{-4} m^2/s (Peters, 2001; Geyer *et al.*, 2008), lateral shear is $\sim 0.2 \times 10^{-3}$ m/s, and the along-channel density gradient is $\sim 3 \times 10^{-4}$ kg/m^4. Equation (5.9) would predict lateral flows of over 1 m/s, which is at least an order of magnitude larger than observed lateral flows in this system (Chant and Wilson, 1997; Lerczak and Geyer, 2004). Lateral flows are reduced because, as will be discussed later, they are shut down by density stratification and this mechanism is not included in the underlying physics contained in equations (5.7) and (5.9).

5.5. Flow curvature

It has long been recognized that a vertically sheared flow with curvature will develop a secondary flow with the lower-layer flow directed toward the inside of the bend and the upper-layer flow directed away from the bend (Rozovski, 1957). The dynamics that drive this lateral flow are easily conceptualized by first considering how flow rounds a bend. Consider a channel with a 90° bend as depicted in Fig. 5.4. Flow must accelerate from a north-westward flow to a south-eastward flow in order for the depth-averaged flow to round the bend. This acceleration is driven by a pressure gradient that sets up by the inertia in the flow. If the pressure gradient is too weak, the depth-averaged flow will not make it around the bend and will cause water to pile up at the outer bend, eventually generating exactly the correct pressure gradient that will steer the flow around the bend. If the flow is vertically sheared, as we would expect a boundary layer flow to be, with swifter flows near the surface, the fluid at the surface has too much momentum and will head toward the outside of the bend. Upon reaching the outer wall it flows downward. In contrast, flow in the lower layer is moving slower and thus the acceleration will drive the flow to the inside of the bend where it will upwell at the inner wall. This results in the classic helical flow pattern shown in Fig. 5.4b.

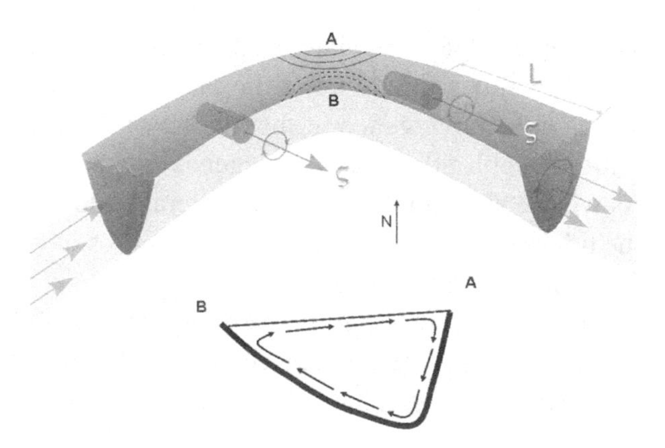

Figure 5.4. Upper panel: schematic showing the set-up of secondary flows associated with flow curvature. Flow entering the channel is vertically sheared and has a relative vorticity vector ζ pointed to the south-east. Lines at channel bends depict sea level, with solid lines elevated levels and dashed lines lower levels. L corresponds to the length scale of the spin-down of secondary flows. Lower panel: structure of secondary flows in and downstream of channel bend. The channel cross-section in the lower panel is drawn asymmetrically to depict the tendency for secondary flows in channel bends to deepen the channel on the outside of the bend.

A second way to conceptualize the effects of flow curvature is by using a vorticity argument. In the upstream flow depicted in Fig. 5.4a, the flow is vertically sheared and the relative vorticity vector is pointed to the south-east. This vorticity is depicted as a cylinder in the flow that rotates clockwise (looking from the south-east) with the vertical shear. As this fluid rounds the bend, this south-easterly pointing vorticity vector corresponds to a secondary flow toward the inside of the bend at depth and toward the outside of the bend at the surface (Fig. 5.4b). Eventually, bottom friction will generate vertical shear and a vorticity vector that points toward the west (Fig. 5.4a).

These dynamics are more precisely described, of course, with the use of a momentum equation written in streamwise coordinates [analogous to equation (5.1)] for a homogenous fluid and neglecting the Earth's rotation:

$$\frac{\partial u_n}{\partial t} + u_s \frac{\partial u_n}{\partial s} - \frac{u_s^2}{R} + g\frac{\partial \eta}{\partial n} - \frac{\partial}{\partial z}\left(A_v \frac{\partial u_n}{\partial z}\right) = 0. \tag{5.10}$$

Taking the depth-average of this equation and neglecting stress at the sea surface results in:

$$\overline{u_s \frac{\partial u_n}{\partial s}} - \frac{\overline{u_s^2}}{R} + g\frac{\partial \eta}{\partial n} - \frac{\tau_b}{\rho H} = 0, \tag{5.11}$$

where τ_b is the bottom stress. Note that the first term in equation (5.10) vanishes upon vertical integration because the depth-average of the cross-stream flow (u_n) is exactly zero. Taking the difference between equations (5.10) and (5.11) and neglecting time dependence and the first term in equation (5.11), which is generally small, yields:

$$u_s \frac{\partial u_n}{\partial s} - \frac{\partial}{\partial z}\left(A_v \frac{\partial u_n}{\partial z}\right) + \frac{\tau_b}{\rho H} = \frac{u_s^2 - \overline{u_s^2}}{R}. \tag{5.12}$$

The first term in equation (5.12) is the streamwise advection of lateral flow, the second and third terms are frictional forces. The term on the right-hand side is the shear forcing and drives the secondary flows. Clearly, equation (5.12) indicates that the forcing increases quadratically with the shear.

Kalkwijk and Booij (1986) provided analytical solutions of equation (5.11) for a case with a logarithmic velocity profile and a parabolic eddy viscosity profile. Given a non-dimensional quadratic drag coefficient, $C_D = 3 \times 10^{-3}$, the maximum strength of secondary flows as in Kalkwijk and Booij (1986) is $6\,|\overline{u_s}|H/R$ (Geyer, 1993) (Fig. 5.5). However, observations of secondary flows around a headland by Geyer (1993) clearly demonstrated that lateral flows were twice as large as predicted by Kalkwijk and Booij (1986). Geyer (1993) attributed the discrepancy

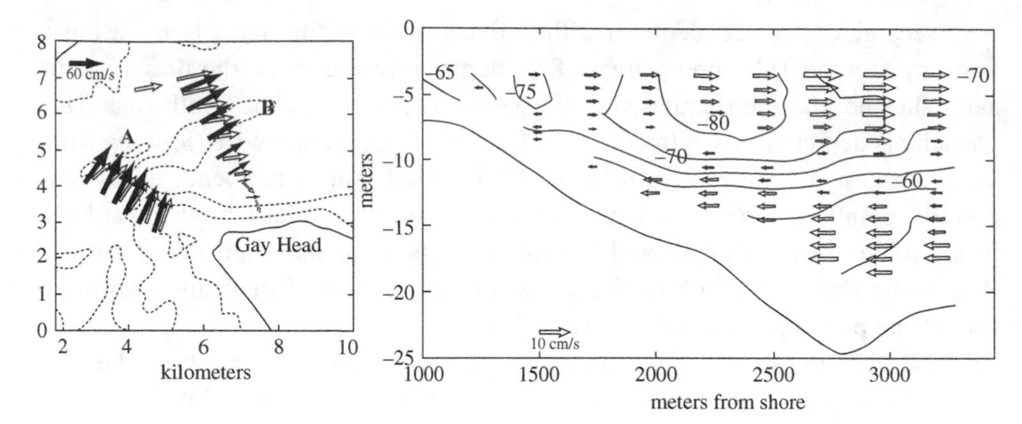

Figure 5.5. Composite from Geyer (1993) showing secondary flows from shipboard surveys off Martha's Vineyard. In left panel, filled arrows are near-surface vectors while arrows are near-bottom vectors. Veering angle for transect B is over 30°. Right panel shows streamwise flows (contours) and secondary flows (arrows) for transect B.

between the theory and observations to the effects of stratification and assessed the role of stratification by numerically solving equation (5.12), neglecting the first term. Geyer (1993) noted that stratification allowed the vertical shear in the streamwise flow to be stronger than it would be in a logarithmic layer, and thus the forcing term to the secondary flows, $\left(u_s^2 - \overline{u_s^2}\right)/R$, is stronger. In addition, because this term is balanced by friction, $\frac{\partial}{\partial z}\left[A_v \frac{\partial u_n}{\partial z}\right]$, reduced eddy viscosity in the presence of stratification requires stronger shear in the secondary flow for frictional forces to balance the (increased) shear forcing. Thus, Geyer (1993) argued that the effects of stratification were twofold in augmenting secondary circulation for it (a) increased the shear in the streamwise flow that forces the secondary flows and (b) suppressed vertical eddy viscosity, thus requiring stronger lateral shears (i.e., secondary flows) to balance the shear forcing.

Similar conclusions were reached by Chant (2002) based on a long-term mooring deployment in Newark Bay, NJ, near a region of strong flow curvature. During times of low river flow a strong secondary flow was observed as linearly proportional to tidal current amplitude (Fig. 5.6a,b). Chant (2002) argued that the shear forcing would increase quadratically with tidal current speed and this would be balanced by a quadratic increase in the cross-channel frictional term. The quadratic increase in the frictional term occurs because eddy viscosity increases linearly with tidal current speed, as suggested by Bowden and Fairbairn (1952a,b). Thus, a linear increase in vertical eddy viscosity and in the vertical shear (secondary flows) will cause the stream-normal stress term to increase quadratically to match the quadratic

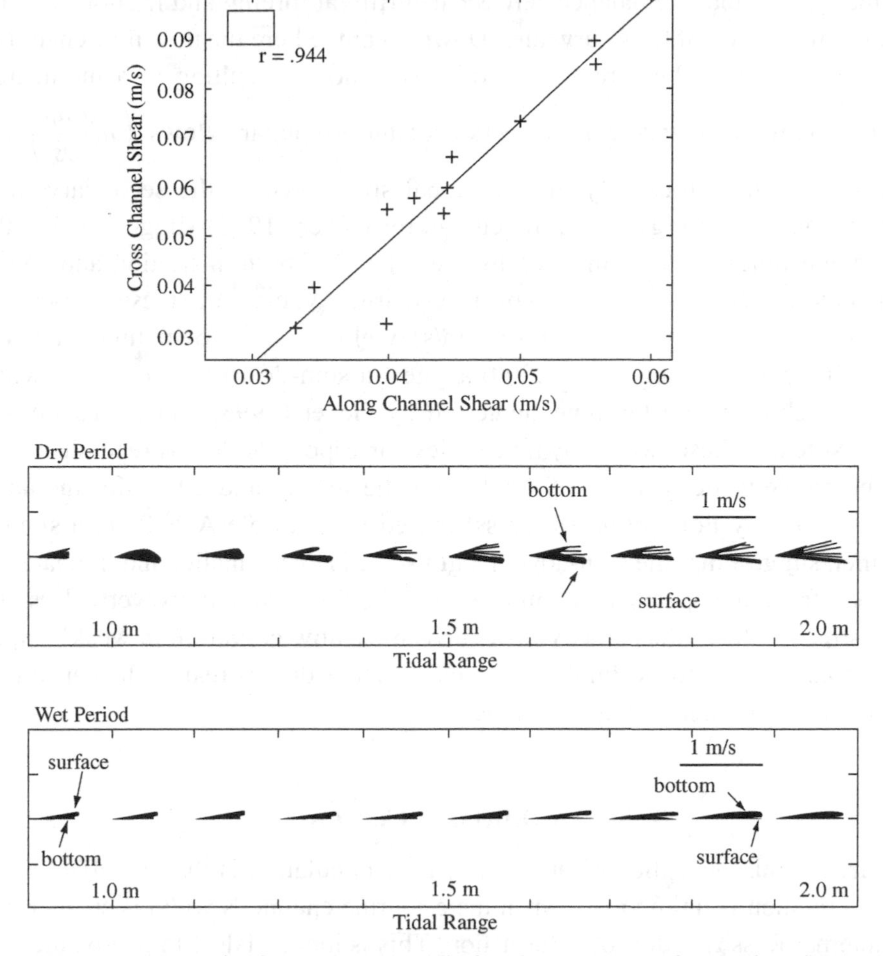

Figure 5.6. Composite from Chant (2002) showing response of secondary flow to variations in tidal range and stratification from a mooring in Newark Bay. Upper panel shows linear relationship between along-channel shear and cross-channel shear (i.e., secondary flows). Middle panel shows veering of current vector during low river discharge during ebb tide in a region of counter-clockwise flow curvature. Flow toward the bottom of the page is toward the outside of the bend, while flow toward the upper side of the page represents flow toward the inside of the bend. Data is binned as a function of tidal range. Bottom panel shows lack of veering during times of high river discharge, indicating that secondary flows are shut down by stratification.

increase in the forcing to the secondary flow. Therefore, secondary flows increase linearly with tidal current speed, despite the quadratic increase in forcing. However, in the same record, during times of high river discharge, Chant (2002) noted that secondary flows were shut down due to buoyancy effects, and this is discussed in more detail later (Fig. 5.6c).

Finally, note that the balance between centrifugal forcing and friction can only occur in the region of flow curvature. Downstream, where the flow may encounter a straight channel, the forcing would be shut down resulting in a momentum balance between downstream advection of the secondary flows $\left(u_s \frac{\partial u_n}{\partial s}\right)$ and friction, resulting ultimately in a frictional spin-down of the secondary flows downstream of the region of flow curvature. Geyer (1993) suggested that this spin-down timescale is dominated by the effects of bottom friction and can be estimated as $H/C_D u_s$. Given typical estuarine values of these parameters ($H = 10$ m, $C_D = 3 \times 10^{-3}$, and $u_s = 0.5$ m/s) yields a spin-down time of a few hours. In contrast, Fong *et al.* (2009) suggest a spin-down time of H/u^*, which is $(C_d)^{1/2}$ shorter than the time suggested by Geyer (1993) and corresponds to 400 s. Note that these two decay time scales correspond to downstream distances, for the above velocity scale, of 4000 m for the former and 200 m for the latter. Observations by Fong *et al.* (in press), based on multiple ADCPs in a sinuous channel, suggest that the spin-down length scale is even smaller and indicate that internal friction appears to dominate over bottom friction. Nevertheless, the downstream effect of secondary flows driven by flow curvature in weakly stratified cases appears to be limited. The case of the downstream effect in highly stratified environments is discussed next.

5.6. Effects of Coriolis

A common misconception regarding estuarine circulation is that the effect of the Earth's rotation is unimportant when the estuarine channel's width is smaller than the internal Rossby radius of deformation. This is indeed false! In reality, the first-order momentum balance in the cross-channel direction is often geostrophic. Consider, for example, the depth-averaged momentum equation in the cross-channel direction for a homogenous fluid and (momentarily) neglecting the advective term:

$$\frac{\partial v}{\partial t} = -fu - \frac{1}{\rho}\frac{\partial P}{\partial x} + \frac{\partial}{\partial z}\left(\frac{\tau_y}{\rho}\right). \tag{5.13}$$

An along-channel flow (u) will accelerate the cross-channel velocity, which tends to tilt the sea level upward to the right side of the channel (looking downstream in the northern hemisphere) and produces a lateral pressure gradient. If the along-channel flow has no shear, then no cross-channel flows will develop and the flow will be geostrophic. However, as is always the case, there will be shear in the along-channel flow due to frictional and density effects and thus there will be an imbalance between the depth-independent cross-channel barotropic pressure gradient and the depth-dependent Coriolis acceleration. This produces an ageostrophic cross-channel flow.

In a similar approach to that used to investigate the effects of curvature, the effects of Earth's rotation on driving secondary flows are studied by subtracting the vertical average of equation (5.13) from equation (5.13) itself. This analysis, for now, assumes a homogenous fluid and continues to neglect the advective terms. With these assumptions, the vertical average of equation (5.13) is:

$$\frac{\partial \overline{v}}{\partial t} = -f\overline{u} - g\frac{\partial \eta}{\partial y} + \frac{\tau_{by}}{\rho H}. \tag{5.14}$$

Subtracting equation (5.13) from equation (5.14) results in:

$$\frac{\partial(\overline{v} - v)}{\partial t} = -f(\overline{u} - u) - \frac{\tau_{by}}{\rho H} + \frac{\partial}{\partial z}\left(\frac{\tau_y}{\rho}\right). \tag{5.15}$$

For a steady-state case the shear forcing associated with Coriolis accelerations is balanced by friction, consistent with Ekman dynamics. Thus, in the case of a river flowing seaward the bottom layer flow will be directed to the left of the depth-averaged flow, as expected in a bottom Ekman layer. Using analytical solutions of Kalkwijk and Booij (1986), Geyer (1993) finds that for a quadratic bottom drag coefficient of 3×10^{-3} the magnitude of the lateral circulation forced by Coriolis accelerations is $3fH$. Interestingly, the magnitude of the secondary flow is *not* proportional to the flow speed. This is in contrast to curvature-induced secondary flows, which are proportional to flow speed. A second difference between curvature-forced secondary flows and those associated with Coriolis is that the direction of lateral circulation associated with Coriolis forcing changes sign with the sign of the along-channel flow, whereas in the case of curvature-forced secondary flows the sign of the secondary flow is independent of the sign of the along-channel or streamwise flow.

The strength of the Coriolis forcing, however, does not have a simple relationship with boundary layer thickness β as given by $\beta = \sqrt{2A_v/\sigma}$ (where σ is the tidal frequency; Lerczak and Geyer, 2004). Yet, given the range of vertical eddy viscosities A_v considered in Lerczak and Geyer ($3\text{–}22 \times 10^{-4}$ m^2/s), the strength of secondary flows is independent of β and has a tidally varying amplitude of $\frac{1}{8}\frac{f}{\sigma}U_o$, where U_o is the tidal current amplitude. Thus, for a mid-latitude estuary with tidal currents of 1 m/s, Coriolis-forced secondary flows will have an amplitude of 9 cm/s. The effects of rotation can be important, even in narrow estuaries, because of the lateral flows produced.

The vertical structure of secondary flows that are forced by the Earth's rotation varies with the Ekman number (β/H). When the boundary layer occupies the entire water column, secondary flows are characterized by a single cell, while for small Ekman numbers (thin boundary layers) a more complex lateral flow structure develops that is comprised of multiple cells that vary over the tidal cycle (Lerczak and Geyer, 2004).

A major impact of Coriolis-forced motion on estuarine circulation is that it generates a cross-channel asymmetry in the structure of along-channel exchange flow. This asymmetry is characterized by the inflow (when looking toward the ocean in the Northern Hemisphere) tending to the right side of the channel and the outflow to the left (Valle-Levinson *et al.*, 2000, 2007; Lerczak and Geyer, 2004; Valle-Levinson, 2008). Lerczak and Geyer (2004) suggest this asymmetry would tend to accumulate sediment on the right side of Northern Hemisphere channels and produce a morphologically laterally asymmetric channel, such as that found in the Hudson River and other estuarine channels. This asymmetry could feed back by driving lateral flows associated with differential advection, which would tend to augment the trapping of sediment on the right flank. The feedback between geomorphology and lateral flow processes has been observed to occur in estuarine systems (Geyer *et al.*, 2001; Fugate *et al.*, 2007) and treated analytically by Huijts *et al.* (2006).

5.7. Effects of stratification

While Geyer (1993) suggested that stratification tends to strengthen secondary flows, others (Chant and Wilson, 1997; Seim and Gregg, 1997; Chant, 2002; Lerczak and Geyer, 2004) show clear evidence that secondary flows are shut down by the buoyancy effects of vertical density stratification. This discrepancy occurs because, while under weak stratification, the dynamics are consistent with Geyer's analysis (discussed earlier), under stronger stratification the tilting of the halocline by secondary flows produces a baroclinic pressure gradient that opposes the shear forcing. This tilting occurs via the advective term $\frac{\partial w}{\partial y}\frac{\partial s}{\partial z}$, which appears as the second term on the right-hand side of equation (5.6). The tendency for vertical stratification to suppress secondary flows is analogous to the shutting down of Ekman transport during upwelling conditions on sloping continental shelves, as described by MacCready and Rhines (1991) and Garrett *et al.* (1993).

Chant and Wilson (1997) presented data from the Hudson River estuary around a region of strong flow curvature, where secondary flows were relatively weak (similar to the situations depicted in Fig. 5.6c). Chant and Wilson added baroclinic

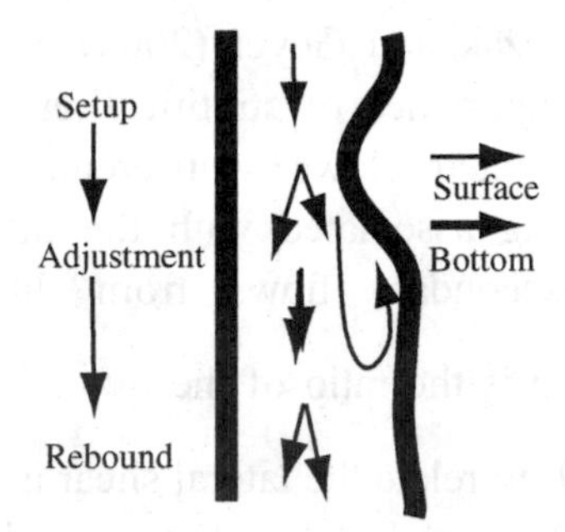

Figure 5.7. Schematic from Chant and Wilson (1997) showing setup of cross-channel baroclinicity, down-stream adjustment and oscillatory rebound down stream.

pressure gradient to equation (5.12) and compared it to estimates of the shear forcing. The cross-stream baroclinic pressure gradient was defined as

$$\frac{g}{\rho}\frac{\partial}{\partial n}\int_0^z \rho(z')dz' - \overline{\frac{g}{\rho}\frac{\partial}{\partial n}\int_0^h \rho(z')dz'}, \tag{5.16}$$

where the overbar represents depth average. The second term then represents the depth-averaged value of the baroclinic pressure gradient. Chant and Wilson (1997) computed equations (5.12) and (5.16) using CTD data and with the forcing term $\left(u_s^2 - \overline{u_s^2}\right)/R$ based on shipboard ADCP data. They found them to be nearly in balance.

As with weakly stratified flows, the momentum balance between pressure gradient and centrifugal accelerations must change away from the region of flow curvature as the shear-forcing term goes to zero. However, rather than simply spinning down, the cross-stream density gradient (set up in the region of flow curvature) will adjust and result in a lateral sloshing of the secondary flows (Fig. 5.7). On the one hand this could potentially extend the downstream influence of curvature-induced secondary flows because of the reduced internal friction associated with stratification, allowing the lateral sloshing to potentially extend far downstream. On the other hand, the oscillatory nature of a lateral sloshing will limit the cross-stream excursion of a parcel of fluid and thus limit the cross-stream mixing. Chant (2002) provided a scaling of the lateral mixing associated with a cross-stream seiche and suggested that it was generally a weak mechanism driving cross-stream mixing.

As alluded to earlier, stratification also constrains lateral flows generated by differential advection. If the channel is sufficiently stratified, the tendency for lateral flows to tilt isopycnals will produce a baroclinic pressure gradient that can shut down secondary flows. Lerczak and Geyer (2004) introduced a parameter that characterizes the relative importance of stratification on suppressing secondary flows to the generation of secondary flows by differential advection. The parameter, γ, is the ratio of the forcing associated with the tilting of the isopycnals to the forcing that drives secondary flows from differential advection, i.e., $\gamma = \frac{(\partial\rho/\partial z)(\partial w/\partial y)}{(\partial\rho/\partial x)(\partial u/\partial y)}$, which is the ratio of the two right-hand terms in equation (5.6). Using expression (5.9) to relate the lateral shear in the along-channel flow to the scale of the secondary flows and scaling the cross-channel gradient in vertical velocity as $\frac{\partial w}{\partial y} = \frac{Hu}{B^2}$, Lerczak and Geyer (2004) obtained:

$$\gamma \sim \frac{1}{24}\frac{N^2H^2}{B^2}\frac{H^2}{A_v}\frac{1}{\sigma} = \frac{1}{24}\frac{T_{fr}T_T}{T_{Iw}}, \tag{5.17}$$

where N is the buoyancy frequency and T_T, T_{fr} and T_{Iw} are the time scales associated with the tide, friction and internal waves, respectively. The internal wave time scale is set by the cross-channel travel time for an internal wave, and is equal to the ratio of the internal wave speed squared (N^2H^2) to the width of the channel squared (B^2). The frictional time scale (H^2/A_v) is the time scale required to mix momentum completely in the vertical. The tidal time scale is, of course, the tidal period. Lerczak and Geyer (2004) argue that when $\gamma \ll 1$, stratification is unable to suppress secondary circulation. However, as γ approaches unity, tilting of the pycnocline by the secondary flow significantly suppresses the lateral motion. As will be discussed later, the variability in stratification at both tidal and fortnightly time scales has been shown to play a central role in the modulation of the dynamics that govern estuarine circulation over the spring/neap cycle.

5.8. Diffusive boundary layer

The final process that can drive lateral circulation consists of mixing along the sloping boundaries. The no-flux boundary condition requires that the vertical density gradient vanishes near the bottom and results in a cross-channel baroclinic pressure gradient that draws fluid up from the lower layer toward the shoaling flank (Fig. 5.4d). For a fluid with a constant eddy viscosity and a Prandtl number (ratio of eddy viscosity to eddy diffusivity of salt) of 1, Garrett *et al.* (1993) find that the thickness of the diffusive boundary layer, δ, is

$$\delta^{-4} = \frac{1}{4A_v^2}(f^2 + N^2\sin^2\theta), \tag{5.18}$$

where θ is the slope of the bottom relative to the horizontal. The maximum cross-channel flow that develops from diffusive boundary layers v_{BL} is given by (Garrett *et al.*, 1993; Lerczak and Geyer, 2004):

$$v_{BL} = \frac{A_v}{\sqrt{2\delta}} \cot \theta. \tag{5.19}$$

In a series of idealized numerical simulations inspired by conditions in the Hudson River estuary, lateral flows associated with diffusive boundary layers as scaled by equation (5.19) were typically 2 cm/s and nearly an order of magnitude smaller than lateral circulation driven by other processes (Lerczak and Geyer, 2004). Subsequently, diffusive boundary layer processes appear to play only a minor role in the dynamics of estuarine secondary circulation. Note, however, that the Lerczak and Geyer (2004) numerical simulations only covered a range of estuarine parameter space occurring in the Hudson River, and thus diffusive boundary layers may be more important in systems that fall outside of this regime. Moreover, Lerczak and Geyer (2004) ran their simulations with a constant eddy visocity. It is likely that more complex flows that develop with more realistic turbulent closures schemes will show that diffusive boundary layers have a greater impact than suggested by Lerczak and Geyer (2004).

5.9. Role of secondary flows on streamwise processes

Thus far this chapter has discussed only the mechanisms that drive lateral circulation and provided field and modeling examples of particular cases. While these studies have advanced our understanding of the dynamics of secondary circulation, they may leave the reader wondering "what is the significance of lateral circulation on estuarine processes?" For example, are secondary flows (as the name may imply) of secondary importance to estuarine dynamics? The answer appears to be a resounding NO! Indeed, analytical (Winant, 2004), numerical (Lerczak and Geyer, 2004), and field observations (Geyer *et al.*, 2008) highlight the important role that secondary flows have on along-channel dispersion and on the very dynamics that drive estuarine exchange, which themselves play a central role in along-channel dispersive processes.

5.9.1. Role in along-channel dynamics

Recent numerical model studies have clearly demonstrated that secondary circulation plays an important role in driving the estuarine exchange flow (Lerczak and Geyer, 2004; Scully *et al.*, 2009). The classic theory of estuarine exchange flow neglects the role of advection and balances the along-channel pressure gradient with frictional forces. In this model, friction is assumed to be equal to the tidally averaged vertical shear (i.e., the exchange flow) times a tidally averaged vertical eddy viscosity. Based on this balance, the estuarine exchange flow Ue is scaled by (Hansen and Rattray, 1966; MacCready, 2004):

$$Ue \sim \frac{1}{48}\frac{gH^3}{A_v\rho_o}\frac{\partial\rho}{\partial y}. \tag{5.20}$$

Equation (5.20) predicts that the estuarine exchange flow will be inversely proportional to the vertical eddy viscosity A_v. Microstructure data from the Hudson River estuary indicate that vertical eddy viscosity varies by an order of magnitude over the spring/neap cycle (Peters, 2001), while hydrographic sections indicate that during low to moderate river discharge the along-channel density gradient is relatively constant over the spring/neap cycle (Lerczak *et al.*, 2006). Thus, equation (5.20) predicts that the exchange flow should vary by an order of magnitude over the spring/neap cycle. However, observations clearly show that the exchange flow varies by a factor of 2–3 over the spring/neap cycle (Geyer *et al.*, 2000), suggesting that important dynamics are missing in the development of classical estuarine theory.

Numerical results from Lerczak and Geyer (2004) describe in detail how secondary flows accelerate exchange flows during spring tides and thus buffer the 10-fold effect predicted by equation (5.20) over the spring/neap cycle. The key to this buffering effect is the interaction between forces that drive the secondary flows due to differential advection, and the time-varying vertical density stratification at both tidal and spring/neap time scales. On the tidal time scale, this asymmetry is characterized by enhanced stratification on the ebb tide and reduced stratification on flood as a result of tidal straining (Simpson *et al.*, 1990). On the neap/spring time scale, modulations in vertical mixing weaken stratification during spring tide and enhance stratification during neap tide. Thus secondary flows are often suppressed during ebb tides, due to the effects of stratification and, in the case of the modeling by Lerczak and Geyer (2004), throughout the tidal cycle during neap tides when the water column remains highly stratified on both flood and ebb. In contrast, during flood phases of the spring tide the model produces the classic two-cell secondary flow field as described by Nunes and Simpson (1985). Lerczak and Geyer (2004) model results (Fig. 5.8a–d) show that during flood, lateral circulation advects low-momentum fluid from the flanks into surface waters in mid-channel, which acts to decelerate the flood – or equivalently, accelerate the upper layer seaward. Meanwhile the strong downward vertical motion in the center of the channel advects the strong landward currents at the surface to the lower layer, which accelerates the lower layer landward. In contrast, during the ebb tide (Fig. 5.8a,b), stratification suppresses the secondary flows and these advective tendencies are not at play.

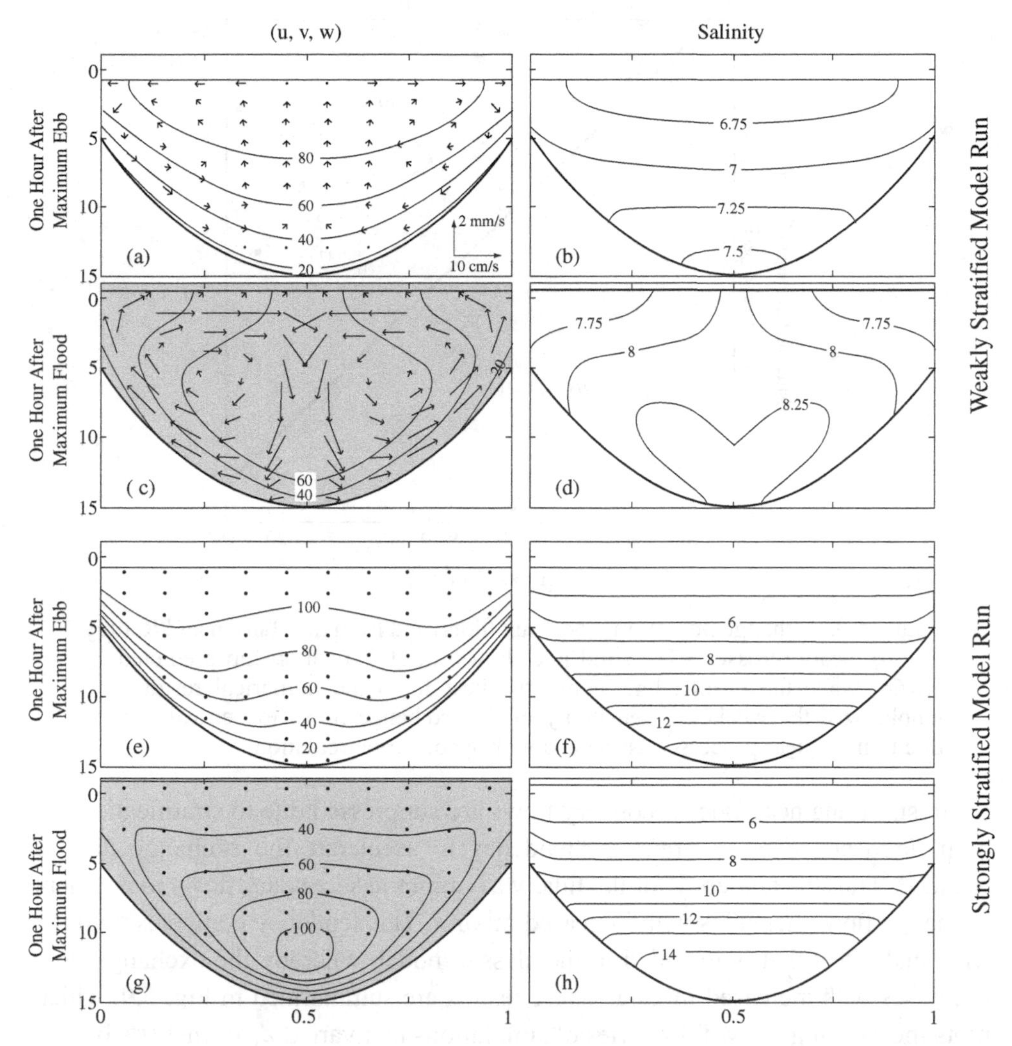

Figure 5.8. Results from Lerczak and Geyer (2004) showing along-channel velocity (contours in left panel), stratification (contours in right panels) and secondary flows (vectors in left panels). Panels ab show "spring tide" conditions during ebb. Panels cd show "spring tide" conditions during the flood. Panels ef show "neap tide" conditions during ebb and panels gh show neap tide conditions during the flood.

Similarly during the neap tide, strong stratification throughout the tidal cycle suppresses secondary flows and the advective momentum exchange is minimal (Fig. 5.8e–h). Consequently, the tidally averaged effects of the secondary flow on the along-channel momentum balance during spring tides resemble the flood tide conditions. This lateral circulation effect, because it accelerates the lower layer landward and the upper layer seaward, tends to augment the exchange flow. In

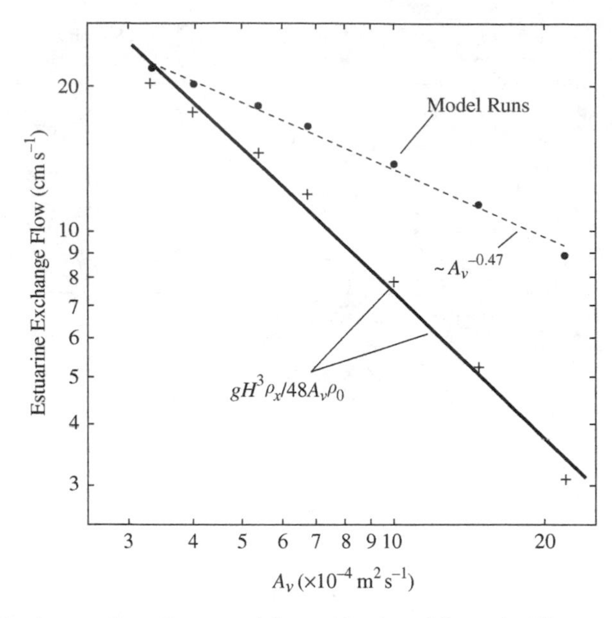

Figure 5.9. Exchange flow from model run (dots) and from the Hansen and Rattray (1966) theory (crosses). The solid line shows the A_v^{-1} relationship predicted by HR66, while the dashed line shows the best fit to the numerical results and emphasized the weakened sensitivity of the exchange flows to vertical mixing due to non-linear processes associated with secondary circulation.

contrast, during neap tides secondary flows are suppressed due to stratification, on both flood and ebb, and thus the tendency to accelerate the exchange flow is reduced. Thus the tendency for the tidally asymmetric secondary flows to augment exchange flow intensifies with increased mixing. This tendency competes with the frictional effect that is included in the classic model, whereby the exchange flow decreases with increased mixing. These results are summarized in Fig. 5.9, which plots the exchange flow for a series of simulations that varied A_v from 3.0×10^{-4} to 24×10^{-4} m^2/s, characteristic of neap tide and spring tide conditions. Results demonstrate that the exchange flow is *less* sensitive to mixing than predicted by equation (5.20). Indeed, rather than falling as A_v^{-1}, the exchange flow in these simulations fell as only $A_v^{-0.47}$. This suggests that the reduced sensitivity of estuarine exchange flow to variations in vertical mixing, relative to that predicted by equation (5.20), is caused by the non-linear effects of secondary flows.

While the results of Lerczak and Geyer (2004) are quite compelling, there are several cautionary notes. First, the results discussed above were run with a constant eddy viscosity, while it is well known that the eddy viscosity varies by orders of magnitude in both space and time. Lerczak and Geyer (2004) did run simulations with the *k–ω* turbulent closure scheme and found that secondary flows did develop during neap tide on flood, but were confined to the bottom boundary layer. While these secondary flows may play an important role in the momentum balance, they do not effectively exchange momentum between the upper and lower layers and thus do not appear to be as effective in augmenting the exchange flow as during weakly stratified conditions. A second cautionary note is that while observations of secondary flows in the Hudson River, and in other estuarine systems, do generally show a reduction in secondary flows during stratified conditions, their structure becomes more complex and characterized by multiple circulation cells in the vertical. Lerczak and Geyer (2004) did present results using the k–ω turbulent closure scheme, which exhibited some of this complexity, most notably the circulation in the lower layer. The model failed to capture the upper layer cell apparent in observations during neap tide floods. Finally, recent numerical results by Scully *et al.* (2009) suggest that the augmentation of the exchange flow during spring tide is largely compensated by increased interfacial stresses. This balance between the exchange flow and interfacial stresses appears to explain why the simple scaling argument proposed by Geyer *et al.* (2000), which neglects both interfacial mixing and secondary flows, accurately predicts the estuarine exchange flow. While it is clear why these two terms (advective effects associated with secondary flows and interfacial stresses) have opposite tendencies, it is unclear why their temporal variability should perfectly compensate for each other over the spring/neap cycle. Is this fortuitous? Does it suggest some dynamical link between secondary flows and mixing? Or is it an artifact of the model?

5.10. Role of secondary flows in dispersion

Secondary circulation is the dominant process driving lateral mixing in estuaries. The interaction between lateral mixing and lateral shears drives along-channel dispersion via a shear dispersion mechanism. This is analogous to the vertical shear dispersion produced by the interaction between vertical shear and vertical mixing. Like vertical shear dispersion, lateral shear dispersion can occur due to both the mean shear and the tidally oscillatory shear (Wilson and Okubo, 1978). In both cases the shear dispersion is orders of magnitude larger than the dispersion driven by small-scale turbulence (Fischer *et al.*, 1979). The mechanism of lateral mixing is itself a vertical shear dispersion whereby the vertical shear in the cross-channel flow, associated with the secondary flow, interacts with vertical mixing. The rate of vertical shear dispersion in the cross-stream direction (D_y) is

$$D_y = \alpha \frac{\Delta v^2 H^2}{K_z}, \tag{5.21}$$

where Δv, H and K_z are the vertical shear in the cross-channel flow (i.e., the secondary circulation), the channel depth and the vertical eddy diffusivity, respectively. The coefficient α depends on the vertical structure of the velocity and diffusivity and in estuarine flows it is ~ 1 to 10×10^{-3} (Geyer *et al.*, 2008). Thus, for typical estuarine flows ($K_z = 1\text{–}10 \times 10^{-3}$ m^2/s, $\alpha = 5 \times 10^{-3}$, $\Delta v = 0.1$ m/s and $H = 10$ m), lateral dispersion associated with vertical shear dispersion will be on the order of 10 m^2/s. The along-channel dispersion associated with lateral shear dispersion is identical to equation (5.21) with the denominator replaced by K_y and the shear by the cross-channel shear in the along-channel flow, i.e.

$$D_x = \alpha \frac{\Delta u^2 H^2}{K_y}. \tag{5.22}$$

Whether lateral shear dispersion is driven by tidally oscillating lateral shear or by the mean shear depends on the lateral mixing time. Unlike steady shear dispersion, which continues to increase with decreasing vertical eddy diffusivity, shear dispersion associated with tidal motion will be maximum when the mixing time is on the order of a tidal cycle. Note that the lateral mixing time T_x is given by $W^2/10K_y$, where W is the oscillatory lateral shear dispersion associated with tides. This lateral mixing may be an important mechanism driving along-channel dispersion in channels that are 100 to 1000 m wide.

Observations of the lateral spread of a dye patch confined to the lower layer in the Hudson River estuary indicated a K_y of ~ 1 m^2/s during the flood tide. This value is consistent with equation (5.21) using the patch thickness of 3–5 m for H of 10 m and vertical eddy diffusivity of $1\text{–}2 \times 10^{-3}$ m^2/s (Geyer *et al.*, 2008). During the ebb tide, weak lateral flows were suppressed by stratification and caused estimates of $K_x \sim 0.2$ m^2/s, which were consistent with equation (5.21). During spring tides, lateral shear dispersion was the dominant process driving dispersion, because vertical shear dispersion was suppressed by strong mixing. However, during neap tides vertical shear dispersion dominated and was significantly larger than the spring tide dispersion rates. Nevertheless, in many narrow estuarine systems, lateral shear dispersion is often the dominant process driving along-channel dispersion (Fischer, 1972; Smith, 1977). Even in estuaries where along-channel dispersion is dominated by vertical shear dispersion, the fact that secondary flows act to augment the estuarine shear suggests that they add to the dispersive nature of the estuary.

5.11. Summary

Secondary circulation has a first-order impact on the along-channel dynamics of an estuary, estuarine dispersion and estuarine geomorphology. There are a number of processes that drive secondary flows, but it appears that in many systems differential advection dominates, with the exception in the vicinity of channel bends where flow curvature dominates. Coriolis acceleration also plays a subtle but significant role in the dynamics of secondary flows, even in very narrow estuaries.

References

Bowden, K. F. and L. A. Fairbairn (1952a) Further observations of the turbulent fluctuations in a tidal current. *Phil. Trans. R. Soc. A* **244**, 335.

Bowden, K. F. and L. A. Fairbairn (1952b) A determination of the friction forces in a tidal current. *Proc. R. Soc. London, Ser. A* **214**, 371–392.

Chant, R. J. (2002) Secondary flows in a region of flow curvature: relationship with tidal forcing and river discharge. *J. Geophys. Res.*, doi:10.1023/2001JC001082.

Chant, R. J. and R. E. Wilson (1997) Secondary circulation in a highly stratified channel. *J. Geophys. Res.* **102**, 23,207–23,215.

Fischer, H. B. (1972) Mass transport mechanisms in partially stratified estuaries. *J. Fluid Mech.* **53**, 671–687.

Fischer, H. B., E. J. List, R. C. Koh, J. Imberger and N. H. Brooks (1979) *Mixing in Inland and Coastal Waters*. Academic Press, New York.

Fong, D. A. *et al.* (2009) Turbulent stresses and secondary currents in a tidally-forced channel with significant curvature and asymmetric bed forms. *J. Hydr. Eng.* **135**(3), 198–208.

Friedrichs, C. T. and D. G. Aubrey (1988) Non-linear tidal distortion in shallow well-mixed estuaries: a synthesis. *Est. Coast. Shelf Sci.* **27**, 521–545.

Fugate, D. C. *et al.* (2007) Lateral dynamics and associated transport of sediment in the upper reaches of a partially mixed estuary, Chesapeake Bay USA. *Cont. Shelf Res.* **27**, 679–698.

Garrett, C. *et al.* (1993) Boundary mixing and arrested Ekman layers: rotating stratified flow near a sloping boundary. *Ann. Rev. Fluid Mech.* **25**, 291–323.

Geyer, W. R. (1993) Three dimensional tidal flow around headlands. *J. Geophys. Res.* **98**, 955–966.

Geyer, W. R., R. Chant and R. Houghton (2008) Tidal and spring–neap variations in horizontal dispersion in a partially mixed estuary. *J. Geophys. Res.* **113**, C07023, doi:10.1029/2007JC004644.

Geyer, W. R. *et al.* (2000) The dynamics of a partially mixed estuary. *J. Phys. Oceanogr.* **30**, 2035–2048.

Geyer, W. R. *et al.* (2001) Sediment transport and trapping in the Hudson River Estuary. *Estuaries* **24**(5), 670–679.

Hansen, D. V. and M. Rattray (1966) New dimensions in estuary classification. *Limnol. Oceanogr.* **11**, 319–326.

Huijts, K. M. H. *et al.* (2006) Lateral entrapement of sediment in tidal estuaries: an idealized model study. *J. Geophys. Res.* **111**, C12016, doi:10.1029/2006JC003615.

Kalkwijk, J. and R. Booij (1986) Adaptation of secondary flow in nearly horizontal flow. *J. Hydr. Res.* **24**, 19–37.

Lerczak, J. A. and W. R. Geyer (2004) Modeling the lateral circulation in straight, stratified estuaries. *J. Phys. Oceanogr.* **34**, 1410–1428.

Lerczak, J. A. *et al.* (2006) Mechanisms driving the time-dependent salt flux in partially stratified estuary. *J. Phys. Oceanogr.* **36**(12), g2283–2298.

MacCready, P. (2004) Toward a unified theory of tidally-averaged estuarine salinity structure. *Estuaries* **27**(4), 561–570.

MacCready, P. and P. B. Rhines (1991) Buoyant inhibition of Ekman transport on a slope and its effect on stratified spin-up. *J. Fluid Mech.* **223**, 631–661.

Nunes, R. A. and J. H. Simpson (1985) Axial convergence in a well mixed estuary. *Est. Coast. Mar. Sci.* **20**, 637–649.

Okubo, A. (1973) Effects of shoreline irregularities on streamwise dispersion in estuarine and other embayments. *Neth. J. Sea Res.* **8**, 213–224.

Peters, H. a. R. B. (2001) Microstructure observations of turbulent mixing in a partially mixed estuary, II: Salt flux and stress. *J. Phys. Oceanogr.* **31**, 1105–1119.

Phillips, O. M., J. M. Shyu and H. Salmun (1986) An experiment on boundary mixing: mean circulation and transport rates. *J. Fluid Mech.* **173**, 473–499.

Pritchard, D. W. (1956) The dynamic structure of a coastal plain estuary. *J. Mar. Res.* **17**, 412–423.

Rozovskii, I. L. (1957) *Flow of Water in Bends of Open Channels*. Ac. Sc. Ukr. SSR; Isr. Progr. Sc. Transl., Jerusalem, 1961.

Scully, M. E., W. R. Geyer and J. A. Lerczack (2009) The influence of lateral advection on the residual circulation: a numerical modeling study of the Hudson River estuary. *J. Phys. Oceanogr.* **39**, 107–124.

Seim, H. E. and M. C. Gregg (1997) The importance of aspiration and channel curvature in producing strong vertical mixing over a sill. *J. Geophys. Res.* **102**, 3451–3472.

Simpson, J. H., J. Brown, J. Matthews and G. Allen (1990) Tidal straining, density currents and stirring in the control of estuarine stratification. *Estuaries* **13**(2), 125–132.

Smith, R. (1977) Long term dispersion of contaminants in small estuaries. *J. Fluid Mech.* **82**, 129–146.

Smith, R. (1978) Longitudinal dispersion of a buoyant contaminant in a shallow channel. *J. Fluid Mech.* **78**, 677–688.

Valle-Levinson, A. (2008) Density-driven exchange flows in terms of the Kelvin and Ekman number. *J. Geophys. Res.* **113**, C04001, doi:10.1029/2007Jc004144.

Valle-Levinson, A., K. Holderied, C. Li and R. J. Chant (2007) Subtidal flow structure at the turning region of a wide outflow plume. *J. Geophys. Res.* **112**, C04004, doi:10.1029/2006JC003746.

Valle-Levinson, A. *et al.* (2000) Convergence of lateral flow along a coastal plain estuary. *J. Geophys. Res.* **17**, 17,045–17,061.

Wilson, R. E. and A. Okubo (1978) Longitudinal dispersion in a partially mixed estuary. *J. Mar. Res.* **36**(3), 427–447.

Winant, C. D. (2004) Three-dimensional wind-driven flow in an elongated, rotating basin. *J. Phys. Oceanogr.* **34**, 462–476.

6

Wind and tidally driven flows in a semienclosed basin

C. WINANT

Scripps Institution of Oceanography, UCSD

6.1. Introduction

To begin, the equations that describe the circulation in an enclosed or semienclosed basin, forced either by winds or tides, are developed, along with the required boundary conditions. The basin, schematically illustrated in Fig. 6.1, can be either enclosed if $-L^* \leq x^* \leq L^*$, or semienclosed, in which case the domain is reduced to $0 \leq x^* \leq L^*$ (starred variables are dimensional). The free surface is located a distance $\eta^*(x, y)$ above $z = 0$, where the free surface is located when the forcing is zero. The maximum depth of the basin is H^* and the local depth, h^*, is a function of x, y but for simplicity is taken to be dependent only on the lateral coordinate: $h^* = h^*(y)$.

6.2. Model equations

The mass conservation equation is:

$$\frac{\partial u^*}{\partial x^*} + \frac{\partial v^*}{\partial y^*} + \frac{\partial w^*}{\partial z^*} = 0. \tag{6.1}$$

For a constant density fluid, the horizontal components of the momentum equation are:

$$\frac{\partial u^*}{\partial t^*} + u^* \frac{\partial u^*}{\partial x^*} + v^* \frac{\partial u^*}{\partial y^*} + w^* \frac{\partial u^*}{\partial z^*} = -g \frac{\partial \eta^*}{\partial x^*} + K \frac{\partial^2 u^*}{\partial z^{*2}}, \tag{6.2}$$

$$\frac{\partial v^*}{\partial t^*} + u^* \frac{\partial v^*}{\partial x^*} + v^* \frac{\partial v^*}{\partial y^*} + w^* \frac{\partial v^*}{\partial z^*} = -g \frac{\partial \eta^*}{\partial y^*} + K \frac{\partial^2 v^*}{\partial z^{*2}}, \tag{6.3}$$

where the vertical momentum balance is assumed to be hydrostatic, $p^* = \rho g(\eta^* - z^*)$. The vertical eddy diffusivity, K, is taken to be constant. Lateral mixing is ignored because the ratio $K_h H^{*2}/K_z^* B^{*2}$ (K_h represents the horizontal eddy viscosity) is assumed to be small.

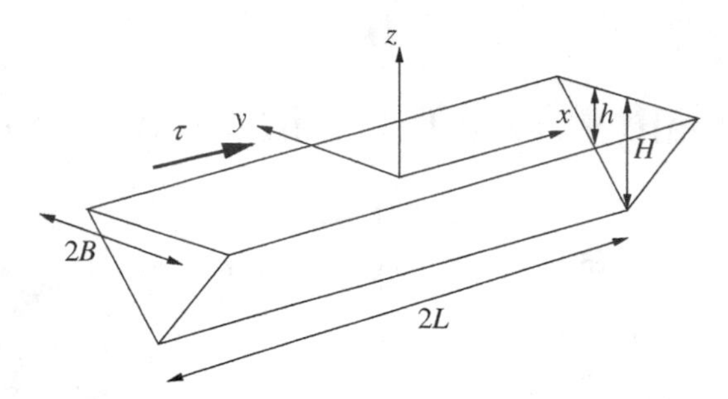

Figure 6.1. Closed basin sketch.

6.3. Boundary conditions

Boundary conditions are required at the bottom ($z = 0$), at the surface ($z^* = \eta$), as well as at the lateral boundaries ($x^* = \pm L$ and $y^* = \pm B$). Of these the simplest is the no-slip bottom boundary condition:

$$u^* = v^* = w^* = 0 \quad \text{at} \quad z^* = -h^*. \tag{6.4}$$

At the surface ($z^* = \eta^*$) the kinematic boundary condition is:

$$w^* = \frac{\partial \eta^*}{\partial t^*} + u^* \frac{\partial \eta^*}{\partial x^*} + v^* \frac{\partial \eta^*}{\partial y^*} \tag{6.5}$$

and the dynamic boundary conditions state that the pressure is constant and the shear stress is equal to whatever stress is imposed at the surface by the wind,

$$\frac{\partial u^*}{\partial z^*} = \frac{\tau_s^x}{\rho K} \quad \frac{\partial v^*}{\partial z^*} = \frac{\tau^x y_s}{\rho K}, \tag{6.6}$$

where τ_s^x and τ_s^y are the wind stress components applied at the surface.

Because the lateral diffusion terms have been omitted in the momentum equations, there is no need for no-slip and shear stress boundary conditions at the lateral boundaries. Instead, the transport normal to closed lateral boundaries must be set to zero. On an open boundary either the transport or the sea level has to be specified.

6.4. Wind-driven viscous flow in a closed shallow basin

As a first example of three-dimensional circulation in an enclosed basin, consider the circulation driven at the surface by a wind stress, constant in time and space, and aligned with the x-axis: $\tau_s^x = \tau, \tau_s^y = 0$. L^* and B^* are used to non-dimensionalize horizontal spatial coordinates, and H^* plays a similar role in the vertical. It is assumed that $L^* > B^* \gg H^*$. Here and in the remainder, unstarred variables are non-dimensional.

$$x = x^*/L^*, \quad y = y^*/B^* \quad \text{and} \quad z = z^*/H^*. \tag{6.7}$$

The development in this section follows the work of Csanady (1973), Wong (1994), and focuses on what happens near the middle of the basin. Mathieu *et al.* (2002) extend the theory to the full basin, including the turning regions (where $x \approx \pm 1$). Winant (2004) further extends the analysis to include the effect of the Earth's rotation. The boundary conditions at the surface suggest a scale for the horizontal velocities:

$$u = u^* \frac{\rho K}{\tau H^*} \quad \text{and} \quad v = v^* \frac{\rho K}{\tau H^*}. \tag{6.8}$$

The mass conservation equation then suggests a scale for the vertical velocity:

$$\frac{\tau H^*}{\rho K L^*} \frac{\partial u}{\partial x} + \frac{\tau H^*}{\rho K B^*} \frac{\partial v}{\partial y} + \frac{1}{H^*} \frac{\partial w^*}{\partial z} = 0. \tag{6.9}$$

This, in turn, suggests:

$$w = w^* \frac{\rho K}{\tau H^*} \frac{L^*}{H^*}. \tag{6.10}$$

The mass conservation equation for the non-dimensional variables is:

$$\frac{\partial u}{\partial x} + \frac{1}{\alpha} \frac{\partial v}{\partial y} + \frac{\partial w}{\partial z} = 0, \tag{6.11}$$

where $\alpha = B^*/L^*$ is the horizontal aspect ratio of the basin.

If the sea level is non-dimensionalized as follows:

$$\eta = \eta^* \frac{\rho g H^*}{\tau L^*}, \tag{6.12}$$

the x-component of the momentum equation takes on the simple form

$$0 = -\frac{\partial \eta}{\partial x} + \frac{\partial^2 u}{\partial z^2}, \tag{6.13}$$

reflecting the balance between the pressure gradient and vertical stress divergence. The lateral equation is:

$$0 = -\frac{\partial \eta}{\partial y} + \frac{\partial^2 v}{\partial z^2}. \tag{6.14}$$

The boundary conditions in z are:

$$u = v = w = 0 \quad \text{at} \quad z = -h, \tag{6.15}$$

while at the surface, $z = \eta$,

$$\frac{\partial u}{\partial z} = 1; \quad \frac{\partial v}{\partial z} = 0. \tag{6.16}$$

6.4.1. The horizontal velocities in wind-driven flow

The general solution to equation (6.13) is:

$$u = \frac{\partial \eta}{\partial x}\frac{z^2}{2} + Az + B. \tag{6.17}$$

The boundary condition at the surface ($z = \eta$) determines A:

$$A = 1 - \frac{\partial \eta}{\partial x}\eta, \tag{6.18}$$

then the boundary condition at $z^* = -h^*$ gives for B:

$$B = h - \frac{\partial \eta}{\partial x}\left(\frac{h^2}{2} + \eta h\right), \tag{6.19}$$

so that:

$$u = \frac{\partial \eta}{\partial x}\left[\frac{z^2 - h^2}{2} - \eta(z + h)\right] + z + h \tag{6.20}$$

or, equivalently, with $\tilde{z} = z/h$:

$$\frac{u}{h} = h\frac{\partial \eta}{\partial x}\left[\frac{\tilde{z}^2 - 1}{2} - \eta(\tilde{z} + 1)\right] + \tilde{z} + 1. \tag{6.21}$$

The first term on the right is the velocity forced by the pressure gradient (negative because $\tilde{z}^2$ is always less than 1) and the second term is the velocity driven by the wind stress. A similar expression can be derived for the lateral velocity v:

$$\frac{v}{h} = h\frac{\partial \eta}{\partial y}\left[\frac{\tilde{z}^2 - 1}{2} - \eta(\tilde{z} + 1)\right]. \tag{6.22}$$

The difference between equations (6.21) and (6.22) is due to the fact that the wind stress acts only in the x-direction. A positive pressure gradient (sea level increasing downwind) drives the flow toward negative x. Note also that the velocity profile depends on h. If h is very small, then the second term in the expression for u dominates over the first.

The ratio u/h is illustrated as a function of $\tilde{z}$ in Fig. 6.2 when $\eta = 0$, as for instance near the center of the basin ($x \approx 0$). When the water depth h is very small, the velocity increases linearly from the bottom to the surface, as in Couette flow. As the

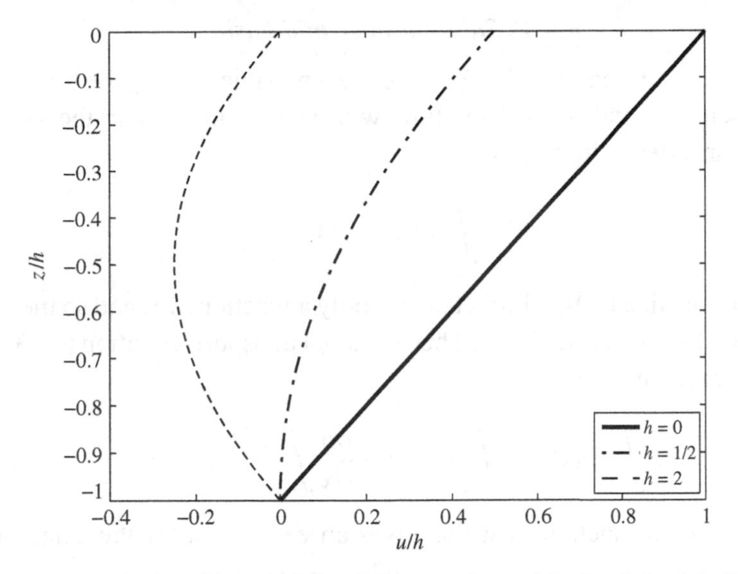

Figure 6.2. Vertical profiles of u/h for different values of the local depth h. The solid line is the same solution as Couette flow.

water depth increases, the first term in equation (6.21) increases, and the velocity becomes more negative, corresponding to upwind flow driven by the pressure gradient. Note that in all cases, $\partial u/\partial z = 1$ at the surface, because of the surface boundary condition.

6.4.2. *Wind-driven transport*

Equation (6.21) seems to imply that the axial velocity is a function of *both* the wind stress (implicitly, through the non-dimensionalization) *and* the sea level gradient. Intuitively, it is expected that these two variables depend on each other, and that if the wind stress increases, the sea level gradient will increase as well. To determine this relationship, it is useful to define the axial transport as

$$[u] = \int_{-h}^{\eta} u\,dz \approx \int_{-h}^{0} u\,dz = \frac{h^2}{2} - \frac{\partial \eta}{\partial x}\frac{h^3}{3}, \tag{6.23}$$

where the assumption that η is small justifies the approximation. Just how small η is will be discussed later. The first term on the right shows that wind stress forces transport downwind, whereas a positive pressure gradient forces transport toward negative x. The above follows from the two definite integrals:

$$\int_{-h}^{0} (z^2 - h^2)dz = -\frac{2}{3}h^3, \quad \int_{-h}^{0} (z+h)dz = \frac{h^2}{2}.$$

6.4.3. Solution near mid-basin

Near the middle of the basin, if $L^* > B^*$, we expect two things: first, the flow will be in the x-direction only and second, the flow will have no net flux in the x-direction, since the basin extends from $y = \pm 1$:

$$\int_{-1}^{1} [u]\, dy = 0.$$

If $v = 0$, then equation (6.14) shows that η is only a function of x, and so the axial sea level gradient does not depend on y. Then the axial transport [equation (6.23)] can be integrated across the basin:

$$\int_{-1}^{1} [u]\, dy = \int_{-1}^{1} \frac{h^2}{2}\, dy - \frac{\partial \eta}{\partial x} \int_{-1}^{1} \frac{h^3}{3}\, dy = 0. \tag{6.24}$$

Since h is a known function of y, this gives an expression for the axial elevation gradient:

$$\frac{\partial \eta}{\partial x} = \frac{3}{2} \frac{\int_{-1}^{1} h^2\, dy}{\int_{-1}^{1} h^3\, dy} = \frac{3}{2} \frac{<h^2>}{<h^3>}. \tag{6.25}$$

If, for instance, the bottom profile is a linear trough, that is $h = 1 - |y|$, then $<h^2> = 1/3$ and $<h^3> = 1/4$, and $\partial \eta / \partial x = 2$. In dimensional terms this corresponds to:

$$\frac{\partial \eta^*}{\partial x^*} = 2 \frac{\tau}{\rho g H^*}. \tag{6.26}$$

How big is this? A $10\,\mathrm{m\,s^{-1}}$ wind corresponds to $\tau = 0.1$ Pa. If the basin is 10 m deep and 10 km long:

$$\frac{\partial \eta^*}{\partial x^*} = 2 \frac{0.1}{1000 \times 10 \times 10} = 2 \times 10^{-6} \tag{6.27}$$

so that at the end of the basin, 5 km away from the midpoint where $\eta = 0$, $\eta^* = 2 \times 10^{-6} \times 5 \times 10^3$ m or about 1 cm.

Returning to non-dimensional variables, and knowing the axial sea elevation gradient, equation (6.21) gives the axial velocity u, illustrated in Fig. 6.3. Near the surface the flow is downwind, as in Couette flow, while at depth the flow responds to the pressure gradient required to force the total axial transport to zero. The solution given here is valid only near the central part of the basin, where the lateral sea level gradient is zero. Near the ends ($|x| \approx 1$), the downwind flow turns to connect with the upwind return flow as described by Mathieu *et al.* (2002). If the Earth's rotation is included in the problem (Winant, 2004; Sanay and Valle-Levinson, 2005), the

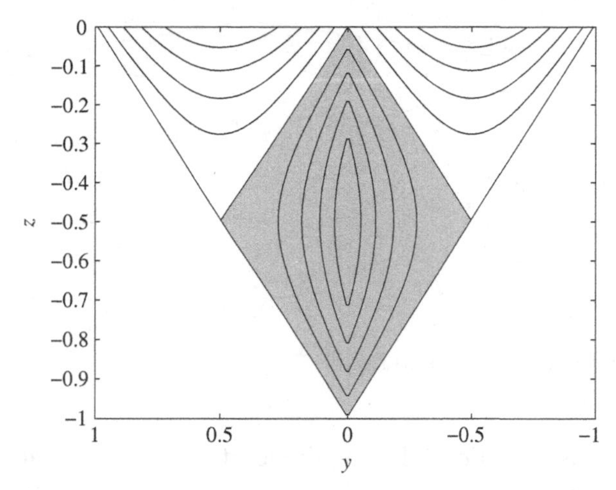

Figure 6.3. Sectional view of the axial flow near mid-basin ($x = 0$). The view is downwind, toward positive x. Shaded areas are upwind. The contour interval is 0.25. Near the surface, the axial flow is downwind and at depth, near the center of the basin, the flow is upwind.

lateral velocity is non-zero even near mid-basin, with surface flow to the right of the wind and to the left of the wind at depth.

6.5. Tidal fluctuations in a semienclosed basin

Now we consider a semienclosed basin, forced along the open ($x = 0$) plane. As before, the origin of z is the location of the free surface at rest. $\eta^*(x, y, t)$ again represents the location of the free surface above the $z = 0$ level (Fig. 6.4). As in the previous section, rotation of the Earth is neglected for simplicity. Winant (2007) has shown that unless friction is large, rotation fundamentally alters the magnitude and structure of the fluctuating lateral flow. The basin length is L^*, width $2B^*$, and maximum depth H^*. At the entrance the tide forces the sea level to vary as $\eta^*(x = 0, t) = C^* \cos \omega^* t^*$, where C^* is the tidal amplitude and ω^* is the frequency (in radians per second) of the tide.

6.5.1. Equations of motion and boundary conditions

Non-dimensional (unstarred) variables are introduced as follows:

$$t = \omega^* t^*, \quad x = x^*/L^*, \quad y = y^*/B^*, \quad (z, h) = (z^*, h^*)/H^* \tag{6.28}$$

and

$$u = u^*/\epsilon\omega^* L^*, \quad v = v^*/\epsilon\omega^* B^*, \quad w = w^*/\epsilon\omega^* H^*, \quad \eta = \eta^*/C^*, \tag{6.29}$$

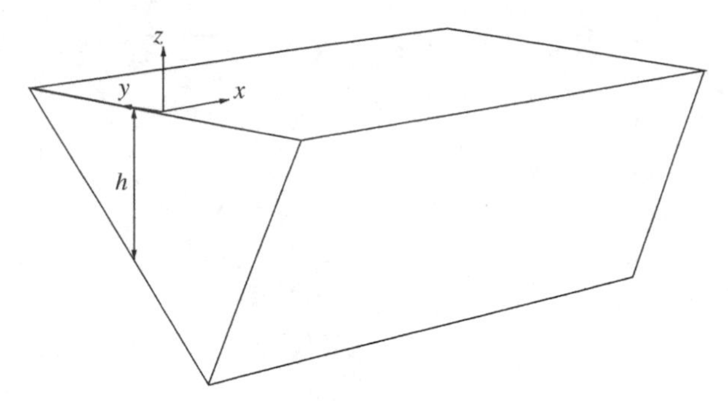

Figure 6.4. Generic semienclosed basin.

where $\epsilon = C^*/H^*$ is the ratio of the amplitude of the tidal wave at the open end to the maximum depth. In this non-dimensional system, the surface is located at $z = \epsilon\eta$. The continuity equation is:

$$\frac{\partial u}{\partial x} + \frac{\partial v}{\partial y} + \frac{\partial w}{\partial z} = 0 \tag{6.30}$$

and the vertically integrated continuity equation, combined with the surface boundary condition, is:

$$\frac{\partial \eta}{\partial t} + \frac{\partial}{\partial x}\left(\int_{-h}^{\epsilon\eta} udz\right) + \frac{\partial}{\partial y}\left(\int_{-h}^{\epsilon\eta} vdz\right) = 0. \tag{6.31}$$

As with any free-surface problem, the fact that surface dynamic and kinematic boundary conditions are applied at a location that is part of the solution increases the complexity considerably.

The horizontal components of the momentum equation are:

$$\frac{\partial u}{\partial t} + \epsilon\left(u\frac{\partial u}{\partial x} + v\frac{\partial u}{\partial y} + w\frac{\partial u}{\partial z}\right) = -\frac{1}{\kappa^2}\frac{\partial \eta}{\partial x} + \frac{\delta^2}{2}\frac{\partial^2 u}{\partial z^2}, \tag{6.32}$$

$$\frac{\partial v}{\partial t} + \epsilon\left(u\frac{\partial v}{\partial x} + v\frac{\partial v}{\partial y} + w\frac{\partial v}{\partial z}\right) = -\frac{1}{\alpha^2\kappa^2}\frac{\partial \eta}{\partial y} + \frac{\delta^2}{2}\frac{\partial^2 v}{\partial z^2}, \tag{6.33}$$

where

$$\delta = \sqrt{2K_z^*/\omega^* H^{*2}}, \quad \kappa = \omega^* L^*/\sqrt{g^* H^*}. \tag{6.34}$$

Here δ, the relative amplitude of the periodic boundary layer to the maximum depth, measures the relative importance of friction to local acceleration. κ is a relative measure of the length of the basin to the wavelength of the tidal wave.

At the surface ($z = \epsilon\eta$) the kinematic boundary condition is:

$$w = \frac{\partial \eta}{\partial t} + \epsilon\left(u\frac{\partial \eta}{\partial x} + v\frac{\partial \eta}{\partial y}\right) \tag{6.35}$$

and the dynamic boundary conditions state that the pressure is constant and the shear stress is zero, $u_z = v_z = 0$. At the bottom $(z = -h), u = v = w = 0$. On the closed sides $(x = 1$ and $y = \pm 1)$ the velocity normal to the boundary must be zero.

Perturbation and Taylor series expansion

If ϵ is small, the original non-linear problem can be broken down into an ordered set of linear problems by introducing perturbation expansions for the dependent variables:

$$(u, v, w, \eta) = (u_0, v_0, w_0, \eta_0) + \epsilon(u_1, v_1, w_1, \eta_1) + \cdots \tag{6.36}$$

into the equations. Neglecting terms of order ϵ^2 or greater gives:

$$\frac{\partial u_0}{\partial x} + \frac{\partial v_0}{\partial y} + \frac{\partial w_0}{\partial z} + \epsilon\left(\frac{\partial u_1}{\partial x} + \frac{\partial v_1}{\partial y} + \frac{\partial w_1}{\partial z}\right) = 0. \tag{6.37}$$

Introducing the perturbation expansion equation (6.36) into the vertically integrated continuity equation (6.31) and taking the limit as $\epsilon \to 0$ gives two equations. To lowest order:

$$\frac{\partial \eta_0}{\partial t} + \left(\int_{-h}^{0} u_0 dz\right)_x + \left(\int_{-h}^{0} v_0 dz\right)_y = 0 \tag{6.38}$$

and to order ϵ:

$$\frac{\partial \eta_1}{\partial t} + \frac{\partial}{\partial x}\left(\int_{-h}^{0} u_1 dz + \eta_0 u_0|_{z=0}\right) + \frac{\partial}{\partial y}\left(\int_{-h}^{0} v_1 dz + \eta_0 v_0|_{z=0}\right) = 0. \tag{6.39}$$

The horizontal components of the local momentum equations become:

$$\frac{\partial u_0}{\partial t} - f\alpha v_0 + \frac{1}{\kappa^2}\frac{\partial \eta_0}{\partial x} - \frac{\delta^2}{2}\frac{\partial^2 u_0}{\partial^2 z} + \epsilon\left(u_0\frac{\partial u_0}{\partial x} + v_0\frac{\partial u_0}{\partial y} + w_0\frac{\partial u_0}{\partial z} + \frac{\partial u_1}{\partial t} - f\alpha v_1 + \frac{1}{\kappa^2}\frac{\partial \eta_1}{\partial x} - \frac{\delta^2}{2}\frac{\partial^2 u_1}{\partial^2 z}\right) = 0, \tag{6.40}$$

$$\frac{\partial v_0}{\partial t} + \frac{f}{\alpha}u_0 + \frac{1}{\alpha^2\kappa^2}\frac{\partial \eta_0}{\partial y} - \frac{\delta^2}{2}\frac{\partial^2 v_0}{\partial^2 z} + \epsilon\left(u_0\frac{\partial v_0}{\partial x} + v_0\frac{\partial v_0}{\partial y} + w_0\frac{\partial v_0}{\partial z} + \frac{\partial v_1}{\partial t} + \frac{f}{\alpha}u_1 + \frac{1}{\alpha^2\kappa^2}\frac{\partial \eta_1}{\partial y} - \frac{\delta^2}{2}\frac{\partial^2 v_1}{\partial^2 z}\right) = 0. \tag{6.41}$$

With the perturbation expansion, the surface boundary conditions become:

$$u_{0z} + \epsilon u_{1z} = 0, \quad v_{0z} + \epsilon v_{1z} = 0, \tag{6.42}$$

$$w_0 + \epsilon w_1 = \frac{\partial \eta_0}{\partial t} + \epsilon \frac{\partial \eta_1}{\partial t} + \epsilon \left(u_0 \frac{\partial \eta_0}{\partial x} + v_0 \frac{\partial \eta_0}{\partial y} \right) \tag{6.43}$$

at $z = \epsilon\eta$, where terms of order ϵ^2 or greater have been neglected.

A Taylor series expansion is used to transform the boundary condition at $z = \epsilon\eta$ into a condition at $z = 0$:

$$u_z|_{z=\epsilon\eta} = u_z|_{z=0} + \epsilon\eta u_{zz}|_{z=0} + \cdots = 0 \tag{6.44}$$

or, with the perturbation expansion:

$$u_{0z}|_{z=0} + \epsilon u_{1z}|_{z=0} + \epsilon\eta_0 u_{0zz}|_{z=0} + \cdots = 0, \tag{6.45}$$

$$w_0|_{z=0} + \epsilon\eta \left(\frac{\partial w_0}{\partial z} \right)_{z=0} + \epsilon w_1|_{z=0} = \frac{\partial \eta_0}{\partial t} + \epsilon \left(\frac{\partial \eta_1}{\partial t} + u_0|_{z=0} \frac{\partial \eta_0}{\partial x} + v_0|_{z=0} \frac{\partial \eta_0}{\partial y} \right). \tag{6.46}$$

Using mass conservation, the last expression can we rewritten as:

$$w_0|_{z=0} + \epsilon w_1|_{z=0} = \frac{\partial \eta_0}{\partial t} + \epsilon \left(\frac{\partial \eta_1}{\partial t} + \frac{\partial \eta_0 u_0|_{z=0}}{\partial x} + \frac{\partial \eta_0 v_0|_{z=0}}{\partial y} \right). \tag{6.47}$$

Because lateral friction is ignored, only the velocity normal to the sides of the basin is required to go to zero. At the open end the transport or, equivalently, the sea level is specified.

Lowest-order problem

In the limit as $\epsilon \to 0$, the momentum equations and boundary conditions become:

$$\frac{\partial u_0}{\partial t} + \frac{1}{\kappa^2} \frac{\partial \eta_0}{\partial x} - \frac{\delta^2}{2} \frac{\partial^2 u_0}{\partial^2 z} = 0, \tag{6.48}$$

$$\frac{\partial v_0}{\partial t} + \frac{1}{\alpha^2 \kappa^2} \frac{\partial \eta_0}{\partial y} - \frac{\delta^2}{2} \frac{\partial^2 v_0}{\partial^2 z} = 0. \tag{6.49}$$

Solutions are sought that are harmonic in time, for example:

$$u_0(x, y, z, t) = \mathrm{Re}(U(x, y, z)e^{-it}) = |U(x, y, z)| \cos(t - \phi_U(x, y, z)), \tag{6.50}$$

with equivalent forms for the sea level and the other velocity components. Solutions for the complex amplitudes of the horizontal velocities in terms of the complex amplitude of the sea level are given by Lamb (1932):

$$U = \frac{iN_x}{\kappa^2}\left[\frac{\cos((1+i)z/\delta)}{\cos((1+i)h/\delta)} - 1\right], \tag{6.51}$$

$$V = \frac{iN_y}{\alpha^2\kappa^2}\left[\frac{\cos((1+i)z/\delta)}{\cos((1+i)h/\delta)} - 1\right]. \tag{6.52}$$

The vertical velocity is obtained by integrating the continuity equation from the bottom up:

$$W = -\int_{-h}^{z} (U_x + V_y) dz'. \tag{6.53}$$

The horizontal transports are given by:

$$[U] = \frac{iN_x P}{\kappa^2} \quad \text{where} \quad P = \frac{\delta \tan((1+i)h/\delta)}{(1+i)} - h \tag{6.54}$$

and

$$[V] = \frac{iN_y P}{\alpha^2\kappa^2}. \tag{6.55}$$

6.5.2. *Fluctuating sea level*

Assuming harmonic solutions, the lowest-order vertically integrated mass balance [equation (6.38)] becomes

$$-iN + [U]_x + [V]_y = 0. \tag{6.56}$$

If equations (6.54) and (6.55) are introduced into equation (6.56), a second-order elliptic partial differential equation for N is obtained:

$$(PN_y)_y + \alpha^2[(PN_x)_x - \kappa^2 N] = 0, \tag{6.57}$$

where the subscripted independent variables denote partial differentiation. At the basin entrance, the sea level is specified, and on the closed boundaries, the transport vanishes:

$$N = 1 \text{ at } x = 0 \quad \text{and} \quad N_x = 0 \text{ at } x = 1, \tag{6.58}$$

$$N_y = 0 \text{ at } y = \pm 1. \tag{6.59}$$

In the limit as $\alpha \to 0$, equation (6.57) shows that $N_y = 0$ everywhere, and an equation for N can be obtained by integrating the same equation across the basin:

$$\frac{d^2N}{dx^2} + \kappa^2\mu^2 N = 0, \tag{6.60}$$

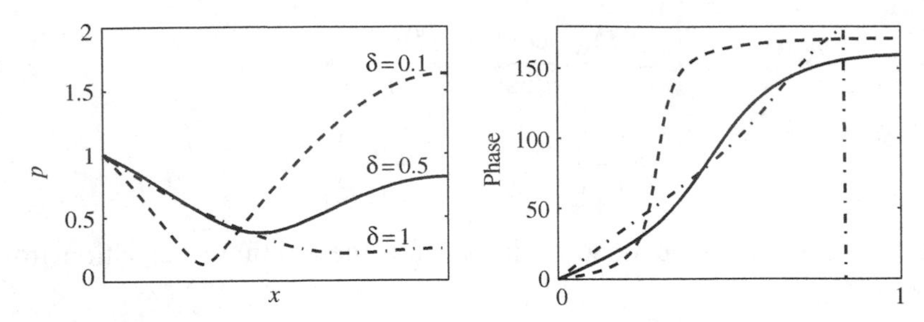

Figure 6.5. Amplitude and phase of $N^{(0)}$ for three different values of δ. For this calculation, $\kappa = 1.5$ and the basin depth is a linear function of lateral position. A positive phase is associated with a signal that is delayed relative to zero phase.

where $\mu = 1/\sqrt{< -P >}$ and the angle brackets stand for the lateral average. A solution that satisfies the boundary condition is:

$$N = \frac{\cos \kappa\mu(1-x)}{\cos \kappa\mu}. \tag{6.61}$$

This solution is discussed in more detail by Ianniello (1977), Li and Valle-Levinson (1999), and Winant (2007). It is illustrated in Fig. 6.5 for a triangular basin with $\kappa = 1.5$, corresponding to the basin length being slightly longer than the quarter wavelength of the tide, and three values of δ. κ is taken as 1.5, chosen so that for a maximum basin depth of 10 m, the corresponding length for a semidiurnal tide would be just over 100 km. Three values of δ corresponding to low, moderate and high friction are explored. If δ is small (0.1), the real part of μ is much greater than the imaginary part, and the tidal wave is nearly a standing gravity wave. There is a node (where the amplitude is minimum) in elevation at one quarter of the gravity wavelength from the closed end. The phase of the sea level changes by π in this vicinity. For small friction, the amplitude is always maximum at the closed end. Friction also determines the minimum amplitude of the wave, and the distance over which the phase changes from 0 to π.

As friction increases to $\delta = 0.5$, the apparent wavelength shortens (as illustrated by the movement of the node toward $x = 1$), because the real part of μ increases. The increasing imaginary part of μ results in lower amplitude at the closed end, and a more progressive change in the phase over the length of the basin, compared to the low friction case. For larger friction, $\delta = 1$, the real and imaginary parts of μ are close to equal in magnitude, and the node is no longer apparent. For very large friction the wave is completely damped before reaching the closed end. In this case, the boundary condition at the closed end must be replaced by a condition that $N^{(0)}$ vanishes for large x and the solution is

$$N^{(0)} = e^{-(1+i)\kappa\delta\gamma x}, \tag{6.62}$$

where γ is a parameter of order 1 that depends weakly on the geometry of the basin.

6.5.3. Fluctuating velocities

Knowing N, the full three-dimensional velocity field can be evaluated. The local velocities (u_0, v_0 and w_0) are described, rather than their complex amplitudes. Further, the velocities are scaled by the local amplitude of the fluctuating sea level. Because of the kinematic surface boundary condition on the vertical velocity, the scaled vertical velocity is always minus one at the surface one quarter period after local high water, independent of the axial location or the other parameters in the problem. While the magnitude of the velocities varies considerably with friction, the scaled velocities have comparable magnitude. They are illustrated in Fig. 6.6 for three different values of δ, at six different times over one half cycle, beginning at high water. Velocities in the following half period are minus the velocities illustrated. The section is located at mid-basin ($x = 0.5$).

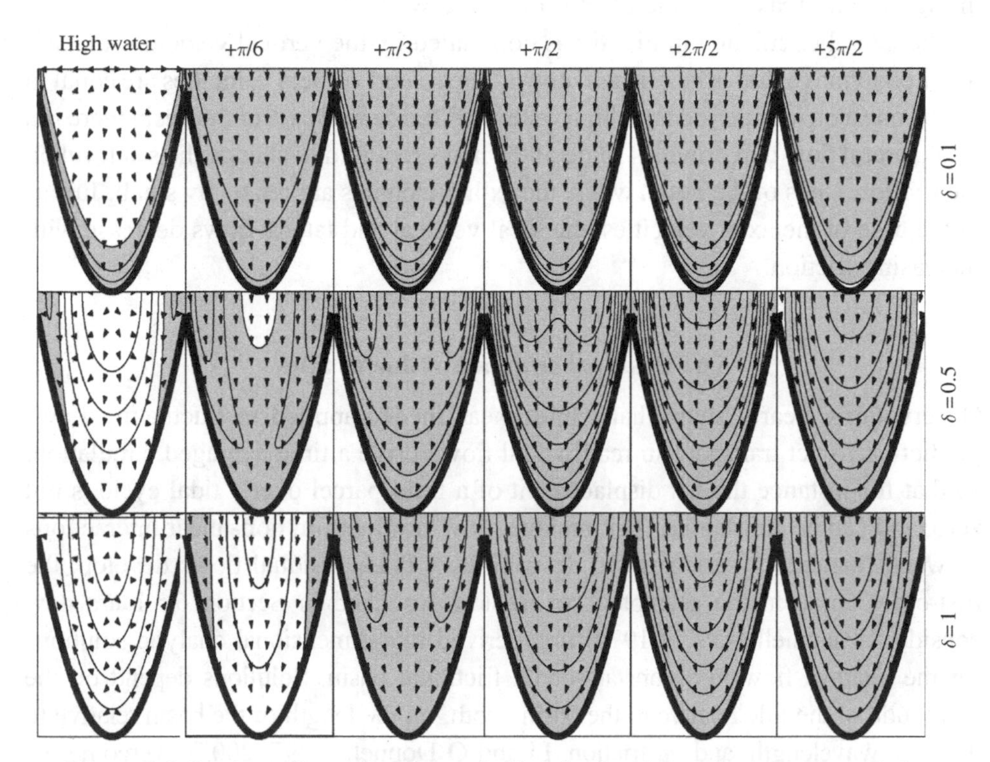

Figure 6.6. Sections illustrating the scaled local velocities at six times through the tidal cycle, and for three different values of δ. The viewer is looking toward the closed end of the basin. The section is located at $x = 0.5$. The leftmost column corresponds to local high water. The calculation is for $f = 0$, $\alpha = 0.025$, and $\kappa = 1.5$. The axial velocity is negative in the shaded area, and the axial velocity contour interval is 0.25. Lateral and vertical velocities are represented by arrows, the scaled vertical velocities at the surface for $\pi/2$ after high water are equal to -1. Reproduced from Winant (2007).

When friction is relatively small ($\delta = 0.1$), and the tidal wave is close to standing, the axial velocity is nearly constant over the section except in the bottom boundary layer where it decreases to satisfy the no-slip condition at the bottom. At high water, the axial velocities in the section are weak and change sign, with little net flow: a slack flow condition. As time progresses, the axial flow becomes negative throughout the section, corresponding to an ebb condition. The largest ebb flow leads low water by a phase of $\pi/2$. Increasing friction changes the axial flow pattern in two ways. First, the timing of the axial flow changes relative to local sea level; at the highest value of friction considered here, the maximum ebb current occurs at a phase of $\pi/4$ before low water, as expected for a frictional system. Second, the shear in the axial velocity is distributed throughout the section, rather than being confined near the bottom. Since the amplitude of the sea level variation decreases with increasing friction (Fig. 6.5), the local velocities (as opposed to the scaled velocities illustrated in Fig. 6.6) decrease with increasing friction as well.

The lateral circulation in Fig. 6.6 is dominated by the vertical velocities that are required to move the surface up and down. The scaled vertical velocities are equal to minus one a quarter period after high water. Off the central axis of the basin there is a weak lateral flow that provides some of the mass required to change the sea level on the shallow sides of the basin, where the axial velocities are relatively small. Just as in the case of the axial velocities, the local vertical and lateral flows decrease with increasing friction.

6.6. Residual circulation due to tides

Observations clearly show that while linear theory applied to fluctuating flows predicts zero net transport, in reality tidal flows drive a time-averaged circulation, so that for instance the net displacement of a fluid parcel over a tidal cycle is not zero, and can be substantial. This residual flow results from non-linear interactions between fluctuating variables. To describe how these residual flows develop, the first-order terms in the asymptotic expansion described in Section 6.5.1 are now considered. Ianniello (1977, 1979) first derived two-dimensional analytic solutions for the residual flow in a constant-depth frictional basin. Solutions depend on the amplitude of the tide relative to the basin depth, on the length of the basin relative to the tidal wavelength, and on friction. Li and O'Donnell (1997, 2005) overcome the constant-depth limitation by developing a two-dimensional model of the vertically integrated flow that includes laterally variable bathymetry. Their analysis shows that the vertically integrated residual circulation varies laterally, and depends on the basin shape as well as the parameters identified by Ianniello (1977). Following Longuet-Higgins (1969), Ianniello (1977) distinguishes between the Eulerian mean velocity, the time average of the local velocity at some point, and the Lagrangian velocity, the time-averaged velocity of a fluid parcel. Although the mass transported in an unsteady flow is more readily described in terms of the Lagrangian velocity, the following treatment is limited to the Eulerian mean flow for simplicity. The Lagrangian component is described by Ianniello (1977), as well as by Winant (2008) for the case with rotation.

6.6.1. Equations of motion and boundary conditions

First-order problem

The solution for the lowest-order fluctuating flow was described in the previous section. To find the residual flow, the time-averaged solution to the first-order problem is sought. The governing equations are obtained by subtracting the lowest-order equations from the full expansion [equations (6.37) through (6.41)]. If the equations and boundary conditions governing the fluctuating flow are subtracted from the full equations, the resulting problem in the limit is:

$$\frac{\partial u_1}{\partial x} + \frac{\partial v_1}{\partial y} + \frac{\partial w_1}{\partial z} = 0, \tag{6.63}$$

$$\frac{\delta^2}{2}\frac{\partial^2 u_1}{\partial z^2} - \frac{\partial u_1}{\partial t} = \left(u_0\frac{\partial u_0}{\partial x} + v_0\frac{\partial u_0}{\partial y} + w_0\frac{\partial u_0}{\partial z}\right) + \frac{1}{\kappa^2}\frac{\partial \eta_1}{\partial x}, \tag{6.64}$$

$$\frac{\delta^2}{2}\frac{\partial^2 v_1}{\partial z^2} - \frac{\partial v_1}{\partial t} = \left(u_0\frac{\partial v_0}{\partial x} + v_0\frac{\partial v_0}{\partial y} + w_0\frac{\partial v_0}{\partial z}\right) + \frac{1}{\alpha^2\kappa^2}\frac{\partial \eta_1}{\partial y}. \tag{6.65}$$

The surface boundary conditions, applied at $z = 0$, are:

$$u_{1z} = -\eta_0 u_{0zz}|_{z=0}, \quad v_{0z} = -\eta_0 v_{0zz}|_{z=0}, \quad w_1 = \eta_{1t} + \frac{\partial \eta_0 u_0}{\partial x} + \frac{\partial \eta_0 v_0}{\partial y}, \tag{6.66}$$

while at the bottom, $z = -h$:

$$u_1 = v_1 = w_1 = 0. \tag{6.67}$$

The Eulerian mean flow

The residual flow equations are obtained by averaging the equations and boundary conditions governing the first-order solutions [equations (6.63) through (6.67)] over a tidal cycle. If the subscript E denotes the average over one period of first-order variables (e.g., $u_E = \overline{u_1}$ is the Eulerian average axial velocity):

$$\frac{\partial u_E}{\partial x} + \frac{\partial v_E}{\partial y} + \frac{\partial w_E}{\partial z} = 0, \tag{6.68}$$

$$\frac{\delta^2}{2}\frac{\partial^2 u_E}{\partial z^2} + f\alpha v_E - \frac{1}{\kappa^2}\frac{\partial \eta_E}{\partial x} = \overline{\left(u_0\frac{\partial u_0}{\partial x} + v_0\frac{\partial u_0}{\partial y} + w_0\frac{\partial u_0}{\partial z}\right)} = R_x, \tag{6.69}$$

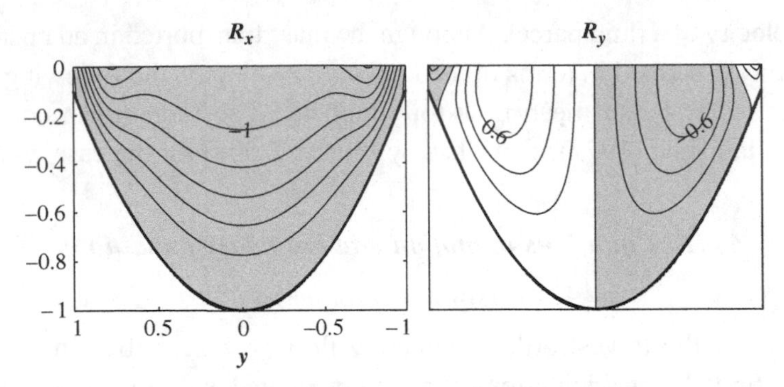

Figure 6.7. The axial and lateral component of the Reynolds stress at a section located mid-channel ($x = 0.5$), for $\alpha = 0.025$, $\kappa = 1.5$. The view is toward the closed end of the basin. f and δ are noted above each pair of frames. Contour intervals are evenly spaced. Shaded areas correspond to negative values.

$$\frac{\delta^2}{2}\frac{\partial^2 v_E}{\partial z^2} - \frac{f}{\alpha}u_E - \frac{1}{\alpha^2\kappa^2}\frac{\partial \eta_E}{\partial y} = \overline{\left(u_0\frac{\partial v_0}{\partial x} + v_0\frac{\partial v_0}{\partial y} + w_0\frac{\partial v_0}{\partial z}\right)} = R_y. \tag{6.70}$$

The surface boundary conditions, applied at $z = 0$, are:

$$\frac{\partial u_E}{\partial z} = -\overline{\eta_0 u_{0zz}|_{z=0}} = T_x, \qquad \frac{\partial v_E}{\partial z} = -\overline{\eta_0 v_{0zz}|_{z=0}} = T_y, \tag{6.71}$$

$$w_E = \frac{\partial}{\partial x}\overline{\eta_0 u_0|_{z=0}} + \frac{\partial}{\partial y}\overline{\eta_0 v_0|_{z=0}} = \frac{\partial S_x}{\partial x} + \frac{\partial S_y}{\partial y}, \tag{6.72}$$

where $S_x = \overline{\eta_0 u_0|_{z=0}}$ and $S_y = \overline{\eta_0 v_0|_{z=0}}$ are the two components of the Stokes transport. At the bottom, $z = -h$:

$$u_E = v_E = w_E = 0. \tag{6.73}$$

Forcing

The Eulerian flow is forced in part by the divergence of the tidal Reynolds stresses (R_x and R_y), by the Stokes transport (S_x and S_y), and by the boundary condition at $z = 0$ (T_x and T_y). In the Li and O'Donnell (1997, 2005) models, the last two contributions add up to twice the Stokes transport.

The Reynolds stress divergence terms are determined by the zeroth-order flow, described in Section 6.5.3. The axial and lateral components are illustrated in Fig. 6.7 at the mid-point ($x = 1/2$) of the basin. The axial Reynolds stress divergence (R_x) is dominated by $\overline{u_0 u_{0x}}$. The vertical shear in the fluctuating axial velocity for this intermediate value of friction is responsible for the vertical and horizontal structure of R_x. When friction is small, R_x is relatively uniform over the entire section, the amplitude decreasing to zero in a thin boundary layer. The leading term in R_y is

$\overline{v_0 u_{0y}}$. R_y has the opposite sign on either side of the basin, because v_0 changes sign on either side of the basin. R_x and R_y have comparable amplitudes.

6.6.2. Residual sea level

The time average of equation (6.39) is:

$$\frac{\partial}{\partial x}\left(\int_{-h}^{0} u_E dz + \overline{\eta_0 u_0|_{z=0}}\right) + \frac{\partial}{\partial y}\left(\int_{-h}^{0} v_E dz + \overline{\eta_0 v_0|_{z=0}}\right) = 0, \tag{6.74}$$

where the terms in parentheses represent the total time-averaged transports in the axial and lateral directions, respectively. With the convention that an integral from the bottom to $z = 0$ can be represented by square brackets, equation (6.74) becomes:

$$\frac{\partial}{\partial x}([u_E] + S_x) + \frac{\partial}{\partial y}([v_E] + S_y) = 0, \tag{6.75}$$

where u_E includes the response to the Reynolds stress, the surface boundary condition (T_x and T_y), and the sea level gradients. It is convenient to think of all these transports as consisting of a part driven by the three components of forcing, $[u_F]$, and a second part, $[u_\eta]$, that represents the transport due to the sea elevation gradients, set up to satisfy mass conservation for the residual flow:

$$\frac{\partial}{\partial x}([u_\eta] + [u_F]) + \frac{\partial}{\partial y}([v_\eta] + [v_F]) = 0. \tag{6.76}$$

The meaning of equation (6.76) is that in a steady state, the sea level must be distributed in such a way that the divergence of the total transport is zero. Winant (2008) has shown how the transport driven by sea level gradients can be related to these gradients and the forcing, so that equation (6.76) can be transformed into an elliptic partial differential equation for the residual sea level, η_E:

$$\frac{\partial}{\partial x}\left(P_1 \frac{\partial \eta_E}{\partial x}\right) + \frac{1}{\alpha^2}\frac{\partial}{\partial y}\left(P_1 \frac{\partial \eta_E}{\partial y}\right) = -\frac{\partial [u_F]}{\partial x} - \frac{\partial [v_F]}{\partial y}, \tag{6.77}$$

where

$$P_1 = -\frac{2h^3}{3\delta^2\kappa^2}. \tag{6.78}$$

The two components of the sea elevation gradients computed for $\alpha = 0.025$, $\kappa = 1.5$, $\delta = 0.5$ are illustrated in Fig. 6.8. The axial sea elevation gradient, $(\eta_E)_x$ is of order one, and $(\eta_E)_x$ is independent of y. The lateral sea elevation gradient, $(\eta_E)_y$ is of order α^2, in the sense to balance the bottom friction associated with the lateral mean current in the inner part of the basin, and in balance with the lateral Stokes transport near the open end.

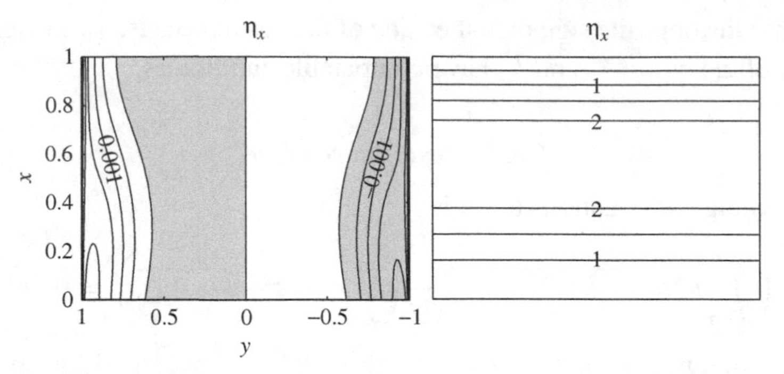

Figure 6.8. Plan view of the axial and lateral component of the sea-level gradient for $\alpha = 0.025$, $\kappa = 1.5$, and $\delta = 0.5$. The closed end of the basin is at the top of each frame. Contour intervals are evenly spaced. Shaded areas correspond to negative values.

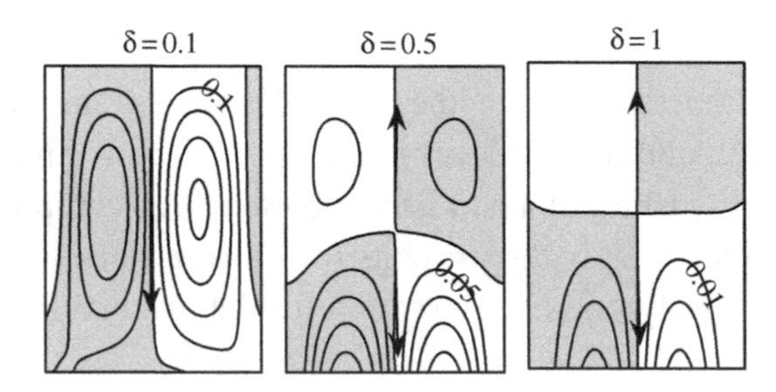

Figure 6.9. Plan view of the transport streamfunction for three different values of δ. The basin length ($\kappa = 1.5$) and the width ($\alpha = 0.025$) are the same for all three cases. The closed end of the basin is at the top. The arrows illustrate the direction of the flow. Contour intervals are evenly spaced. Shaded areas correspond to negative values.

6.6.3. Transport and Eulerian local velocities

Transport streamfunction

Equation (6.74) guarantees the existence of a transport streamfunction defined as:

$$\psi_x = \int_{-h}^{0} v_E dz + S_y, \quad \psi_y = -\int_{-h}^{0} u_E dz - S_x. \tag{6.79}$$

The transport streamfunction computed for different values of δ is illustrated in Fig. 6.9. The streamfunction patterns for moderate and large friction are similar, although the amplitude decreases with increasing friction. This pattern is very similar to the two-cell pattern described by Li and O'Donnell (2005). The lateral distribution of the forcing results in overall outflow near the center of the basin. The integrated residual transport across the width of the basin is zero since the residual flow is in steady state. The direction of the axial component, $-\psi_y$ depends on the relative magnitudes of the different components as a function of lateral position. The direction of the residual transport is determined by small differences in the lateral distributions of the different transport components, subject to the condition that the residual axial transport, integrated across the basin width, is zero. Winant (2008) provides a detailed discussion of these results.

Three-dimensional Eulerian mean flow

The local Eulerian velocities are illustrated in Fig. 6.10 for $f = 0$ and for three different values of δ. For high friction ($\delta = 1$), the shifting pattern of axial velocity is explained by the relative magnitude of the different forcings. Near the entrance the forcing due to correlations with fluctuating sea level ($[u_T]$ and $\overline{u_0\eta_0}$) are larger than $[u_R]$. As already discussed above in the context of the streamfunction, this forcing overcomes the pressure gradient forcing in shallower depth, resulting in positive Eulerian mean velocities at the sides of the basin. Near the closed end the Reynolds forcing overwhelms the sea level forcing, with the result that positive Eulerian mean velocities are found near

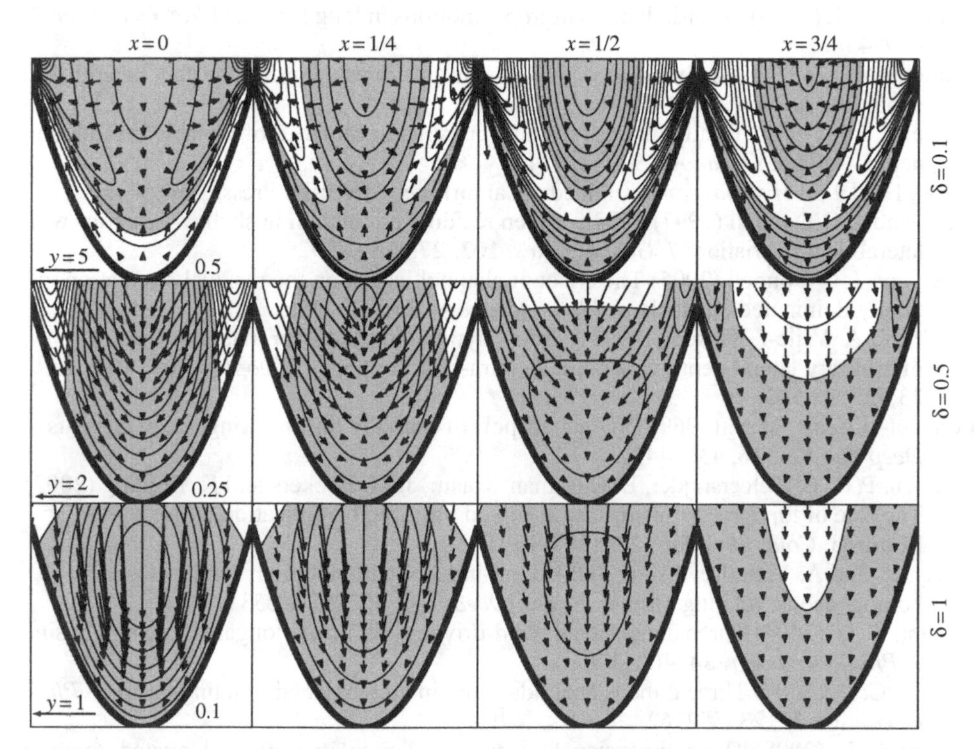

Figure 6.10. Sections illustrating the Eulerian mean velocities at different axial positions in the basin, for $\alpha = 0.025$, $\kappa = 1.5$ and $f = 0$, for three different values of δ. The viewer is looking toward the closed end of the basin. The axial velocity is negative in the shaded area, and the axial velocity contour interval is 0.25. Lateral and vertical velocities are represented by arrows.

the middle of the basin. For $\delta = 1$ the transition between the two patterns occurs near $x = 1/2$, where the Eulerian mean is everywhere negative, illustrating the point that the Eulerian velocity does not integrate to zero over any section. The inflow and outflow connect through the mostly vertical flow in the cross-section (y–z-plane). Patterns of mean Eulerian velocities for moderate friction ($\delta = 1/2$) are qualitatively similar to the case of high friction, the main difference being the lateral velocities that have a larger horizontal component. When $\delta = 0.1$ friction is confined to narrow boundary layers near the bottom, and the Reynolds forcing is uniform everywhere else. The pattern of the flow driven by the Reynolds forcing is thus almost exactly the same as the flow driven by the sea level gradient (which also acts uniformly with depth), with the result that the pattern of the Eulerian mean is mostly determined by the Stokes forcing. In this case, as with the cases described above, the lateral flow is mostly in the sense to connect inflow and outflow.

References

Csanady, G. T. (1973) Wind-induced barotropic motions in long lakes. *J. Phys. Oceanogr.* **3**, 429–438.

Ianniello, J. P. (1977) Tidally induced residual currents in estuaries of constant breadth and depth. *J. Mar. Res.* **35**, 755–786.

Ianniello, J. P. (1979) Tidally induced residual currents in estuaries of variable breadth and depth. *J. Phys. Oceanogr.* **9**, 962–974.

Lamb, H. (1932) *Hydrodynamics*, 6th edn. Cambridge University Press, New York.

Li, C. and J. O'Donnell (1997) Tidally driven residual circulation in shallow estuaries with lateral depth variation. *J. Geophys. Res.* **102**, 27,915–27,929.

Li, C. and J. O'Donnell (2005) The effect of channel length on the residual circulation in tidally dominated channels. *J. Phys. Oceanogr.* **35**, 123–456.

Li, C. and A. Valle-Levinson (1999) A 2-d analytic tidal model for a narrow estuary of arbitrary lateral depth variation: the intra-tidal motion. *J. Geophys. Res.* **104**, 23,525–23,543.

Longuet-Higgins, M. S. (1969) On the transport of mass by time-varying ocean currents. *Deep Sea Res.* **16**, 431–447.

Mathieu, P. P., E. Deleersnijder, B. Cushman-Roisin, J. M. Beckers and K. Bolding (2002) The role of topography in small well-mixed bays, with application to the lagoon of Mururoa. *Cont. Shelf Res.* **22**, 1379–1395.

Sanay, R. and A. Valle-Levinson (2005) Wind-induced circulation in semienclosed homogeneous, rotating basins. *J. Phys. Oceanogr.* **35**, 2520–2531.

Winant, C. D. (2004) Three dimensional wind-driven flow in an elongated, rotating basin. *J. Phys. Oceanogr.* **34**, 462–476.

Winant, C. D. (2007) Three dimensional tidal flow in an elongated, rotating basin. *J. Phys. Oceanogr.* **37**, 2345–2362.

Winant, C. D. (2008) Three dimensional residual tidal circulation in an elongated, rotating basin. *J. Phys. Oceanogr.* **36**, 1278–1295.

Wong, K. C. (1994) On the nature of transverse variability in a coastal plain estuary. *J. Geophys. Res.* **99**, 14,209–14,222.

7

Mixing in estuaries

STEPHEN G. MONISMITH

Stanford University

7.1. Overview

This chapter is intended to serve as an overview of mixing processes operant in estuaries. At the smallest scales this means turbulent mixing that ultimately leads to irreversible changes in property distributions, while at the largest scales, e.g., the width or length of an estuary, this means dispersion by some form of spatially and temporally varying flow. At times it can be important to differentiate between dispersion and diffusion in that the former really refers to changes in time of the separation of marked particles or fluid elements, whereas the latter refers to changes in macroscopic fluid properties, e.g., salinity, by molecular motions, e.g., diffusion. These distinctions are clearly laid out in Fischer *et al.* (1979), the best reference on mixing in the opinion of the author, and the text for a graduate course at Stanford University from which this chapter is largely drawn. Much of what follows is described in Fischer *et al.* (1979), and the reader of this chapter is encouraged to look there for further details.

7.2. Basics of mixing

7.2.1. Fickian diffusion

The fundamental model for the mixing of fluids is that of Fickian diffusion for which it is postulated that the flux of some material, $\overrightarrow{q}$, is proportional to the spatial gradient of its concentration, C, i.e.:

$$\overrightarrow{q} = -K\nabla C. \tag{7.1}$$

Here K is the diffusivity or diffusion coefficient and has units of m^2/s. Most importantly, it describes down-gradient fluxes, i.e., diffusion acts to eliminate spatial variations in concentration. For molecular diffusion, equation (7.1) is exact, and can be derived on the basis of the random motions of a large number of molecules. As a model of dispersion, it is only approximate, and is often used primarily because of analytical convenience.

Including advection, the Reynolds transport theorem including Fickian diffusion results in the fundamental advection–diffusion equation

$$\frac{\partial C}{\partial t} + \nabla \cdot \left(\vec{U}C\right) = -\nabla \cdot \left(\vec{q}\right) = \nabla \cdot (K\nabla C). \tag{7.2}$$

In the absence of advection, and when C only varies in (x), equation (7.2) reduces to "the diffusion equation":

$$\frac{\partial C}{\partial t} = \frac{\partial}{\partial x}\left(K\frac{\partial C}{\partial x}\right) = K\frac{\partial^2 C}{\partial x^2}. \tag{7.3}$$

The fundamental solution to equation (7.3), i.e., one that describes the diffusion of a mass M placed instantaneously at the origin, is

$$C(x,t) = \frac{M}{\sqrt{4\pi Kt}}\exp\left(\frac{-x^2}{4Kt}\right), \tag{7.4}$$

a solution which has the property that the variance of the concentration distribution grows linearly in time:

$$\sigma^2 = \frac{1}{M}\int_{-\infty}^{+\infty} x^2 C(x,t)\,dx = 2Kt. \tag{7.5}$$

Note that equation (7.5) is a prescription for computing dispersion coefficients from the observed evolution of the concentration of diffusing tracers. If the fluid velocity is non-zero but uniform, equation (7.4) is modified by the use of the Galilean transformation,[1] which for a velocity U in the x-direction is $x' = x - Ut$, such that x in equation (7.4) is replaced by x'.

The fundamental behavior revealed by equation (7.4) is that for timescale T, diffusion acts to spread materials of a spatial scale

$$\delta \sim \sqrt{KT}. \tag{7.6}$$

This relationship between time and space is really no more than dimensional analysis, since δ has units of length while K has units of length^2 time^{-1}. Nonetheless, this fundamental behavior significantly affects a wide range of estuarine transport processes. In the presence of forcing that oscillates with frequency Ω, the effects of that forcing are confined to a diffusive layer that has a thickness $O\left(\sqrt{K/\Omega}\right)$. For example, Banas *et al.* (2004) use this scaling to estimate the distance that changes in estuarine salinity due to coastal upwelling offshore of an estuary penetrate into the estuary.

These simple solutions can easily be extended to multiple space dimensions, at least for the case where diffusion coefficients are constant, e.g., diffusion of a point mass in 3D and assuming anisotropic mixing, i.e., different diffusivities for x-, y-, and z-directions, viz:

$$C(x, y, z, t) = \frac{M}{8\sqrt{\pi^3 K_x K_y K_z t^3}} \exp\left(-\frac{x^2}{4K_x t} - \frac{y^2}{4K_y t} - \frac{z^2}{4K_z t}\right). \tag{7.7}$$

This form of fundamental solution is sometimes useful in estuaries, since often $K_z \ll K_x, K_y$ due to the effects of stratification as well as the fact that scales of turbulence (see below) are often much smaller in the vertical direction than they are in the horizontal direction.

Equations (7.4) and (7.7) apply in the absence of boundaries. The presence of a boundary at some value of x generally requires that the flux $= -K\partial C/\partial x = 0$ at that point. On the other hand, a concentration field computed using equation (7.4) gives $\partial C/\partial x \neq 0$, and thus violates the no-flux condition.

The effects of boundaries are incorporated using the method of image sources, an approach that takes advantage of the linearity of the governing equations (Crank, 1975). The method of image sources works as follows. Consider a source placed at $x = 0$ in a fluid in which there is a no-flux boundary at $x = -H$. If a second source is also placed at $x = -2H$, the concentration field becomes:

$$C(x, t) = \frac{M}{\sqrt{4\pi K t}} \left\{ \exp\left(\frac{-x^2}{4Kt}\right) + \exp\left(\frac{-(x + 2H)^2}{4Kt}\right) \right\}. \tag{7.8}$$

C now satisfies the boundary condition at $x = -H$, since the flux from the so-called image source will cancel that from the real source at this place in space. The presence of a second boundary, say at $x = H$, somewhat complicates matters since while addition of a second image source at $x = 2H$ cancels the flux from the real source at $x = H$, it now creates a non-zero flux at $x = -H$. Likewise, the image source at $x = -2H$ has a non-zero flux at $x = H$. These in turn can be cancelled by additional images at $x = \pm 4H$, giving rise ultimately to an infinite series of image sources. Fortunately, the exponential decay implicit to equation (7.8) means that these series generally converge rapidly. In a like fashion, the effects of time-varying sources can also be computed using linear superposition.

As an aside, the fundamental solution can also be derived through consideration of the random walk of a large number of particles taking uncorrelated steps (a.k.a. the drunkard's walk). This correspondence is the basis for the random walk approach used to represent mixing in particle-tracking models that are commonly used in population models of larval organisms (see, e.g., Simons *et al.*, 2007). For these models, advection can be handled explicitly whereas diffusion is represented by a random step, resulting in a particle step (in 1D) for timestep Δt of

$$\Delta x = U\Delta t + \sqrt{2K\Delta t}Z, \tag{7.9}$$

where Z is a random number that is either 1 or -1. If K varies spatially then equation (7.9) must be modified to include an additional velocity dK/dx in order to properly model the combined effects of advection and diffusion (Visser, 1997). An important practical aspect of random walk models is that for large numbers of particles (a requirement for particle-tracking results to properly replicate continuous diffusion models), one must ensure that the random number generator used to construct Z is truly random (Hunter *et al.*, 1993).

The discussion above supposes unsheared flows. If the flow is sheared, a new dispersion mechanism, shear flow dispersion, first described by Taylor (1922), can arise. Consider a linear shear flow shown in Fig. 7.1, in which both shear and dispersion operate. Since the cross-sectional average velocity can be removed by a Galilean transformation, it can be assumed that there is no mean advection. If we mark a line of particles at $t = 0$ and then follow its motion and deformation afterwards, first we would see it tilt. This tilt creates transverse variations in concentration that can be operated on by mixing across the cross-section. As the flow proceeds, tilting by the shear can come to be balanced by transverse diffusion, a state reached in a time $T_m \sim B^2/K$, the time required for mixing to be completed across the channel, such that for longer times, diffusion and shear are in balance.

To see how this proceeds, we can use the random walk approach to compute the diffusion of a large number of particles in a linear shear (Fig. 7.2). As seen in Fig. 7.2, the variance in the lateral direction quickly becomes constant, indicating that mixing across the channel has been completed, whereas after transverse mixing is complete, the variance in the streamwise direction grows linearly in time, exactly as expected for Fickian diffusion. However, for the parameters used for the random walk, the diffusion coefficient should be $K = 0.125$ as per equation (7.9),[2] whereas the effective longitudinal dispersion coefficient inferred from equation (7.5) using

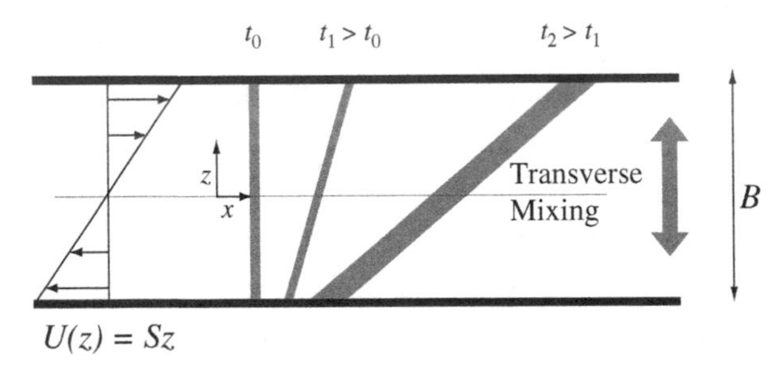

Figure 7.1. Sketch of shear flow dispersion.

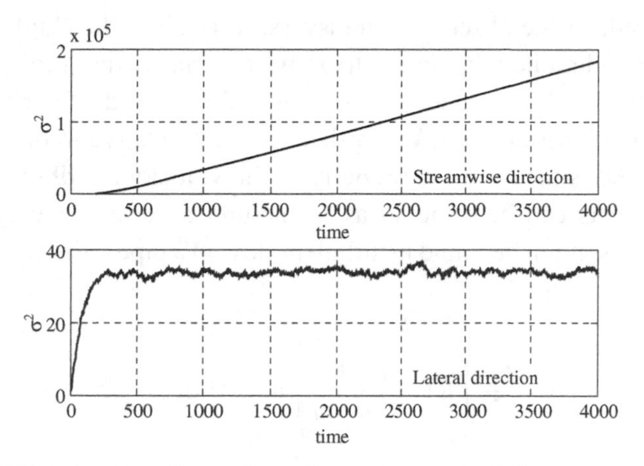

Figure 7.2. Diffusion in a linear shear flow. These calculations were made using 1000 particles, $S = 0.05$, $\Delta t = 1$, $\Delta x = 0.5$, and $B = 2$.

the computed rate of change with time of the variance is 26 (in this case dimensionless). This is the fundamental characteristic of shear flow dispersion – greatly enhanced longitudinal mixing, in this case by a factor of approximately 200.

Detailed analysis of this process, first carried out by G. I. Taylor (1953), can be described as follows. A transverse variation in concentration $C'(z)$ is created by a sheared velocity $u'(z)$ acting on a mean longitudinal gradient of the concentration averaged across the cross-section, $\overline{C}(x)$; this produces a source which is balanced by diffusion:

$$-u'\frac{d\overline{C}}{dx} = D\frac{dC'^2}{dz^2}, \tag{7.10}$$

where u' is the deviation of the velocity from the cross-sectional average and D is the diffusion coefficient for lateral mixing. Equation (7.10) can be integrated to give C'. Thus, the average total flux is:

$$\frac{F}{B} = \frac{1}{B}\int_0^B u'C'\,dz \propto \frac{d\overline{C}}{dx}. \tag{7.11}$$

Thus, the enhanced diffusion could be calculated from the observed velocity profile through simple integration:

$$K = \frac{1}{BD}\int_0^H u'(z)\int_0^z\int_0^{z'} u'(z'')\,dz''dz'dz. \tag{7.12}$$

Taylor's result is remarkable in that it shows that (a) shear flow dispersion is stronger for wider channels, and (b) shear flow dispersion is weakened by transverse diffusion. Indeed, in the absence of transverse diffusion, the fluid line described above stretches indefinitely, giving a streamwise variance that grows like t^2, i.e., a diffusion coefficient that depends on the scale of the cloud of particles, since the length of the cloud grows like t. When shear is weak relative to diffusion, $K \rightarrow 0$. Overall, K expresses the diffusive property of the velocity distribution.

Taylor's analysis can be done so as to include a vertically varying diffusion coefficient such as might be found in turbulent flow in a pipe or in a tidal channel, for which case:

$$K = \frac{1}{H} \int_0^H u'(z) \int_0^z \frac{1}{\kappa_t(z)} \int_0^{z'} u'(z'')\, dz'' dz' dz, \tag{7.13}$$

where $\kappa_t(z)$ is the laterally varying (turbulent) diffusivity analogous to D in equation (7.13) and we are considering mixing over the fluid depth, H. For the case of pipe flow, Taylor's analysis gives $K = 10au_*$ (here a is the pipe diameter and u_* is the shear velocity), whereas for channel flow that obeys the law of the wall, equation (7.39), $K = 5.9Hu_*$.

7.3. Turbulence in estuaries[3]

Most flows in estuaries are turbulent, i.e., they:

(a) Vary chaotically in time, albeit around a well-defined mean value. Figure 7.3a, a short record of velocity measurements made in Elkhorn Slough, a small estuary in Northern California,[4] shows how all three velocity components fluctuate around a generally non-zero average value.
(b) Exhibit a wide, continuous range of scale variations. This is illustrated in Fig. 7.3b, a plot of power spectra derived from the velocities shown in Fig. 7.3a.
(c) Cascade energy from large to small spatial scales, extracting energy from the mean flow at large scale and dissipating it as heat at the smallest scales.

We start here by discussing (b) and (c) and will then address the implications of (a).

7.3.1. The cascade

The cascade of energy from large to small scales is fundamental to turbulence. As expressed by L. F. Richardson ca. 1920:

> "Big whorls have little whorls
> Which feed on their velocity;

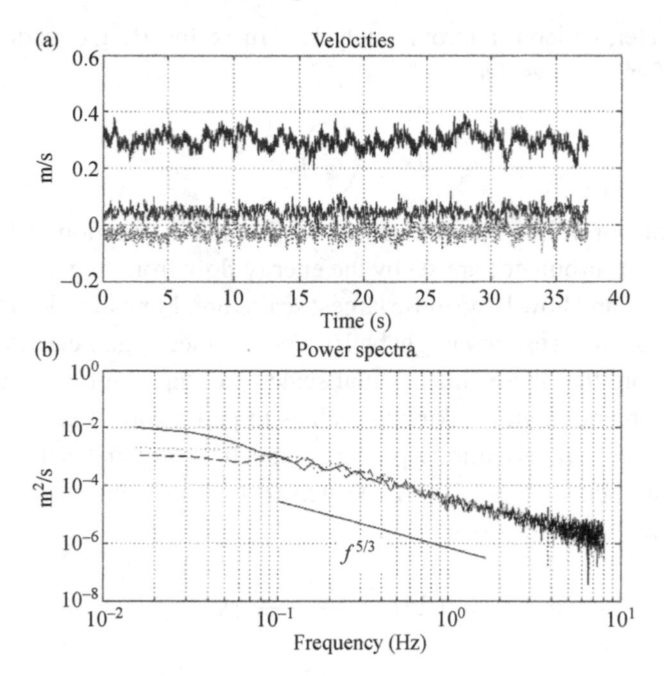

Figure 7.3. (a) Turbulent velocities measured in Elkhorn Slough. The components are streamwise (—), transverse (- - -) and vertical (···). (b) Power spectra for the velocities shown in (a) – the 5/3 slope expected for inertial subrange of turbulence is also shown. Unpublished data provided by N. Nidzieko.

And little whorls have lesser whorls,
And so on to viscosity"

This small poem (based on a similar poem by Swift) expresses the idea that smaller scales extract energy non-linearly from larger scales. Mechanistically, this is accomplished by vortex stretching and instability. The largest scales of turbulence motion, the energy-containing scales, are set by the geometry of the flow, e.g., the depth of a tidal channel or the distance from a solid boundary like the bed. Motions at this scale are essentially independent of viscosity. The rate at which they transfer energy to smaller scales can be estimated by a simple scaling argument. If u is the velocity scale of the eddies which have length scale l, then they will have kinetic energy u^2 and a characteristic turnover time of l/u, such that their rate of energy transfer will be $u^2/(l/u) = u^3/l \sim \varepsilon$, where ε, is the rate of energy dissipation. This means that at any intermediate scale l,

$$u(l) \sim (\varepsilon l)^{1/3}. \tag{7.14}$$

However, at the smallest scales, known as the Kolmogorov scale, l_K, where the actual dissipation of turbulence takes place, $l_K \sim \nu u^2/l^2$, where ν is the molecular viscosity of water, which is approximately 10^{-6} m^2/s. Inserting equation (7.14) into this estimate for l_K shows that

$$l_K = \left(\frac{\nu^3}{\varepsilon}\right)^{1/4}. \tag{7.15}$$

The fundamental premise of this cascade, as embodied in equation (7.15), is that the smallest scales of turbulence are set by the energy flow from larger scales, and thus by how energetic and how large those largest scales are. For example, a typical value of ε in a tidal flow in an estuary might be 10^{-6} W m^{-3} (see, e.g., Peters, 1997), so that $l_K \sim 10^{-3}$ m. Note that these small spatial scales are important to small organisms like zooplankton and phytoplankton as well as to particle aggregation.

The intermediate scales exhibit a particular spectral form that can be used to infer ε from measurements (Pope, 2000). Written in terms of the wavenumber in the streamwise direction (the x-axis):

$$\phi_{vv}(k) = 0.25\left(\frac{4 - \cos(\phi)}{3}\right)k^{-5/3}\varepsilon^{2/3}, \tag{7.16}$$

where

$$\int_0^\infty \phi_{vv} dk = \overline{v^2} \tag{7.17}$$

and ϕ is the angle between the velocity component and the x-axis. Most commonly, spectra are computed from temporal variations and advection in the streamwise direction at velocity U using the Taylor frozen turbulence hypothesis which states that temporal variations are simply the result of spatial variations being carried past the sensor without other temporal changes. This means that

$$k = \frac{2\pi}{U}f, \tag{7.18}$$

with f the frequency in Hz. For example, for the spectrum shown in Fig. 7.3, fitting the measured streamwise velocity spectrum ($\phi = 0$) to equation (7.16) gives $\varepsilon \simeq 2 \times 10^{-5}$ m^2/s^3.

7.3.2. *Mixing (dispersion) in turbulent flows*

One consequence of the spatial and temporal variability of turbulent flows is that they are remarkably dispersive. Under certain conditions, this can be represented by a Fickian diffusion coefficient, albeit one that is orders of magnitude larger than molecular diffusion coefficients. In reality, mixing by turbulence is really a two-step affair: stirring by turbulent velocity shear creates property variations on scales that are small enough for molecular diffusion to smear them out, and molecular diffusion then acts to make property distributions uniform. While the discussion below focuses on the stirring aspects of turbulence, it is important to recognize that in some applications, e.g., external fertilization of benthic invertebrates, the distinction can be important.

The theoretical basis for turbulent dispersion was originally given by Taylor (1922), who carried out a set of elegantly simple calculations using statistical properties of the velocity field. His arguments for dispersion in one dimension (x) are as follows.[5] Consider a set of particles, all initially located at $x = 0$. If the Lagrangian velocity of any particle is u, then the location of that particle is $X(t) = \int_0^t u\,dt$. Thus the average of the squared displacement of the cloud of all the particles, i.e., the standard deviation of particle position, is:

$$\langle X^2(t)\rangle = \int_0^t\int_0^t \langle u(\tau_1)u(\tau_2)\rangle\, d\tau_1\, d\tau_2, \tag{7.19}$$

where the angle brackets are used to indicate an average over all the particles. Note that for a Fickian diffusion process $\langle X^2\rangle$ would increase linearly in time.

Now, define the Lagrangian autocorrelation coefficient to be

$$R_x(\tau_2 - \tau_1) = \frac{\langle u(\tau_1)u(\tau_2)\rangle}{\langle u^2\rangle}. \tag{7.20}$$

R_x is difficult to determine since it refers to velocities measured following a particle, rather than being measured at a fixed point as is more easily done (Squires and Eaton, 1991). In terms of R_x, equation (7.19) can be written as

$$\langle X^2(t)\rangle = \langle u^2\rangle \int_0^t\int_0^t R_x(\tau_2 - \tau_1)\, d\tau_1 d\tau_2 == \langle u^2\rangle \int_0^t\int_0^t R_x(s)\, d\tau_1 d\tau_2, \tag{7.21}$$

where the integral is written in terms of the difference in time $s = \tau_2 - \tau_1$. A *very* deft mathematical step by Taylor enabled him to convert this double integral into a single integral:

$$\langle X^2(t)\rangle = 2\langle u^2\rangle \int_0^t (t - s)R_x(s)\, ds. \tag{7.22}$$

This general relation has two limits. For small times, $R_x \approx 1$, which implies that

$$\langle X^2(t)\rangle \approx 2\langle u^2\rangle \int_0^t (t-s)\, ds = \langle u^2\rangle t^2, \tag{7.23}$$

i.e., a diffusion coefficient that increases linearly in time. For large time, $R_x \to 0$, so that the integral in equation (7.22) is finite and can be split into two parts:

$$\begin{aligned}\langle X^2(t)\rangle &= 2\langle u^2\rangle \left(\int_0^t tR_x(s)\, ds - \int_0^t sR_x(s)\, ds \right) \\ &= 2\langle u^2\rangle t \int_0^t R_x(s)\, ds - 2\langle u^2\rangle \int_0^t sR_x(s)\, ds.\end{aligned} \tag{7.24}$$

The first term on the right of equation (7.24) is proportional to t, while the second term becomes constant as $t \to \infty$. Thus, equation (7.24) can be written as:

$$\langle X^2\rangle \simeq 2\langle u^2\rangle T_x t + \text{constant}, \tag{7.25}$$

where the Lagrangian timescale is

$$T_x = \int_0^\infty R_x(s)\, ds. \tag{7.26}$$

In effect, T_x is the time taken for a particle to forget its initial velocity. Thus, asymptotically, turbulent dispersion can be described by a Fickian diffusion coefficient, K_x:

$$\frac{d}{dt}\langle X^2\rangle = 2\langle u^2\rangle T_x = K_x. \tag{7.27}$$

Thus, Taylor's theory supports the notion that turbulence can be represented by an effective diffusion coefficient, albeit if one waits long enough, i.e., for $t \geq T_x$. Since the scale of the dispersing cloud of particles is $\langle u^2\rangle^{1/2} t$, this statement can be viewed as applying to spatial scales that are larger than

$$L_x = \langle u^2\rangle^{1/2} T_x. \tag{7.28}$$

For example, in tidal flow in a channel, $\langle u^2\rangle \approx u_*{}^2$, $T_x \sim z/u_* \sim H/u_*$, and thus $L_x \sim H$, so that strictly speaking, turbulent diffusion coefficients applied to compute flow and mixing for scales less than the depth (as will be presented below) can only be viewed as useful approximations.

7.3.3. Reynolds averaging

Given that in estuarine flows the largest scales might be $O(10 \text{ m})$, whereas the smallest scales are $O(10^{-3} \text{ m})$, there is generally a large separation of scales between largest and smallest scales. If one wanted to carry out a full direct numerical solution to the governing equations of motion (see Pope, 2000), one would need 10^{12} grid points to cover this range of scales in three dimensions. This is approximately a factor of 100 larger than the very largest of such calculations being carried out in 2008.

As a consequence of this scale separation, prediction of turbulent flows requires some form of averaging to be done. The most common averaging used in engineering, oceanography, and meteorology is known as Reynolds averaging, an averaging in time that essentially amounts to low-pass filtering. In Reynolds averaging, any variable of interest, say φ, is decomposed into its average and fluctuating (primed) part, i.e.:

$$\varphi(x, y, z, t) = \Phi(x, y, z) + \widetilde{\varphi}(x, y, z, t) + \varphi'(x, y, z, t), \tag{7.29}$$

where averaged variables (denoted as upper case)[6]

$$\Phi = \frac{1}{T_a} \int_0^{T_a} \varphi \, dt. \tag{7.30}$$

Ideally, T_a should be infinite in which case Φ is not a function of time (as shown above). Practically this is not viable; thus, often in estuarine flows adequate averaging can be had by setting T_a =10 min (see Stacey *et al.*, 1999b). This definition works well except in the presence of waves, where things become more complicated since some of what might be described as turbulence, e.g., velocity variance, is not turbulence as described above. For surface waves as might be found on an estuarine shoal, the Reynolds decomposition can be extended to include a wave variable component, $\widetilde{\varphi}$, determined in some way from measured free surface deformations (see Jones and Monismith, 2008). Even this approach isn't entirely successful because turbulence variations that are coherent with the waves are often considered to be waves. This is not surprising since these waves exert a phase-dependent stretching of turbulent vortices that can significantly affect their dynamics (Texeira and Belcher, 2003), even possibly leading to the formation of Langmuir circulations, depth-scale secondary flows that can effect rapid water column mixing in estuarine shallows (Gargett and Wells, 2007). The degree to which this separation works can be attributed to the fact that the wave motions tend to occur at frequencies that are well separated from both the energy-containing and dissipative scales. In contrast, there is at present no clear way to separate internal waves and turbulence.

Applying Reynolds averaging to the Navier–Stokes equations gives rise to what are known as the Reynolds averaged Navier–Stokes equations (RANS). Carrying out this procedure shows that the mean velocities and pressures obey the same governing equations as before except that new terms, the Reynolds stresses, appear:

$$\frac{DU_i}{Dt} = -\frac{1}{\rho_0}\frac{\partial P}{\partial x_i} - \frac{\rho}{\rho_0} g\delta_{i3} - \frac{\partial}{\partial x_i}\left(\overline{u_i' u_j'}\right). \tag{7.31}$$

Here we have used the Einstein summation notation[7] to compactly represent all three momentum equations and have made use of the Boussinesq approximation. Per usual convention, ρ_0 is the reference density of seawater (ca. 1000 kg m^3), and $i = 3$ is assigned to the vertical direction. The new terms are often written as stresses, i.e.:

$$\tau_{ij} = -\left(\overline{u_i' u_j'}\right), \tag{7.32}$$

although they really represent the average momentum flux associated with correlations of turbulent motions, i.e., the mixing of momentum by turbulent eddies. Often these are written in terms of an eddy viscosity, ν_t, e.g., for the usually dominant vertical mixing of horizontal momentum:

$$\tau_{13} = \nu_t \frac{\partial U_1}{\partial x_3} = \nu_t \frac{\partial U}{\partial z}. \tag{7.33}$$

Because viscous stresses are essentially negligible compared to the Reynolds stresses, they have not been included in equation (7.31). The Reynolds stresses also serve to define the largest scales of motion through Prandtl's concept of a mixing length, which for a one-dimensional shear flow, $U(z)$, can be written as:

$$L_m = \sqrt{\frac{-\overline{u'w'}}{\partial U/\partial z|\partial U/\partial z|}}. \tag{7.34}$$

In a similar fashion, when Reynolds averaging is applied to the conservation of a scalar [equation (7.2)], e.g., salt, additional scalar fluxes, $-\overline{u_i' C'}$, appear. Again by analogy to molecular diffusion, these are written in terms of eddy diffusivities, κ_t; for estuaries, the vertical flux

$$-\overline{w'c'} = \kappa_t \frac{\partial C}{\partial z} \tag{7.35}$$

is usually the most important.

Strictly speaking, equations (7.33) and (7.35) are definitions of ν_t and κ_t, not predictive models, although in virtually all circulation models used to predict flows in estuaries (see, e.g., Warner *et al.*, 2005), separate models for ν_t and κ_t

are essential! The reason for this is that the Reynolds stresses are unknown; indeed, the RANS equations for a variable-density fluid provide five equations to solve for 14 unknowns. Attempts to develop models for the Reynolds stresses based on straightforward manipulation of the Navier–Stokes equations lead to what is known as the closure problem, e.g., if one writes equations for τ_{ij}, the imbalance between equations and unknowns only grows larger, including terms that involve triple correlations and products of the fluctuating velocities and pressures.

Several sets of field measurements of ν_t in estuaries have recently been reported (see, e.g., Scully *et al.*, 2005). All make use of acoustic Doppler current profilers (ADCPs) to measure the Reynolds stresses following an approach outlined by Lohrman *et al.* (1990) and analyzed in detail by Stacey *et al.* (1999a). ADCPs transmit short pings of sound to travel through the water column along a set of beams (usually four). The Doppler shift of the sound produced when the outgoing ping is scattered from material in the water column is used to compute velocities in range-gated bins extending outward from the ADCP. The fluctuating velocities at the same height but along oppositely directed ADCP beams can be combined to produce an estimate of the Reynolds stress:

$$\overline{u'w'} = \frac{\overline{v_1'^2} - \overline{v_2'^2}}{4\sin(\theta)\cos(\theta)}, \tag{7.36}$$

where v_i' are the along-beam velocities and 2θ is the angle between the two beams. Reynolds stresses determined this way will be shown below.

Fortunately, some simple prescriptions for eddy viscosity exist, notably for turbulent channel flow:

$$\nu_t = \kappa_t = \kappa u_* z(1 - z/H), \tag{7.37}$$

where $\kappa \simeq 0.41$ is the von Karman constant (Nezu and Nakagawa, 1993). This is a generalization of the linearly varying law of the wall eddy viscosity valid for small values of z/H. This eddy viscosity distribution leads to the well-known law of the wall (Pope, 2000) for homogeneous, turbulent flows driven by free surface pressure, e.g., unstratified tidal flows. The basic momentum balance for such a 1D flow would be

$$g\frac{\partial \zeta}{\partial x} = \frac{\partial}{\partial z}\left(\nu_t \frac{\partial \overline{u}}{\partial z}\right) = -\frac{u_*^2}{H}, \tag{7.38}$$

where ζ is the deviation of the free surface from its at-rest position. Integration of equation (7.38) using equation (7.37) and assuming that $U(z = z_0) = 0$ gives:

$$U = \frac{u_*}{\kappa} \ln\left(\frac{z}{z_0}\right). \tag{7.39}$$

Here z_0 is the roughness height, a length scale that characterizes the roughness of the bed. Commonly, u_* is related to the velocity at 1 m above the bed (1 mab) through the drag coefficient, C_D:

$$u_*{}^2 = C_D U|U|(z = 1\,\text{mab}). \tag{7.40}$$

For typical smooth estuarine beds $C_D \approx 0.003$, although it can be higher depending on bed forms, vegetation, or other roughness elements like corals. Indeed, equations (7.39) and (7.40) can be combined to connect C_D and z_0:

$$C_D = \frac{\kappa}{\ln(1\,\text{m}/z_0)}. \tag{7.41}$$

Values of C_D and z_0 are generally found by least-squares fitting to observed velocity profiles (see, e.g., Gross and Nowell, 1985; Lueck and Lu, 1997), usually with good success. However, while most circulation models assume that C_D is constant, there is abundant evidence from field measurements that it is not constant (see, e.g., Cheng *et al.*, 1999; Fong *et al.*, 2009).

Scalar mixing coefficients (κ_t) are generally assumed to be equal to eddy viscosities, i.e., the turbulent Prandtl number

$$Pr_t = \frac{\nu_t}{\kappa_t} \tag{7.42}$$

is equal to 1. This is generally taken to be reasonable in unstratified flows, but is known to be flawed in the presence of stratification (see below). Moreover, scale dependence, as suggested by Taylor's analysis for early times, appears to be a common feature of horizontal mixing in geophysical flows like those in estuaries (see Okubo, 1971; Fong and Stacey, 2003).

7.3.4. Turbulent kinetic energy

A relatively simple (and later useful) example, the turbulent kinetic energy equation, suffices to illustrate the problem. The turbulent kinetic energy is defined as half the sum of the three component velocity variances:

$$TKE = \frac{1}{2}\left(\overline{u'^2} + \overline{v'^2} + \overline{w'^2}\right) = \frac{1}{2}\left(\overline{u_i'^2}\right). \tag{7.43}$$

An evolution equation for the *TKE* is:

$$
\begin{aligned}
&\frac{D}{Dt}(TKE) = T + P + B - \varepsilon \\
&T = -\frac{\partial}{\partial x_j}\left(\frac{1}{\rho_0}\overline{p' u_j'} + \frac{1}{2}\overline{u_i' u_i' u_j'} - 2\nu\overline{u_i' e_{ij}'}\right) \\
&P = -\overline{u_i' u_j'}\frac{\partial U_i}{\partial x_j} \\
&B = -g\overline{u_3'(\alpha\theta' + \beta s')} \\
&\varepsilon = 2\nu\overline{e_{ij}' e_{ij}'}.
\end{aligned}
\tag{7.44}
$$

Here we define each of the individual terms in the *TKE* equation, i.e., T for transport, P for production, B for buoyancy flux, and ε for dissipation. Other symbols are: thermal expansivity, α; temperature, θ; saline expansivity, β; salinity, S; rate of strain tensor, $e_{ij} = \frac{1}{2}(\partial u_i/\partial x_j + \partial u_j/\partial x_i)$.

P represents the production of turbulence by the interaction of the mean shear and the Reynolds stresses. Physically, this is the result of eddies intensifying by being strained (or stretched) by the mean shear. Normally $P > 0$ and a result one obtains (for example) if the eddy viscosity is positive in a one-dimensional shear flow is $P = \nu_t(\partial U/\partial z)^2$. For example, near the bed in a flow described by equation (7.39),

$$P \simeq \frac{u_*^3}{\kappa z}. \tag{7.45}$$

T serves to redistribute *TKE*, generally transferring *TKE* from regions of high *TKE* to low *TKE*. The term involving the fluctuating pressure represents pressure work, whereas the other terms represent advection of *TKE* by turbulence and its diffusion by viscosity. T is thought to be important for breaking surface waves where it transfers downward the intense turbulence produced by breaking (Craig and Banner, 1994).

B is the production or consumption of *TKE* by buoyancy forces produced by fluctuating fluid densities. In the presence of stable stratification, B will be a sink since turbulent mixing will move heavy fluid elements up and light fluid elements down, making B negative. In contrast, B can be positive in the presence of convection produced by surface cooling or whenever heavier fluid is carried over lighter fluid. In this case heavy fluid moves down and light fluid moves up. In terms of the Brunt Vaisala or buoyancy frequency

$$N^2 = -\frac{1}{\rho}\frac{\partial \rho}{\partial z}g, \tag{7.46}$$

B can be written as

$$B = -\kappa_t N^2. \tag{7.47}$$

For stably stratified flows, the ratio of B to P is known as the flux Richardson number, R_f.

ε is always a sink for *TKE*; it is the rate at which energy transferred to small scales by the turbulence cascades and is converted into heat. ε is often used in oceanography as the chief descriptor of turbulence, since it can readily be measured with shear probes, devices that can measure near-dissipation-scale strain rates and so almost directly compute ε (see, e.g., Peters, 1997). Moreover, in many estuarine flows, the *TKE* equation often reduces to the near-equilibrium balance of $P \approx \varepsilon$, enabling measurements of ε and vertical shear in the mean velocity to be used to estimate Reynolds stresses (Peters, 1997). Values of ε observed in estuaries range from 10^{-9} to 10^{-5} W m^{-3}.

7.3.4. Effects of stratification

The effects of stable density stratification on turbulence and mixing can be profound. For the case where the stratification is sufficiently strong, this effect can be summarized as the simple model that "stratification kills turbulence". Figure 7.4 shows an example of this behavior, a laboratory experiment in which a turbulent layer is confined to the region near the turbulence source (an oscillating grid) by the density stratification underneath. The source of this effect is the simple physics that work must be done against gravity to move fluid particles vertically in a stratified fluid.

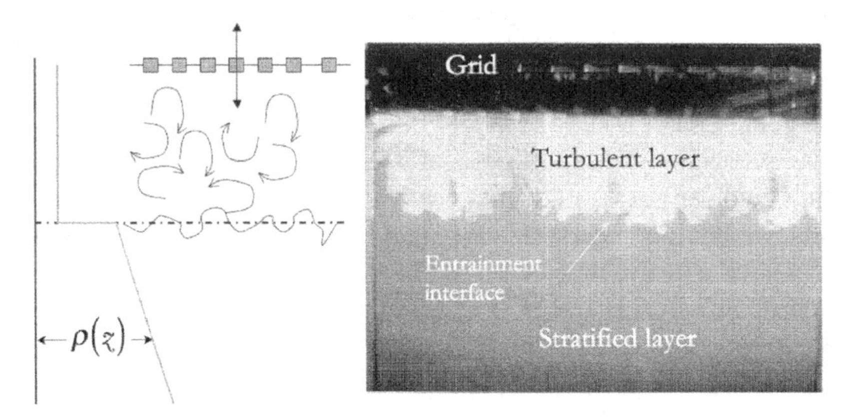

Figure 7.4. The penetration of a turbulent mixing region into a quiescent stratified fluid. The experiment itself is described in Maxworthy and Monismith (1988). The photograph was taken by T. Shay.

As initially developed by Richardson (1920), the competing effects of shear and stratification can be assessed as follows. Consider a 1D shear flow with $U = U_0 + Sz$ and stratification $\rho = \rho_0 + \Gamma z$. Here S refers to the shear, *not* to salinity.[8] Suppose an eddy mixes fluid over height δ. The kinetic and potential energies of this region are:

$$PE = \int_{-\delta/2}^{\delta/2} \rho g z \, dz, \qquad KE = \tfrac{1}{2} \int_{-\delta/2}^{\delta/2} \rho_0 U^2 \, dz. \tag{7.48}$$

Before mixing:

$$PE_i = \int_{-\delta/2}^{\delta/2} (\rho_0 + \Gamma z) g z \, dz = \frac{\Gamma g \delta^3}{12}, \qquad KE_i = \tfrac{1}{2} \int_{-\delta/2}^{\delta/2} \rho_0 (U_0 + Sz)^2 \, dz = \tfrac{1}{2}\rho_0 U_0^2 \delta + \tfrac{1}{2}\rho_0 \left(\frac{S^2 \delta^3}{24} \right), \tag{7.49}$$

while afterwards:

$$PE_f = \int_{-\delta/2}^{\delta/2} \rho_0 g z \, dz = 0, \qquad KE_f = \tfrac{1}{2} \int_{-\delta/2}^{\delta/2} \rho_0 {U_0}^2 \, dz = \tfrac{1}{2}\rho_0 U_0^2 \delta. \tag{7.50}$$

Thus, the changes in potential and kinetic energy are:

$$\Delta PE = PE_f - PE_i = \frac{-\Gamma g \delta^3}{12}, \qquad \Delta KE = KE_f - KE_i = -\rho_0 \frac{S^2 \delta^3}{48}. \tag{7.51}$$

However, since $\Delta PE + \Delta KE \leq 0$:

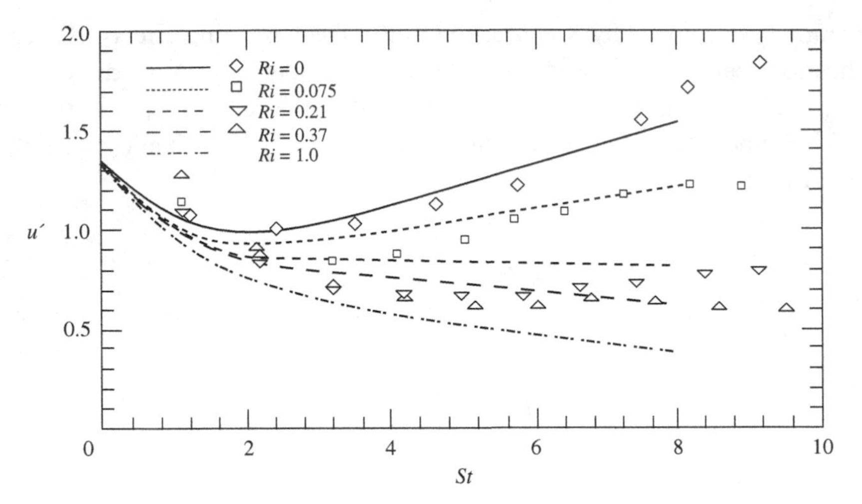

Figure 7.5. The behavior of the fluctuating streamwise velocity in the presence of uniform shear and stratification – taken from Holt *et al.* (1992) with permission. The symbols represent laboratory results of Rohr *et al.* (1988). Here, the x-axis is the shear time, i.e., the time made dimensionless by the characteristic time scale of the mean velocity shear, i.e., S^{-1}, and the y-axis represents the fluctuating streamwise velocity.

$$\frac{-\Gamma g}{\rho_0 S^2} = \frac{N^2}{S^2} = Ri \leq \frac{1}{4}, \tag{7.52}$$

where Ri, the Richardson number, determines the overall stability of the flow. Remarkably, linear stability theory shows that $Ri \leq 1/4$ is a necessary condition for instability (Turner, 1973).

The effect of Ri on turbulence has been explored parametrically in a series of laboratory and numerical experiments examining how turbulence evolves in the presence of shear and stratification (summarized in Ivey *et al.*, 2008). In Fig. 7.5, the results of DNS[9] computations by Holt *et al.* (1992) show this behavior: for $Ri < 0.25$, turbulence grows with time, for $Ri > 0.25$,[10] it decays. In the higher Ri cases, it is thought that this decay process is enhanced by the transfer of energy from turbulent motions to internal waves, waves that owe their existence to the presence of a stable stratification (see Gill, 1982).

Stratification also brings into play a further set of turbulent length scales. The first of these is the Ellison scale:

$$L_E = \frac{\sqrt{\overline{\rho'^2}}}{\partial \rho / \partial z}, \tag{7.53}$$

which is a measure of eddy size, although it can also be significantly altered by internal waves, since for a pure internal wave field, L_E is the local internal wave amplitude. L_E is useful because it is easily measured by a set of fixed sensors. In weakly stratified turbulence, $L_E \sim L_M$, although the constant of proportionality may depend on the parameter Q defined below (Shih *et al.*, 2005). A second, more important scale, is the Ozmidov scale:

$$L_O \equiv \frac{\varepsilon^{1/2}}{N^{3/2}}. \tag{7.54}$$

The physics of L_O are a balance between inertia and buoyancy. Assume an eddy of size L_O has a velocity u, then this force balance means that $u^2/L_O \sim N^2 L_O$, so using the inertial estimate $u \sim (\varepsilon L_O)^{1/3}$, $\varepsilon^{2/3} L_O{}^{-1/3} \sim N^2 L_O$, which can be solved to give equation (7.54).

The Ozmidov scale represents the largest scale turbulence can attain, much like the Kolmogorov scale defines the smallest scale. The ratio of these two scales, a measure of the separation of the largest and smallest length scales, is an important determinant of how stratified turbulence behaves. This ratio can be used to define another dimensionless number, sometimes called the activity number, or the strain-based Froude number:[11]

$$Q = Fr_\gamma = \left(\frac{L_O}{L_K}\right)^{4/3} = \frac{\varepsilon}{\nu N^2}. \tag{7.55}$$

Remarkable turbulence observations made in Knight Inlet in British Columbia by Gargett *et al.* (1984) show that when $Q > 200$, small-scale turbulence is unaffected by stratification whereas for $200 > Q > 15$, the effects of stratification extend to smaller and smaller scales. For $Q < 15$, stratification "kills turbulence", and while the fluid may still appear to be turbulent, buoyancy fluxes are essentially zero (Ivey *et al.*, 2008).

A second dynamically important ratio, that of the Ozimdov scale to the turbulent length scale L_T (either L_E, L_M, or L_x), defines the turbulent Froude number:

$$Fr_T = \left(\frac{L_O}{L_T}\right)^{2/3}. \tag{7.56}$$

Fr_T describes how big the turbulent eddies are relative to how big they could be. Large Fr_T corresponds to weak stratification, whereas when $Fr_T < 1$ the turbulence is likely decaying.

A key attribute of stratified turbulence is the buoyancy flux or scalar mixing it produces. Osborn (1980) proposed using ε to determine mixing by assuming that: (a) $R_f = 0.18$ and (b) $B = P - \varepsilon$. This gives the estimate:

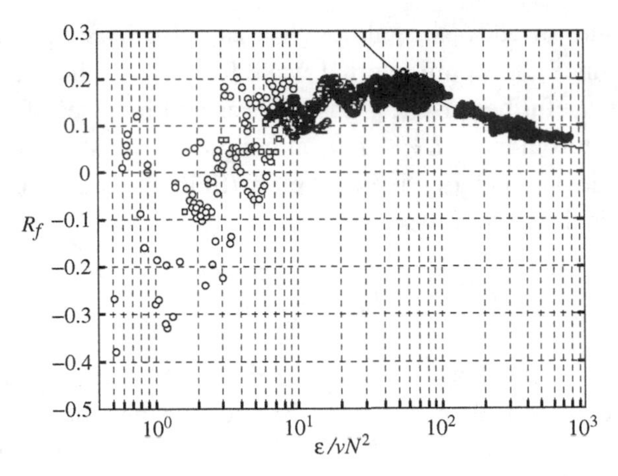

Figure 7.6. The dependence of R_f on the activity parameter Q (taken from Shih *et al.*, 2005 – used with permission).

$$\kappa_t = 0.2\frac{\varepsilon}{N^2}. \tag{7.57}$$

Other approaches acknowledge that Ri_f must depend on turbulence state. For example, Ivey and Imberger (1991) suggest that Ri_f is a function of Fr_T, attaining a maximum of 0.25 at $Fr_T \approx 1$. Shih *et al.* (2005) found that Ri_f could equivalently be written as a function of Q (Fig. 7.6).

7.3.5. Stratified turbulence in San Francisco Bay

To illustrate the ideas discussed above, I will draw on observations made in 1999 in Suisun Cut, a small channel in Northern San Francisco Bay. A basic description of the experiment is given in Brennan (2004). Limited results from this experiment can be found in Stacey and Ralston (2005); others are as yet unpublished. An earlier experiment at this site can be found in Stacey *et al.* (1999b).

Figure 7.7 shows a set of instantaneous fields (velocity, salinity, etc.) measured over a 10-day period that started at a neap tide with strong stratification and ended during a spring tide with relatively weak stratification. Spring–neap variability in the flow is manifest in a significant asymmetry in the bed stress, with ebbs during stratified periods producing much smaller stresses than floods, behavior that is due in part to straining effects in the bottom boundary layer (Stacey and Ralston, 2005). Overall, the lower half of the water column tends to be more turbulent, with values of Q almost always in excess of the transitional value of 200. Likewise, in the lower water column, $Fr_t > 1$, suggesting that while turbulence there may have been affected by stratification, it was not "buoyancy dominated" as would be the case when $Fr_t < 1$.

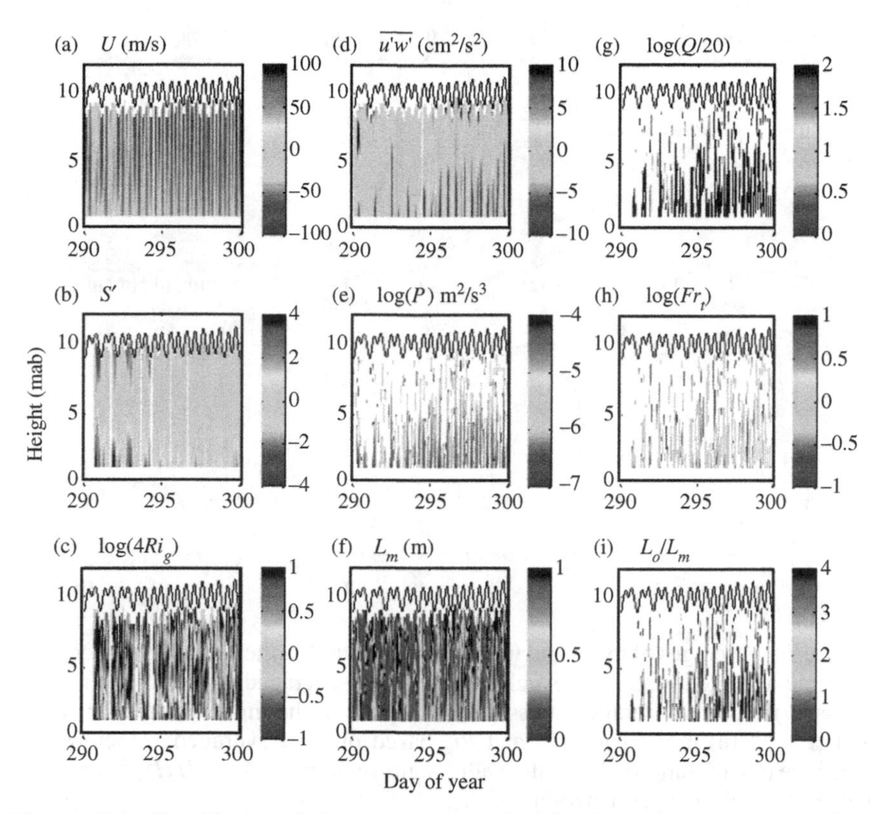

Figure 7.7. Stratified turbulence observed in Northern San Francisco Bay (unpublished data collected by M. Brennan, S. Monismith, M. Stacey, and J. Burau): (a) streamwise velocity; (b) perturbation salinity S' (S_a and S_c have been removed); (c) log(Ri_g/0.25) – this has NANs where N^2<0 ; (d) $\overline{u'w'}$; (e) TKE production, P – white corresponds to NAN in places where the measured production was negative production, a result that may either reflect noise in the stress measurement or stratification effects; (f) turbulence length scale, L_m; (g) $\log\{(P/\nu N^2)/\nu N^2)/20\}$ – with this scaling, darkest shading corresponds to the transition to internal waves; (h) log(Fr_t) – note that log(Fr_t) = 0 corresponds to the transition to strong stratification effects; and (i) length-scale ratio, L_o/L_m – L_o was computed with P rather than ε using the assumption that $P \approx \varepsilon$.

However, the overall effects of stratification can be gauged by how much of several of the fields are colored white in Fig. 7.7. White represents points at which the stress was counter-gradient, i.e., would imply negative values of ν_t. These are places where the stress is either very small, such that the noise inherent to the approach is larger than the stress estimate, as well as stratified regions where momentum fluxes are truly counter-gradient. As expressed by this metric, it is clear that the fraction of the water column that is essentially not turbulent is much larger during the neap than the spring.

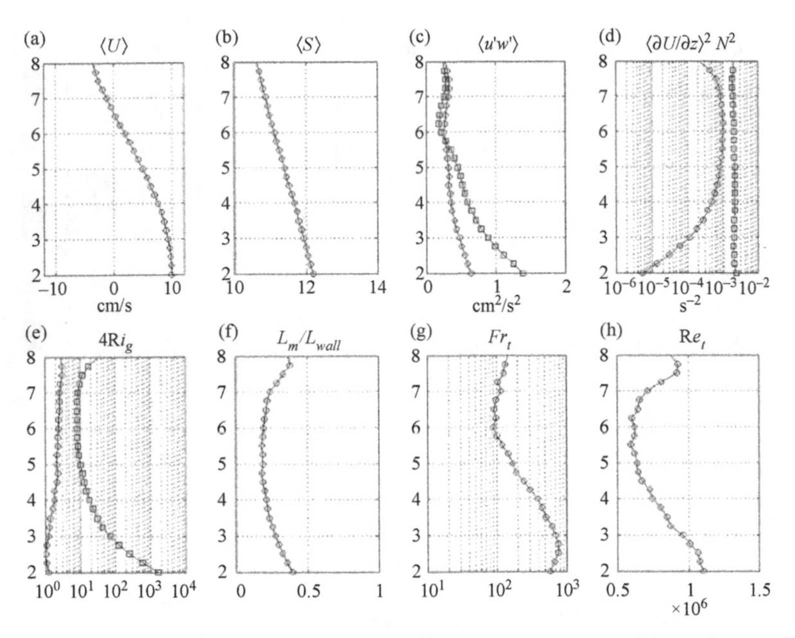

Figure 7.8. Averages of mean and turbulence fields in Northern San Francisco Bay: (a) time-averaged streamwise velocity; (b) time-averaged salinity; (c) median stress (○) and time-averaged stress (□); (d) square of the time-averaged shear (○) and N^2 (□); (e) median Ri_g (○) and Ri_g based on time-averaged velocities and densities (□); (f) ratio of L_m to the wall length scale $\kappa z(1 - z/zHH)$; (g) median value of $\varepsilon/\nu N^2$; (h) median value of Re_t.

Figure 7.8 includes plots of average and median values of several of these variables. The mean velocity distribution shows the form expected for gravitational circulation, as does the mean stress distribution. On average, the water column was stratified and as a consequence, Ri_g using average conditions was much greater than ¼; thus, a model of this flow based on average conditions would predict near-zero values of ν_t. In contrast, the median values of Ri_g are much lower, due to the fact that Ri_g is small for much of the time, time for which the turbulent mixing is energetic. This is reflected in the large values of Q that are common throughout the experiment. On the other hand, the effects of stratification are nonetheless pervasive: L_m, the mixing length, is always smaller than the law of the wall values that would be seen in the equivalent unstratified flow.

The complex behavior seen in Figs 7.7 and 7.8 is challenging to model. For example, Stacey *et al.* (1999b) found that the Mellor Yamada 2.5 closure could be significantly in error when the water column is stratified. While newer models perform better (see, e.g., Umlauf and Burchard, 2005), it is fair to say that we have reached a state with estuarine circulation codes where the turbulence model is a significant factor limiting accuracy of any computations.

7.4. Mechanisms for horizontal mixing in estuaries

At small scales, horizontal mixing in estuaries is the result of turbulent eddies that are thought to be roughly equal in horizontal extent to the local depth. At larger scales, horizontal mixing, which is almost exclusively viewed as being in the axial or longitudinal direction, is produced by vertical and lateral velocity shear produced by tides, winds, and density-driven flows. Most importantly, considering the averaged effects of velocity shear means considering mixing to be the subtidal consequence of tidal processes. Thus, an important conceptual dividing point is that when considering tidally varying mixing, the only mixing mechanism that can be represented (possibly) with a Fickian diffusion coefficient is turbulence. As seen in numerical computations of estuarine transport, temporal and spatial variations in the mean velocity can produce complex stirring and deformation of tracer fields, but only the small-scale turbulent mixing can be considered to be diffusion-like.

Much of what follows will be focused on describing longitudinal variations in subtidal scalar fields, e.g., the subtidal variation in depth-averaged salinity from river to ocean in estuaries like San Francisco Bay. Implicit in this focus is the assumption that these variations tend to play a primary role in system dynamics; in contrast, except in large systems where rotation can be important, lateral variations are assumed to be purely tidal whereas subtidal vertical variations are viewed as being connected to longitudinal variations in scalar field. For example, this is the basis of Hansen and Rattray's (1965) theory of gravitational circulation and estuarine salinity dynamics.

7.4.1. Horizontal turbulent mixing

The details of horizontal mixing remain to be determined, although it appears that in some cases a depth-independent horizontal eddy diffusivity of the form

$$A_x = C_1 u_* H \tag{7.58}$$

may be applicable. Here, u_* is the shear velocity based on the instantaneous Reynolds-averaged velocity, and C_1 is an $O(1)$ constant. Fischer *et al.* (1979) cite values of C_1 between 0.4 and 1.6. It appears, however, that horizontal mixing may also depend on turbulence scales that have horizontal scales larger than the depth, leading to scale-dependent dispersion. Note that at some point, e.g., at the scale of bathymetric variations, perhaps $O(100\,\text{m})$ or less, the distinction between turbulence and mean flows may be difficult to discern. Further discussion of scale-dependent dispersion can be found in Fong and Stacey (2003).

7.4.2. Averaging and its consequences

For many biogeochemical, ecological, or sedimentary processes of interest, what is important is the effect of this tidal mixing, integrated over many tidal cycles. In this case, the net effect of tidal processes can be represented as an effective dispersion coefficient K_x. However, to do so requires some care in defining the averaging. One can average over a tidal cycle at a fixed point in space, producing the Eulerian mean of a given quantity; doing so, one might describe the salinity as having a decomposition similar to that given in equation (7.29):

$$f(x,y,z,t) = \langle f \rangle(x,y,z,t) + \tilde{f}(x,y,z,t) + f'(x,y,z,t), \qquad (7.59)$$

where

$$\langle f \rangle = \frac{1}{T}\int_{t}^{t+T} f\,dt \qquad (7.60)$$

is the average over tidal period T (which still varies in time) and

$$\tilde{f} = \frac{1}{T_a}\int_{t}^{t+T_a} f\,dt - \langle f \rangle \qquad (7.61)$$

is the tidally varying part averaged over period T_a to filter out turbulence f'. If needed f' could be further split into surface wave and turbulence parts. The integrals in equations (7.60) and (7.61) could be considered to be more complicated filtering operations, although as written, they represent simple moving averages. Typically $T \geq 30$ hours, a period that removes the principal semidiurnal and diurnal tidal constituents, while $T_a = 10$ min more often than not.

The temporal decomposition hides an important aspect of the physics of systems with energetic tidal flows that is important to dispersion (Jay, 1991). When this averaging is applied to the velocity field, the result is defined to be the "residual circulation", although as defined, it is a purely Eulerian quantity, the Eulerian mean velocity. On the other hand, if the averaging was done by following a particular particle through a tidal day, in general, a different mean velocity, the Lagrangian mean (residual) velocity, would result. In the absence of tides these would be the same, whereas when the tides are energetic, these can be quite different. The difference between these two quantities is known as the "Stokes drift"; i.e., as described by Longuet-Higgins (1969), one has the conceptual equation relating the two means:

$$\text{Lagrange} = \text{Euler} + \text{Stokes}. \qquad (7.62)$$

A simple example suffices to show the difference. Consider a progressive shallow water wave propagating in the x-direction with velocity

$$\widetilde{U} = \left(\frac{a}{H}\right)\sqrt{gH}\sin\left(\frac{2\pi}{T}\left(x - \sqrt{gH}\,t\right)\right), \tag{7.63}$$

with a corresponding free surface elevation

$$\zeta = a\sin\left(\frac{2\pi}{T}\left(x - \sqrt{gH}\,t\right)\right). \tag{7.64}$$

The tidally averaged net transport through any section is

$$\langle q\rangle = \frac{1}{T}\int_0^T\int_0^{H+\zeta}\widetilde{U\zeta}\,dz\,dt = \frac{a^2}{2H}\sqrt{gH}. \tag{7.65}$$

In contrast, since $\left\langle \widetilde{U}\right\rangle = 0$, the Eulerian net transport $q_E = \langle U\rangle\left\langle H + \widetilde{\zeta}\right\rangle = 0$. The difference between $\langle q\rangle$ and q_E is known as the wave transport and is the depth-integrated Stokes drift for this flow.

As discussed in Jay (1991) and used in practical modeling of Chesapeake Bay by Dortch *et al.* (1992), careful consideration of the averaging involved in removing the tides shows that the subtidal advection–diffusion equation must include the Stokes drift velocity, i.e., the advection term should be:

$$\left(\left\langle \vec{U}\right\rangle + \vec{U}_{Stokes}\right)\cdot\nabla\langle C\rangle. \tag{7.66}$$

For example, in the St. Lawrence Estuary near Ville de Québec, $a \approx 2$ m and $H \approx 10$ m, allowing us to calculate as per equation (7.65) a Stokes drift velocity directed upstream of 0.4 m/s, which is comparable to the net velocity downstream due to river flow! In reality, the situation is even more complicated since the Lagrangian mean flow must satisfy continuity such that the integral over the area of the Lagrangian mean velocity must be equal to the net outflow or inflow to the estuary. If, for example, the net inflow/outflow is zero, this implies that the net mean Eulerian flow is exactly equal and opposite to the net Stokes drift, i.e., if one placed a set of current meters in such a system, one would see, after averaging, a net flow upstream. Such behavior has been documented by Rosman *et al.* (2007) for the related case of surface waves propagating shorewards on a rocky coast.

Finally, while the net advection associated with tides may in reality be zero, lateral variations in either Eulerian or Lagrangian mean velocities, often referred to as "pumping", may lead to significant longitudinal dispersion.

7.4.3. *Advection–diffusion of the subtidal salinity field*

In light of the interest in predicting axial variations in scalar fields, it is also useful to average across the estuarine cross-section; commonly (see, e.g., Fischer, 1972), this is represented with an overbar (like turbulence, unfortunately). That is, both $\langle f \rangle$ and $\tilde{f}$ can be split into a part that varies with position in the cross-section and a part that varies:

$$\langle f \rangle = f_a(x,t) + f_s(x,y,z,t), \tag{7.67}$$

where f_a is the cross-sectional average of $\langle f \rangle$ and f_s is the deviation from that average; and,

$$\tilde{f} = \tilde{f}_a + \tilde{f}_c. \tag{7.68}$$

Jay (1991) discusses the subtleties of the order of averaging, time vs space.

Consider F, the total flux of salinity, S, through a cross-section of area A. Let's suppose that the local (possibly curvilinear) coordinate system has a velocity normal to the cross-section of U, then

$$F = \iint_A US\,dA = A\overline{\langle US \rangle} = A\left\{U_a S_a + \left\langle \tilde{U}_c \tilde{A}_c \right\rangle + \overline{U_s S_s} + \overline{\langle U'S' \rangle}\right\}. \tag{7.69}$$

Thus, the flux has four main components: (a) advection of S_a by the "mean" velocity U_a; (b) flux associated with temporal correlation of $\tilde{U}_c$ and $\tilde{S}_c$, i.e., the rectified flux due to tides; (c) the flux due to spatial correlations of subtidal velocities and salinities, behavior demonstrated, e.g., by gravitational circulation which carries freshwater out at the surface and salty water in at the bottom of an estuary; and (d) the correlations of whatever is left over, for example salt flux associated with tidal variations in gravitational circulation and stratification might fit here. Thus, if filtering and spatially averaging the equation for salt conservation gives an advection–diffusion equation for the cross-sectionally averaged subtidal salinity field:

$$A\frac{\partial S_a}{\partial t} + \frac{\partial}{\partial x}(AU_a S_a) = \frac{\partial}{\partial x}\left(AK_x \frac{\partial S_a}{\partial x}\right), \tag{7.70}$$

where we have lumped together all but the first of the four terms in the averaged flux and written the result as a Fickian diffusion process. Thus, K_x is defined as

$$K_x = -\left\{\left\langle \tilde{U}_c \tilde{A}_c \right\rangle + \overline{U_s S_s} + \overline{\langle U'S' \rangle}\right\}/(\partial S_a/\partial x). \tag{7.71}$$

Like with eddy viscosities and diffusivities, equation (7.71) is more of a definition than a predictive model, although some explicit models of mixing processes that produce K_x are possible.

Following the nomenclature of Fischer (1972), the first term is the flux produced by tidal pumping, a catch-all phrase that refers to coherent tidal variations that produce fluxes. Because they depend on geometry, they tend to be quite site-specific, although Stommel and Farmer presented a brilliantly simple analysis for mixing through a tidal inlet that exemplifies pumping, albeit an analysis that does not yield a prediction of K_x (see chapter 7 of Fischer *et al.*, 1979). Stokes drift effects, which more properly are written as advection, are also included in pumping. Fischer refers to the second term as the effect of residual circulation. The best example of this is Hansen and Rattray's theory of gravitation circulation, which provides a prediction of K_x produced by $\overline{U_s S_s}$. The third term is referred to as the dispersive flux; one model for this term is provided by shear flow dispersion, as extended to include transverse and tidal velocity variations.

7.4.4. Determining $\mathbf{K_x}$ *from observations*

A common approach for determining K_x from observations is to invert the subtidal salt balance equation, viz:

$$K_x = \frac{1}{A\partial S_a/\partial x}\left\{\int_x^{X_r} A\frac{\partial S_a}{\partial t}dx - Q_f S_a\right\}, \tag{7.72}$$

where Q_f is the river flow (assumed constant), X_r is the distance upstream from the mouth of the estuary where the salinity is sensibly zero. In most cases (e.g., Monismith *et al.*, 2002), it is assumed that the salt field is steady (an assumption that Banas *et al.*, 2004 show may be significantly in error in some cases), so that the integral term in equation (7.72) can be omitted. Using this approach and 20 years of salinity data for San Francisco Bay, Monismith *et al.* (2002) found values of K_x that ranged from 10 to 3000 m^2/s, with the largest values occurring near the ocean and for high flows, while the lowest values were found in the upstream low-salinity zone at low river flows. Using the full salt balance, Banas *et al.* (2004) found similar spatial variability with K_x near the mouth of Willapa Bay reaching values of ca. 800 m^2/s, although without the flow dependence seen in the San Francisco Bay case. Finally, applying this approach to the low-salinity zone of the St. Lawrence estuary (e.g., Simons *et al.*, 2007) suggests a value of $K_x = O(1000\ m^2/s)$. Earlier work by Bowden and others summarized in Fischer *et al.* (1979) gives a similar range of values for K_x, with "typical" values of 100–300 m^2/s for many estuaries, e.g. 360 m^2/s for the Mersey according to Bowden and Gilligan (1971). Mechanisms for producing these values of K_x are discussed next.

7.4.5. Shear flow dispersion of tidal currents

The starting point for understanding shear flow dispersion by tidal currents is shear flow dispersion in rivers. Dispersion in rivers can be thought of as describing the steady-state (zero-frequency) limit of a tidal flow, and so provides the basis for understanding shear dispersion in estuaries. Noting that the largest dimension of flow variation is the one likely to give the largest shear flow dispersion coefficient, Fischer (1967) was the first to recognize that Taylor dispersion due to transverse variations in velocity was likely to be much larger than shear dispersion associated with vertical variations. One additional complication in rivers over simpler channel flows is that the local transverse depth variation $H(y)$ must be accounted for. Doing so, Fischer finds that

$$K_x = -\frac{1}{A}\int_0^B \widehat{U}(y)H(y)\left\{\int_0^y \frac{1}{\varepsilon_t H(y')}\left(\int_0^{y'} \widehat{U}(y'')H(y'')\,dy''\right)dy'\right\}dy, \quad (7.73)$$

where B is the width and the local depth-averaged velocity is $\widehat{U}$. In river flows, K_x can be approximated as

$$K_x \approx 0.01\frac{\overline{U}^2 B^2}{\overline{H}u_*}, \quad (7.74)$$

where $\overline{U} = Q/A$ is the average velocity in the river based on the flow, Q. In a steady estuarine flow, $\overline{U} = U_a$. Given that $\overline{U} \sim 20u_*$ and $B > 20\overline{H}$, $K_x/\overline{H}u_* = O(100)$ at least (see table 5.3 in Fischer *et al.*, 1979); i.e., shear flow dispersion associated with transverse mixing is roughly 10 times (or more) larger than shear flow dispersion associated with vertical shear and 100 times (or more) larger than turbulent diffusion.

An important limit to the use of equations (7.73) or (7.74) is that they only apply once the time from initial release is longer than the transverse mixing scale. For a scalar release in the middle of the channel, this mixing time is

$$T_c = 0.1\frac{B^2}{\varepsilon_t}, \quad (7.75)$$

where $\varepsilon_t \approx u_* H$ is the depth-averaged transverse turbulent mixing coefficient. In river flows, equation (7.75) can imply that shear flow dispersion is not applicable until quite a long distance downstream of the release point. In estuarine tidal flows, this mixing requirement has a striking consequence: shear flow dispersion is altered such that for $T_c \ll T$, $K_x \to 0$, whereas for $T_c \gg T$, K_x is the average of the shear flow dispersion coefficient calculated as a function of tidal phase.

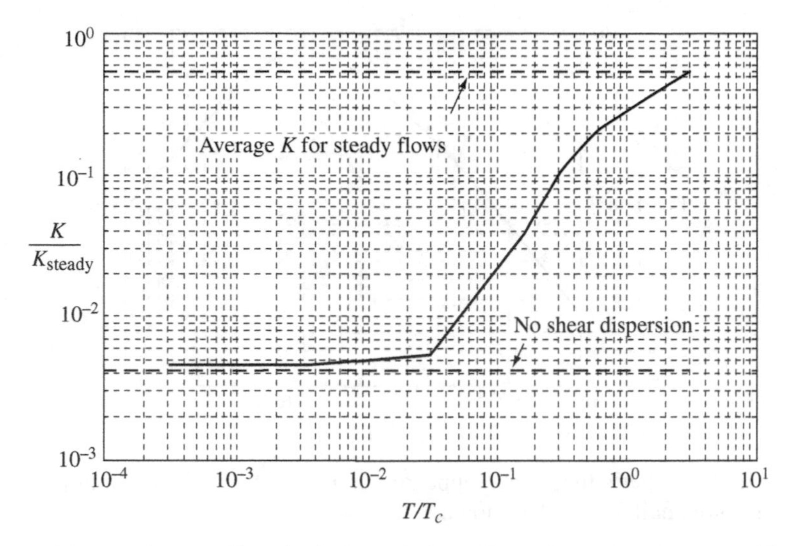

Figure 7.9. Random walk calculation of shear flow dispersion in an oscillating shear flow described by equation (7.76).

The basis of this behavior is that in the limit of no transverse mixing, transverse deviations in scalar concentration produced by shear are eliminated when the shear reverses. For example, the variance of a cloud of particles advecting in the shear flow

$$U(z, t) = Sz \sin(2\pi t/T) \tag{7.76}$$

would vary purely periodically in time if there were no transverse mixing. The results of computations made using the same random walk formulation discussed above (except now with a periodic flow) are given in Fig. 7.9, where the two limiting cases can be discerned and where it is clear that K_x, made non-dimensional by either the transverse mixing coefficient or a value of K_x appropriate to a steady flow of the same strength, is a function of the ratio T/T_c.

Based on an analytical solution to the advection–diffusion equation for the velocity field in equation (7.76) given by Holley *et al.* (1970), Fischer *et al.* (1979) give an approximate expression for K_x that includes this effect of incomplete mixing, as well as the standard approximation connecting the velocity variations in the integrals defining K_x in the steady case with the cross-sectionally averaged velocity:

$$K_x = 0.014\, U_0{}^2 T\, \Psi(T'), \tag{7.77}$$

where $U_s = U_0 \sin(2\pi t/T)$ and $T' = T/T_c$ is used with a constant of 1 rather than 0.1. The function $\Psi(T')$ is plotted in Fig. 7.10 and has the limits $\Psi \to (1/T')$ for $T' \gg 1$ and $\Psi \to 2.6T'$ for $T' \ll 1$.

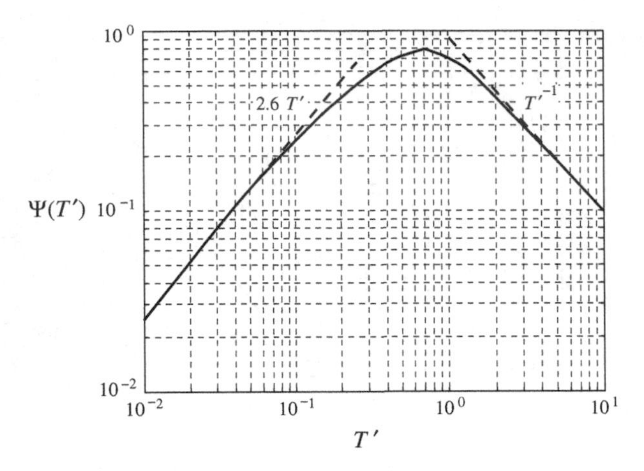

Figure 7.10. Fischer's function Ψ appearing in equation (7.77). Asymptotic limits for large and small T' are also shown.

To see how this works, consider the case of Montezuma Slough, a narrow tidal channel in Northern San Francisco Bay that has $B \approx 30$ m, $H = 6$ m, $U_0 = 0.5$ m/s for the M2 tide ($T = 4.3 \times 10^4$ s). Assuming $C_D = 0.0025$, $u_* = 0.015$ m/s (this is based on the r.m.s. value of U_s) so that by equation (7.58), $\varepsilon_t = 0.06$ m^2/s. Thus, $T_c = 1.5 \times 10^4$ and $T' = 2.7$, so that $\Psi = 0.3$, which finally gives $K_x \approx 25$ m^2/s. However, repeating these calculations for Willapa Bay (Banas *et al.*, 2004) for which $B \approx 8000$ m and $u_* \approx 0.05$ m/s gives $T' = 4 \times 10^{-4}$, so that $\Psi \approx 10^{-3}$, and $K_x \sim 10^{-1}$ m^2/s, i.e., *much* smaller than what is observed. Thus, for narrow estuaries with energetic flows, shear flow dispersion may be important whereas for wide coastal plain estuaries, it is likely irrelevant.

7.4.6. Dispersion by gravitational circulation

A second form of shear flow dispersion for which there is a predictive model is that associated with gravitational circulation. While this topic is dealt with in more detail elsewhere in this volume, a brief presentation is given here for completeness.

The physical basis of gravitational circulation is a balance between the baroclinic pressure gradient and the tidally averaged shear stress divergence (Hansen and Rattray, 1965). From a scaling standpoint, this balance can be written as

$$U_{GC} \sim \frac{\beta \frac{\partial S_a}{\partial x} g H^3}{\nu_t} \sim \frac{\beta \frac{\partial S_a}{\partial x} g H^3}{u_* H} \sim \frac{\beta \frac{\partial S_a}{\partial x} g H^2}{{C_D}^{1/2} U_{r.m.s.}}, \tag{7.78}$$

where U_{GC} is the velocity scale for the vertically sheared gravitational circulation and $U_{r.m.s}$ is the r.m.s. tidal velocity. This straining flow acts to tilt isohalines, producing a vertically variable salinity perturbation whose strength is determined by a balance between vertical mixing and straining, i.e.

$$U_{GC}\frac{\partial S_a}{\partial x} \sim \kappa_t \frac{\partial^2 S'}{\partial z^2}, \tag{7.79}$$

so that

$$S' \sim \frac{U_{GC}\frac{\partial S_a}{\partial x}H^2}{\kappa_t} \sim \frac{\beta\left(\frac{\partial S_a}{\partial x}\right)^2 H^4}{C_D U_{r.m.s.}{}^2}. \tag{7.80}$$

Thus, the total flux integrated over the depth is

$$F \sim U_{GC}S'H \sim \frac{\beta^2\left(\frac{\partial S_a}{\partial x}\right)^3 H^6}{C_D{}^{3/2}U_{r.m.s.}{}^3}, \tag{7.81}$$

implying a diffusion coefficient

$$K_{xGC} \sim \frac{\beta^2\left(\frac{\partial S_a}{\partial x}\right)^2 H^5}{C_D{}^{3/2}U_{r.m.s.}{}^3}. \tag{7.82}$$

As noted by Hansen and Rattray, this relation implies that gravitational circulation acts to resist changes in S_a, since compression of the salinity field by increased outflow, which acts to push salt out, increases the upstream salt flux, which acts in the opposite sense. This fundamental aspect of gravitational circulation explains why the length of salinity intrusion in estuaries shows power law dependencies on Q_f that range from -3 to $-1/7$.

Equation (7.82) can be written in non-dimensional form with the coefficient of proportionality derived by Hansen and Rattray if one assumes that the turbulent viscosities and diffusivities behave as $\nu_t \simeq \kappa_t = \gamma u_* H$, where for homogeneous flows we expect that $\gamma \simeq 0.1$:

$$\frac{K_x}{u_* H} = C_{HR}Ri_x{}^2, \tag{7.83}$$

where $C_{HR} = 5.4 \times 10^{-2}$ and

$$Ri_x = \frac{\beta g \frac{\partial S_a}{\partial x} H^2}{u_*{}^2} \tag{7.84}$$

is the horizontal Richardson number. A key aspect of using this expression is that the coefficient C_{HR} is proportional to γ^{-3}. Given that stratification can dramatically reduce mixing of momentum and salt, it seems reasonable to suppose that $C_{HR} = 5.4 \times 10^{-2}$ is a lower bound and may be much larger depending on the dynamics of the stratification.

An extension of Hansen and Rattray that accounts for tidal stratification dynamics, i.e., strain induced periodic stratification (SIPS), can be carried out numerically following the approach of Simpson and Sharples (1991). The flow is assumed to be locally one-dimensional and to have specified tidal and baroclinic pressure gradients:

$$\frac{\partial U}{\partial t} = -g\left(\frac{\partial \tilde{\zeta}}{\partial x} + \frac{\partial \zeta_a}{\partial x}\right) - g(H - z)\beta\frac{\partial S_a}{\partial x} + \frac{\partial}{\partial z}\left(\nu_t \frac{\partial U}{\partial z}\right), \tag{7.85}$$

where the tidal pressure gradient can be monochromatic, in which case it can be written as

$$g\frac{\partial \tilde{\zeta}}{\partial x} = \frac{2\pi U_0}{T}\sin\left(\frac{2\pi t}{T}\right), \tag{7.86}$$

or can have multiple tidal constituents. Salt conservation implies that

$$\frac{\partial\left(\tilde{S} + S'\right)}{\partial t} = -U\frac{\partial S_a}{\partial x} + \frac{\partial}{\partial z}\left(\kappa_t \frac{\partial\left(\tilde{S} + S'\right)}{\partial z}\right). \tag{7.87}$$

Finally, both ν_t and κ_t are computed using some form of turbulence closure that represents some or all of the processes discussed above, e.g., the popular Mellor–Yamada (1982) level 2.5 turbulence closure that is used in many estuarine circulation models. This model gives well-behaved SIPS flows with gravitational circulation, etc. for $Ri_x < 0.25$, and unphysical runaway stratification for $Ri_x > 0.25$ (Monismith *et al.*, 1996). Using this approach, Monismith *et al.* (2002) found that for $Ri_x < 0.25$, K_x computed as

$$K_x = -\frac{\overline{\langle US \rangle}}{\partial S_a / \partial x} \tag{7.88}$$

obeyed the Hansen and Rattray scaling, but with $C_{HR} = 5 \times 10^2$, i.e., the constant of proportionality was increased by a factor of 10^4 by SIPS stratification over the value given by Hansen and Rattray (1965).

A second, physically appealing extension of Hansen and Rattray's theory was given by Fischer (1972), who examined the gravitational circulation in a triangular, rather than constant-depth estuary, determining that

$$K_x^{\,triangle} = 0.004\left(\frac{B}{H}\right)^2 K_x^{\,Hansen-Rattray}. \tag{7.89}$$

Given that (B/H) might be $O(100)$ or more, Fischer's result shows that including transverse variations in flow and stratification can greatly enhance K_x over what might be computed for a constant-depth cross-section. An idealized analysis including tidal variations in turbulence and stratification for a non-uniform cross-section remains to be done, so that the enhancement of K_x for this case is unknown, but likely comparable to what was found for the uniform-depth case.

7.4.7. *Other processes: dead zones, chaos, etc.*

Spatial flow variations in energetic tidal flows can give rise to more dispersion mechanisms that are not well described by the classical Taylor model. Figure 7.11 shows a series of (computational[12]) clouds dispersing in various parts of the San Francisco Bay Delta. In each series, the first panel shows the bathymetry. To

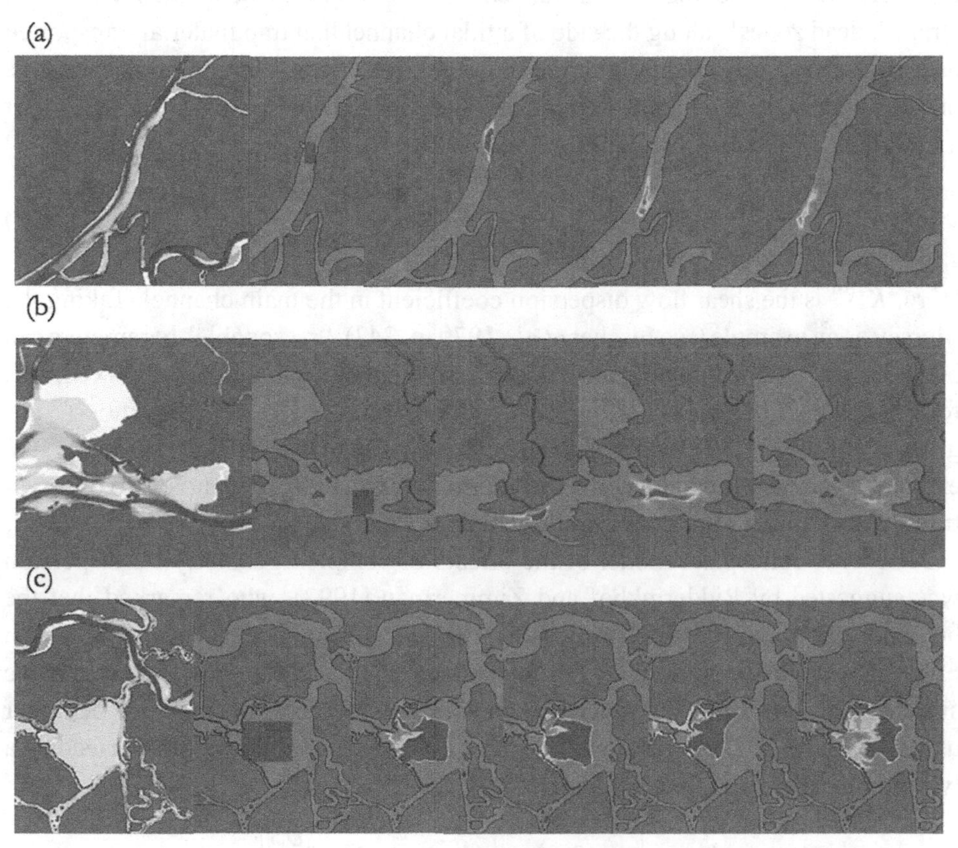

Figure 7.11. Calculated stirring of tracer fields by tidal flows in Northern San Francisco Bay and the Sacramento/San Joaquin Delta: (a) dispersion in the Sacramento River; (b) dispersion in Honker Bay, the eastern part of Suisun Bay; (c) dispersion in Frank's Tract, a large open water embayment in the center of the Delta. The first panel in each sequence is the bathymetry. The computations were done by Dr Nancy Monsen using the circulation code TRIM3D as described in Gross *et al.* (1999).

a first approximation, currents are proportional to the local depth, so transverse shear will be strong where transverse bathymetry gradients are large. Example (a), a cloud dispersing in the tidal Sacramento River, corresponds to the unsteady shear dispersion case treated above. Example (b) shows the results of tidal sloshing of the cloud in Suisun Bay, with the sequence showing how in one tidal cycle, an initial compact tracer field is spread over a large section of the ca. 20 km wide region (what is shown is about 10 km) by the transverse shear in the flow. The final example shows the mixing of a tracer inside (and out) of Frank's Tract, the largest open water portion of the Delta. As the cloud interacts with the complex of channels on the west side of Frank's Tract, it is shredded.

There are few simple models of the complex mixing processes shown in Fig. 7.11. One model that was advanced by Okubo (1973) postulated a series of "traps", dead zones[13] along the side of a tidal channel that trap material transported past and release it back later into the main channel. Parameterizing this exchange process with a trap-to-channel volume of r and a residence time in the trap of k^{-1}, for a tidal flow $U_0 \cos(\sigma t)$, Okubo found that

$$K_x = \frac{K_x{}^{(0)}}{1+r} + \frac{r\,U_0{}^2}{2k(1+r)^2(1+r+\sigma/k)}. \tag{7.90}$$

Here, $K_x{}^{(0)}$ is the shear flow dispersion coefficient in the main channel. Taking the Mersey as an example (see Fischer *et al.*, 1979, p. 242), "reasonable" parameters can be chosen so as to compute a dispersion coefficient close to what is observed; i.e., for $U_0 = 1.5$ m/s, $r = 0.1$, $k^{-1} = 10^4$ s, and $\sigma = 1.4 \times 10^{-4}$ s^{-1}, equation (7.90) gives $K_x = 0.9K_x{}^{(0)} + 360$ m^2/s. While Okubo's model is physically appealing, the exchange parameters cannot really be determined except by comparison to observations, so equation (7.90) is of questionable practical value.

A second, conceptual model of the effects of complex geometry on dispersion was suggested by Ridderinkhof and Zimmerman (1992), who remarked that the behavior of computed dispersion in the Wadden Sea (similar to what is shown in Fig. 7.10) might be viewed as the interaction of tidal flow with a deterministic system of residual eddies produced by the topography. They modeled this behavior using the simple, 2D divergence-free flow [using their notation but correcting a typographical error in their equation (7.2)]:

$$\begin{aligned} U &= \lambda\left[\cos(2\pi t) + \nu(1+\cos(2\pi t))\frac{\partial S}{\partial y}\right], \\ V &= -\lambda\nu[1+\cos(2\pi t)]\frac{\partial S}{\partial x}, \\ S &= \frac{1}{\pi}\sin(\pi x)\sin(\pi y). \end{aligned} \tag{7.91}$$

This dimensionless velocity field produces a set of circulation cells (the eddies) plus a tidal flow; the strength of the tidal flow relative to that of the eddies is measured by v, while λ is effectively the ratio of the tidal excursion to the eddy size (which is 1). Depending on the parameters chosen, the variance of a cloud of particles placed in this velocity field grows much faster than t, behavior referred to as chaotic or anomalous dispersion, although per the discussion above it could also be referred to as scale-dependent dispersion. The basis for this behavior is the existence of hyperbolic points in the flow – places where there are local stagnation points such that on one side of the streamline leading to this stagnation point, flow can go one way, i.e., into one set of eddies or can go another, sampling a different set of eddies. The effect of this behavior is formally chaotic in that a small difference in initial positions between two particles leads to an exponentially increasing separation. Unfortunately, there does not seem to be any simple way of parameterizing this mixing process, something that is hardly surprising since it is decidedly non-Fickian. We note in passing that similar behavior is observed for dispersion produced by tidal flows past headlands (see Signell and Geyer, 1990).

Finally, as a practical matter, it has been suggested that complex dispersion of the type shown above, or as observed in Willapa Bay, might be represented by a dispersion coefficient dependent on the local width and tidal velocity. For example, Banas *et al.* (2004) fit their data to the relation

$$K_x = 0.05\ U_0 B, \tag{7.92}$$

which matches the spatial variation in K_x they observed with an accuracy of ca. 50%. Note that this is well within the accuracy of similar simple models for dispersion in rivers, where a factor of three in accuracy is all that can be expected from relations like equation (7.73).

7.4.8. Synthesis: mixing and salinity intrusion in Northern San Francisco Bay

In any real estuary the net horizontal mixing, i.e., what K_x in equation (7.70) represents, is some combination of all the physical mechanisms discussed above. The example of how these might be combined given in Monismith *et al.* (2002) and used to model salinity intrusion in Northern San Francisco Bay is reprised here. For this system, the length scale for salinity intrusion that is most commonly used is X2, the distance along the channel bottom starting at the Golden Gate (where San Francisco Bay connects to the Pacific Ocean), where the salinity on the bottom is 2. Because the abundance of biota at all trophic estuaries has been found to depend on X2 (Jassby *et al.*, 1995), this metric has been used to formulate regulations governing the diversion of water that would otherwise flow through the Bay and out into the ocean.

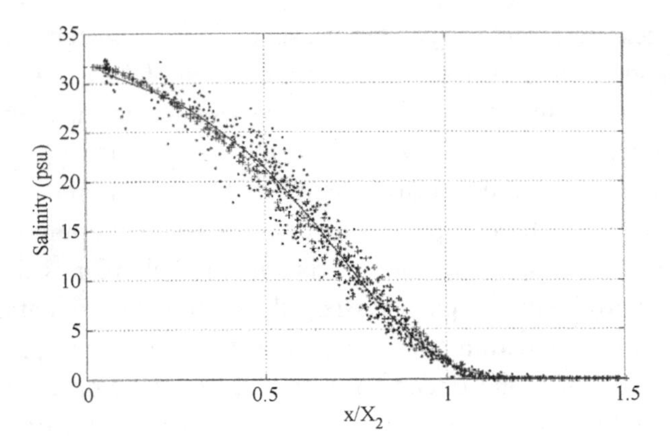

Figure 7.12. Salinity structure in Northern San Francisco Bay plotted as a function of *x*/X2. USGS observations (·) and salinities computed using a flow-dependent stratification correction (+). Taken from Monismith *et al.* (2002) (used with permission of AMS).

X2 has two interesting physical properties: firstly

$$X2 \propto Q^{-1/7}, \tag{7.93}$$

behavior found to describe three decades of flow variation, and secondly, the depth-averaged salinity field tends to take a self-similar form when distance is scaled by X2 (Fig. 7.12). The observed functional dependence of X2 on Q is much weaker than would be predicted by Hansen and Rattray's theory, which suggests $Q^{-1/3}$ at least for estuaries of constant cross-section (MacCready, 1999). On the other hand, the fact that the area and width of San Francisco Bay vary substantially with distance from the Golden Gate ($x = 0$) may explain this behavior.

To examine the effect of width and depth variations on salinity intrusion, consider equation (7.70) parameterized with two dispersion mechanisms: gravitational circulation and tidal shear. The former is represented explicitly as per the Hansen and Rattray formulation, albeit retaining the possibility of adjusting γ, while the latter is considered to be a constant, K_B.[14] Assuming steady state, the salt balance, integrated from some position x to upstream, implies

$$K_B \frac{dS_a}{dx} + \frac{C_{HR}(\beta g)^2 H^5}{\gamma^3 u_*{}^3}\left(\frac{dS_a}{dx}\right)^3 = -\frac{Q}{A} S_a, \tag{7.94}$$

where C_{HR} is the result of Hansen and Rattray's analysis left in terms of ν_t and κ_t. Both K_B and γ are free parameters that can be adjusted to match the computed salinity to the observed salinity distribution at a given flow, i.e., to $S_a = S_0 f\{x\ \mathrm{X2}(Q)\}$. Starting at $x = 0$ where the salinity is set to the ocean value, S_0, equation (7.94) can

Table 7.1. *Flow and mixing reduction*

Flow (m/s)	Ri_E	γ
100	4.8×10^{-3}	0.0041
300	1.5×10^{-2}	0.0040
1000	4.9×10^{-2}	0.0036
3000	1.5×10^{-1}	0.0031
10,000	4.9×10^{-1}	0.0027

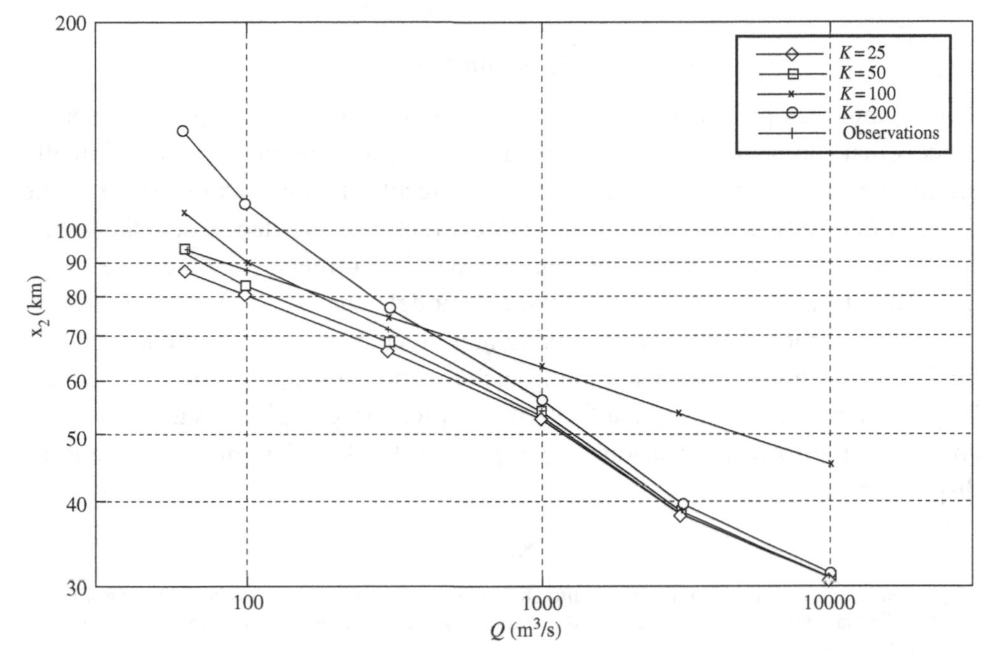

Figure 7.13. X2 (Q): observations (+) and as computed using equation (7.94) with $\gamma = 0.1$ and (◊) $K_B = 25$ m^2/s, (□) $K_B = 50$ m^2/s, (×) $K_B = 100$ m^2/s, and (○) $K_B = 200$ m^2/s. Taken from Monismith *et al.* (2002) (used with permission of AMS).

be solved for dS_a/dx and then integrated to find a new value of S_a and so forth until a position is reached where S_a is suitably close to 0.

Expressed in terms of X2, the computed salinity fields for different values of K_B and $\gamma = 0.1$ are shown in Fig. 7.13, where it is seen that the observed dependence of X2 on Q cannot be replicated using a constant value of γ. In contrast, by choosing $\gamma(Q)$ (Table 7.1) so as to best match X2, the simple model also gives excellent predictions of the spatial structure of the salinity field (also shown in Fig. 7.12). The result that $\gamma < 0.1$ for all flows implies that stratification is always important, even at low flows. The result that γ decreases as Q increases appears to reflect increasing mean stratification at higher flow rates. Behavior that Monismith *et al.* (2002) attributed to the fact that at higher flows Ri_x is likely to be greater than the critical value that determines the transition from SIPS to a sustained stratification state for a larger fraction of the spring–neap cycle than at lower flows. It should be recognized that this interpretation is not unique, as the behavior of X2 vs Q may be obtainable by choosing K_B to suitably depend on width, etc. (Ralston *et al.*, 2008). On the other hand, given the complex tidal mixing operant in Northern San Francisco Bay as seen in Fig. 7.10, any attempt to model mixing there with simple Fickian coefficients may have limited predictive value.

7.5. Summary

This chapter has presented a view of mixing in estuaries that emphasizes simple models that can be used to help understand mixing in estuaries. Overall, virtually all mixing that takes place in estuaries is the result of some form of shear in the velocity field, either at the small scales that characterize turbulence, or at the largest scales that represent horizontal variations in tidal currents. In some cases, e.g., unsteady shear flow dispersion, analytical models can be advanced whereas in others, e.g., chaotic tidal mixing, mixing in a given estuary is determined entirely by the local geometry and tidal forcing such that no simple parameterizations may be possible. At all scales, the effects of stratification can play a significant role, and can serve to make tidal time scale physics the key determinant of subtidal dispersion.

Notes

1. A Galilean transformation means the alteration of the mathematical expression that describes a physical law that happens when a fixed coordinate system is transformed to one moving at a constant velocity.
2. K here is dimensionless.
3. The reader is referred to Tennekes and Lumely (1972) and Pope (2000) for detailed presentations of turbulence theories and observations.
4. This data was taken as part of the experiment discussed in Nidzieko *et al.* (2006).
5. A more detailed version of Taylor's theory can be found in chapter 3 of Fischer *et al.* (1979).
6. All except for fluid density ρ.
7. See Kundu and Cohen (2002).
8. We use S here to denote velocity shear because this notation is widespread in the turbulence literature.
9. Direct numerical simulation.
10. Note that neither the calculations nor the experiments establish the exact boundary for turbulence growth or decay.
11. This should probably be called the Gibson number after Carl Gibson, who was the first to suggest its importance in stratified turbulence.
12. These computations were done by Nancy Monsen of the USGS using Vincenzo Casulli's TRIM code – see Monsen (2001) and Monsen *et al.* (2002).
13. This is the term used in the river mixing literature to describe this process.
14. Alternatively, K_B could be chosen as per equation (7.92).

References

Banas, N. S., B. M. Hickey, P. MacCready and J. A. Newton (2004) Dynamics of Willapa Bay, Washington, a highly unsteady partially mixed estuary. *J. Phys. Oceanogr.* **34**, 2413–2427.

Bowden, K. F. and R. M. Gilligan (1971) Characterisation of features of estuarine circulation as represented in the Mersey Estuary. *Limnol. Oceanogr.* **16**, 490–502.

Brennan, M. L. (2004) *Sediment fluxes in a partially stratified estuary.* PhD thesis, Stanford University.

Cheng, R. T., C. H. Ling, and J. G. Gartner (1999) Estimates of bottom roughness length and bottom shear stress in South San Francisco Bay, California. *J. Geophys. Res.* **104**(C4), 7715–7728.

Craig, P. D. and M. L. Banner (1994) Modeling waves-enhanced turbulence in the ocean surface layer. *J. Phys. Oceanogr.* **24**, 2546–2559.

Crank, J. (1975) *Mathematics of Diffusion*, 2nd edn. Clarendon Press, Oxford.

Dortch, M. S., R. S. Chapman and S. R. Abt (1992) Application of three-dimensional Lagrangian residual transport. *J. Hydr. Div. ASCE* **118**(6), 831–848.

Fischer, H. B. (1967) The mechanics of dispersion in natural streams. *J. Hydr. Div. ASCE* **93**, 187–216.

Fischer, H. B. (1972) Mass transport mechanisms in partially stratified estuaries. *J. Fluid. Mech.* **53**, 671–687.

Fischer, H. B., E. J. List, J. Imberger, R. C. Y. Koh and N. H. Brooks (1979) *Mixing in Inland and Coastal Waters*. Academic Press, New York.

Fong, D. A. and M. T. Stacey, (2003) Horizontal dispersion of a near-bed coastal plume. *J. Fluid Mech.* **489**, 239–267.

Fong, D. A., S. G. Monismith, J. R. Burau and M. T. Stacey (2009) Observations of secondary circulation and bottom stress in a channel with significant curvatures. *J. ASCE Hydr. Div.* **135**(3), 198–208.

Gargett, A. E. and J. R. Wells (2007) Langmuir turbulence in shallow water. Part 1: Observations. *J. Fluid Mech.*, **576**, 27–61.

Gargett, A. E., T. R. Osborn and P. W. Naysmith (1984) Local isotropy and the decay of turbulence in a stratified fluid. *J. Fluid Mech.* **144**, 231–280.

Gill, A. E. (1982) *Atmosphere–Ocean Dynamics*. Academic Press, San Diego.

Gross, E. S., J. R. Koseff and S. G. Monismith (1999) Three-dimensional salinity simulations in South San Francisco Bay. *J. Hydr. Div. ASCE* **125**(11), 1199–1209.

Gross, T. F. and A. R. M. Nowell (1985) Mean flow and turbulence scaling in a tidal boundary layer. *Cont. Shelf Res.* **2**, 109–126.

Hansen, D. V. and M. Rattray, Jr. (1965) Gravitational circulation in straits and estuaries. *J. Mar Res* **23**, 104–122.

Holley, E. R. Jr, D. F. Harleman and H. B. Fischer (1970) Dispersion in homogeneous estuary flow. *J. Hydr. Div. ASCE* **96**, 1691–1709.

Holt, S. E., J. R. Koseff and J. H. Ferziger (1992) A numerical study of the evolution and structure of homogeneous stably stratified sheared turbulence. *J. Fluid Mech.* **237**, 499–539.

Hunter, J. R., P. D. Craig and H. E. Phillips (1993) On the use of random-walk models with spatially-variable diffusivity. *J. Comput. Phys.* **106**(2), 366–376.

Ivey, G. N. and J. Imberger (1991) On the nature of turbulence in a stratified fluid. Part 1: The energetics of mixing. *J. Phys. Oceanogr.* **21**, 650–658.

Ivey, G. N., K. B. Winters and J. R. Koseff (2008) Density stratification, turbulence, but how much mixing? *Ann. Rev. Fluid Mech.* **40**, 169–184.

Jassby, A. D., W. M. Kimmerer, S. G. Monismith, C. Armor, J. E. Cloern, T. M. Powell *et al.* (1995) Isohaline position as a habitat indicator for estuarine resources: San Francisco Bay-Delta, California, U.S.A. *Ecol. Appl.* **5**, 272–289.

Jay, D. A. (1991) Estuarine salt conservation: a Lagrangian approach. *Est. Coast. Shelf Sci.* **32**, 547–565.

Jones, N. L. and S. G. Monismith (2008) The influence of whitecapping waves on the vertical structure of turbulence in a shallow estuarine embayment. *J. Phys. Oceanogr.* **38**(7), 1563–1580.

Kundu, P. K. and I. M. Cohen (2002) *Fluid Mechanics*, 2nd edn. Academic Press, New York.

Lohrmann, A., B. Hackett and L. P. Roed (1990) High resolution measurements of turbulence, velocity and stress using a pulse to pulse coherent sonar. *J. Atmos. Ocean. Tech.* **7**, 19–37.

Longuet-Higgins, M. S. (1969) On the transport of mass by time-varying currents. *Deep-Sea Res.* **16**, 431–447.

Lueck, R. G. and Y. Lu (1997) The logarithmic layer in a tidal channel. *Cont. Shelf Res.* **14**, 1785–1801.

MacCready, P. (1999) Estuarine adjustment to changes in river flow and tidal mixing. *J. Phys. Oceanogr.* **29**, 708–726.

Maxworthy, T. and S. G. Monismith (1988) Differential mixing in stratified fluids. *J. Fluid Mech.* **189**, 571–598.

Mellor, G. L. and T. Yamada (1982) Development of a turbulent closure model for geophysical fluid problems. *Rev. Geophys. Space Phys.* **20**, 851–875.

Monismith, S. G., J. Burau and M. Stacey (1996) Stratification dynamics and gravitational circulation in Northern San Francisco Bay. In T. Hollibaugh (ed.), *San Francisco Bay: The Ecosystem*. AAAS, pp. 123–153.

Monismith, S. G., W. Kimmerer, M. T. Stacey and J. R. Burau (2002) Structure and flow-induced variability of the subtidal salinity field in Northern San Francisco Bay. *J. Phys. Oceanogr.* **32**(11), 3003–3019.

Monsen, N. E. (2001) A study of sub-tidal transport in Suisun Bay and the Sacramento–San Joaquin Delta, *California*. PhD thesis, Stanford University.

Monsen, N. E., J. E. Cloern, L. V. Lucas and S. G. Monismith (2002) A comment on the use of flushing time, residence time, and age as transport time scales. *Limnol. Oceanogr.* **47**(5), 1543–1553.

Nidzieko, N. J., D. A. Fong and J. L. Hench (2006) Comparison of Reynolds stress estimates derived from standard and fast-ping ADCPs. *J. Atmos. Ocean. Tech.* **23**(6), 854–861.

Nezu, I. and H. Nakagawa (1993) *Turbulence in open channel flow*. IAHR monograph, A. A. Balkema, Rotterdam.

Okubo, A. (1971) Oceanic diffusion diagams. *Deep-Sea Res.* **18**, 789–802.

Okubo, A. (1973) Effect of shoreline irregularities on streamwise dispersion. *Neth. J. Sea Res.* **6**, 213–224.

Osborn, T. R. (1980) Estimates of the local rate of vertical diffusion from dissipation measurements. *J. Phys. Oceanogr.* **10**, 83–89.

Peters, H. (1997) Observations of stratified turbulent mixing in an estuary. Neap-to-spring variations during high river run-off. *Est. Coast. Shelf Sci.* **45**, 69–88.

Pope, S. B. (2000) *Turbulent Flows*. Cambridge Press, Cambridge.

Ralston, D. K., W. R. Geyer and J. A. Lerczak (2008) Subtidal salinity and velocity in the Hudson River estuary: observations and modeling. *J. Phys. Oceanogr.* **38**, 753–770.

Richardson, L. F. (1920) The supply of energy to and from atmospheric eddies. *Proc. Roy. Soc. London, Series A* **97**, 354–373.

Ridderinkhof, H. and J. T. F. Zimmerman (1992) Chaotic stirring in a tidal system, *Science* **258**(5085), 1107–1109.

Rohr, J. J., E. C. Itsweire, K. N. Helland and C. W. Van Atta (1988) Growth and decay of turbulence in a stably stratified shear flow. *J. Fluid Mech.* **195**, 77–111.

Rosman, J. H., J. R. Koseff, S. G. Monismith and J. Grover (2007) A field investigation into the effects of a kelp forest (*Macrocystis pyrifera*) on coastal hydrodynamics and transport. *J. Geophys. Res.* **112** C02016, doi:10.1029/2005JC003430.

Scully, M. E., C. Friedrichs and J. Brubaker (2005) Control of estuarine stratification and mixing by wind-induced straining of the estuarine density field *Est. Coasts* **28**(3), 321–326.

Shih, L. H., J. R. Koseff, G. N. Ivey and J. H. Ferziger (2005) Parameterizations of turbulent fluxes and scales using homogenous sheared stably stratified turbulence simulations. *J. Fluid Mech.* **525**, 193–214.

Signell, R. P. and W. R. Geyer (1990) Numerical simulation of tidal dispersion around a coastal headland. In R. Cheng (ed.), *Proceedings of the International Conference on the Physics of Shallow Estuaries and Bays, Asilomar, CA, November 1988*. Springer-Verlag, New York, pp. 210–222.

Simons, R., S. Monismith, F. Saucier, L. Johnson and G. Winkler (2007) Zooplankton retention in the estuarine transition zone of the St. Lawrence estuary. *Limnol. Oceanogr.* **51**(6), 2621–2631.

Simpson, J. H. and J. Sharples (1991) Dynamically-active models in the prediction of estuarine stratification. In D. Prandle (ed.), *Dynamics and Exchanges in Estuaries and the Coastal Zone*. Springer-Verlag, New York, pp. 101–113.

Squires, K. D. and J. K. Eaton (1991) Measurements of particle dispersion obtained from direct numerical simulations of isotropic turbulence. *J. Fluid Mech.* **226**, 1–35.

Stacey, M. T. and D. K. Ralston (2005) The scaling and structure of the estuarine bottom boundary layer. *J. Phys. Oceanogr.* **35**, 55–71.

Stacey, M. T., S. G. Monismith and J. R. Burau (1999a) Measurement of Reynolds stress profiles in unstratified tidal flow. *J. Geophys. Res.* **104**, 10,933–10,949.

Stacey, M. T., S. G. Monismith and J. R. Burau (1999b) Observations of turbulence in a partially stratified estuary. *J. Phys. Oceanogr.* **29**, 1950–1970.

Taylor, G. I. (1922) Diffusion by continuous movements. *Proc. London Math. Soc.*. **20**(1), 196–212.

Taylor, G. I. (1953) Dispersion of soluble matter in solvent flowing slowly through a tube. *Proc. Roy. Soc. London, Series A* **219**, 186–203.

Tennekes, H. and J. L. Lumely (1972) *A First Course in Turbulence*. MIT Press, Boston, MA.

Teixeira, M. A. C. and S. E. Belcher (2003) On the distortion of turbulence by a progressive surface wave. *J. Fluid Mech.* **458**, 229–267.

Turner, J. (1973) *Buoyancy Effects in Fluids*. Cambridge University Press, Cambridge.

Umlauf, L. and H. Burchard (2005) Second-order turbulence closure models for geophysical boundary layers. A review of recent work. *Cont. Shelf Res.* **25**, 795–827.

Visser, A. W. (1997) Using random walk models to simulate the vertical distribution of particles in a turbulent water column. *Mar. Ecol. Progr Series* **158**, 275–281.

Warner, J. C., W. R. Geyer and J. A. Lerczack (2005) Numerical modeling of an estuary: a comprehensive skill assessment. *J. Geophys. Res.* **110**(C05001), doi:10.1029/2004JC002691.

8

The dynamics of estuary plumes and fronts

JAMES O'DONNELL
The University of Connecticut

8.1. Introduction

Motion in the coastal ocean is largely driven by both wind and the tidal oscillation of the adjacent ocean. The complicated shape of the coastal boundary and the irregular bathymetry of estuaries and continental shelves often make the circulation resulting from the interactions between these mechanisms difficult to measure and to understand. When terrestrial runoff intrudes into this dynamic melee, it is unlikely that a simple theory will provide us with quantitative predictability for the motion of particles. Numerical simulations are, therefore, central to practical problems involving the transport of materials in the coastal ocean. Nevertheless, simple and elegant theories for important aspects of buoyancy-influenced coastal currents have been developed and evaluated with careful laboratory and field campaigns. These have become central to a broad understanding of coastal physics and are the focus of this chapter.

8.2. Phenomenology

There are many published reports describing the distribution of salinity at the mouths of rivers and estuaries. A few examples are listed in Table 8.1 to illustrate the diverse range of physical scales that have been examined. Figure 8.1 shows the near-surface salinity distributions from three plumes chosen to illustrate the complexity of behavior that must be understood. The first example, Fig. 8.1a,b, shows the salinity in Long Island Sound at the mouth of the Connecticut River reported by Garvine (1975). The measurements were acquired from a small boat survey in a few hours surrounding low slack water during the spring freshet. The values observed ranged from 5 at the mouth of the river to 28 in the center of the Sound. Figure 8.1b shows an example of the vertical structure of the salinity distribution along the line labeled "A" in Fig. 8.1a. The cross-section shows the same range of variation in

Table 8.1. *Garvine's (1995) plume classification parameters*

Name	K	F	KF	$1/\gamma$	V_E/HU	r/fH
Point Beach	0.1	2	0.2	1	0.1	0.3
Mississippi	0.2	1	0.2	1	0.2	0.7
Amazon	0.6	0.6	0.4	2	0.01	5
Niagara	1	1	1	1	0.08	0.1
Alaska	1	0.3	0.3	6	0.02	0.08
Gaspe	2	0.4	0.8	10	0.01	0.1
Algerian	3	0.3	0.9	8	0.005	0.06
Rhine	3	0.1	0.3	5	0.2	0.7
Delaware	4	0.1	0.4	5	0.3	0.4
SAB	4	0.1	1	5	0.2	0.4

salinity as the surface distribution. It also shows that the brackish water is confined to a shallow surface layer approximately 3 m thick with almost horizontal isohalines. Several different horizontal scales of variation are apparent. The width of the river mouth is approximately 2 km and, as will be discussed later, this is thought to be important in determining the character of the plume. The offshore (north–south) and along-shore extent of the plume are an order of magnitude larger than the source dimension. However, the vertical cross-section suggests that there is a narrow frontal boundary at the offshore edge of the plume at which the salinity variation is rapid. A companion work, Garvine and Monk (1974), focused on the structure of the frontal zone and suggested the scale was 50 m. Recent high-resolution observations by O'Donnell (1997) showed that 5 m was a better estimate of the frontal scale.

A much larger-scale plume is found at the mouth of the Columbia River in the northeast Pacific (Barnes *et al.*, 1972), and this has recently been the focus of extensive field programs summarized by Hickey *et al.* (1998, 2005). Figure 8.1c displays an example of the distribution of salinity 5 m below the surface in the winter of 1991, a period of northerly (upwelling favorable) winds. The river effluent clearly dilutes the ocean water 40 km from shore and at least 100 km north of the river mouth. Figure 8.1d shows a vertical cross-section of the density (σ_T) field along a seaward transect 30 km north of the river mouth. The plume appears to be a surface structure at this time, with the $\sigma_T = 22$ isopycnal confined to the upper 15 m of the water column. The spatial scales apparent in these observations contrast with those observed in the Connecticut plume, with horizontal extent of the plume at least an order of magnitude larger. The small-scale structure of the bounding front has recently been resolved by Nash and Moum (2005) and Orton and Jay (2005).

An early account of an estuary plume was provided by Wright and Coleman (1971), who surveyed the hydrography in the Gulf of Mexico at the mouth of a

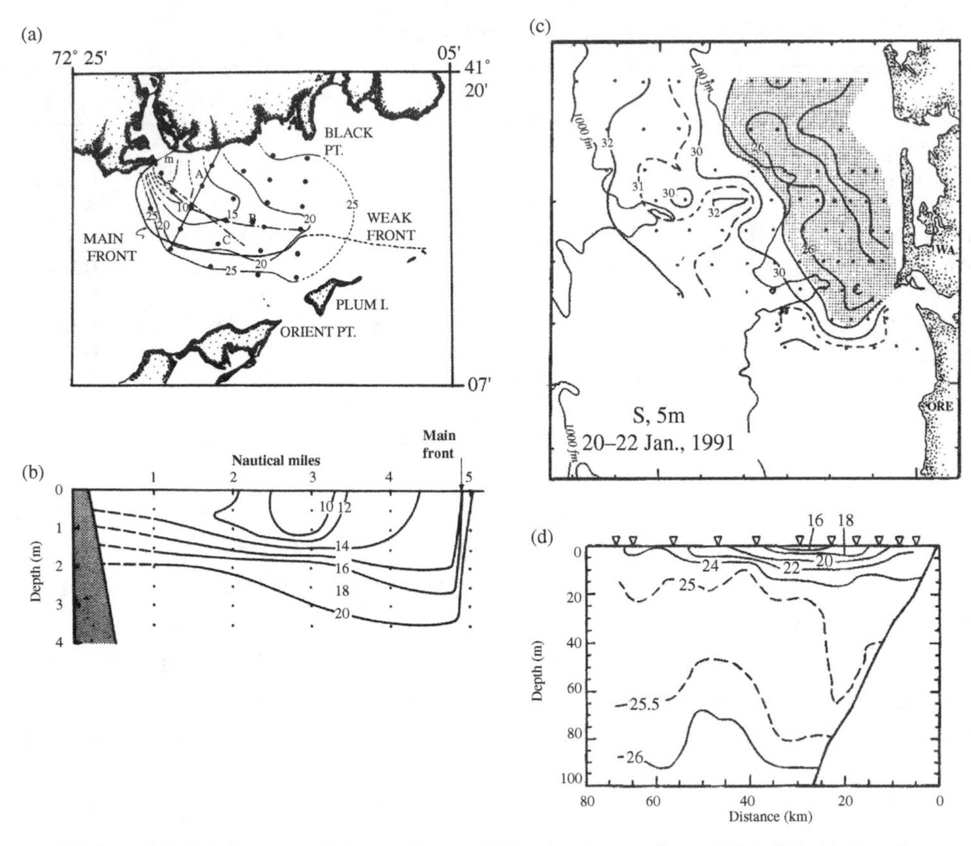

Figure 8.1. (a) A map of the surface salinity in eastern Long Island Sound at the mouth of the Connecticut River. The large dots show the locations where water samples were obtained and the line labeled "A" indicates the location of the vertical cross-section of salinity shown in (b). The curved line labeled "Front" is the approximate location of a foam line observed at the surface. (b) A vertical cross-section of salinity through the plume along the line labeled "A" in surface map shown in (a). The large dots show the location of samples, though higher resolution was employed near the front and the locations not shown. Both figures were presented by Garvine (1974). (c) A map of the salinity distribution 5 m below the surface at the mouth of the Columbia River on 20–22 January, 1991. Areas where the salinity is less than 28 are shaded. (d) A vertical cross-section of the density (σ_T) field along an east–west transect 30 km to the north of the river mouth. Both frames (c) and (d) are from Hickey *et al.* (1998). (e) A schematic diagram of the pattern of surface density (σ_T) in the Gulf of Mexico at the mouth of the Mississippi River during the flood phase of the tide, derived from a combination of airborne radiometer observations and water samples. (f) The ebb distribution comparison to (e). (g) A vertical cross-section of density (σ_T) along the seaward extension of the channel direction during the flood tide on 22 October, 1969. Frames (e)–(g) are the work of Wright and Coleman (1971). (h) The salinity distribution 3 m below the surface in the northern Gulf of Mexico showing that the inner shelf waters along the coast of Louisiana and Texas are measurably freshened for a distance of order 1000 km west of the Mississippi River delta. From Sahl *et al.* (1997).

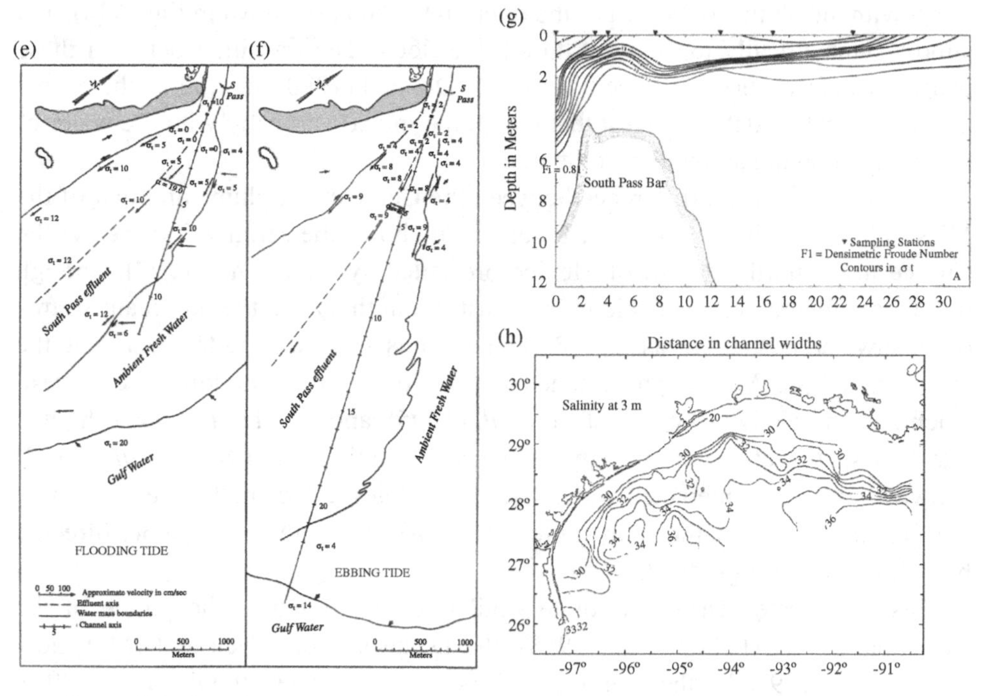

Figure 8.1. (cont.)

major distributary of the Mississippi River, South Pass. Figure 8.1e,f shows schematics that combine their interpretations of the airborne, high spatial resolution, infrared surface ocean temperature measurements and spatially coarse salinometer measurements from ship surveys. The temperature images showed distinct frontal boundaries that are represented in the schematics by the solid curves. The surface σ_T values are labeled at the measurement locations. The solid straight line shows the direction of the channel axis, with tick marks indicating distance from the source scaled by the width of the channel. The dashed line shows the centerline of the discharge plume, inferred from the temperature measurements. Within a kilometer of the river channel the σ_T values are less than 2 and increase to 10 two or three kilometers further from the source. Figure 8.1g shows a characteristic density cross-section along the centerline of the plume during the flood. The magnitude of the stratification and the depth of the pycnocline is very similar to that found at the mouth of the Connecticut. The horizontal variation is also similar. There are important differences, however. Figure 8.1e,f contrasts the structure at flood and ebb phases. There is a slight difference in the angle between the channel axis and the center of the plume at the ebb and flood, however, the data are probably inadequate to distinguish between the two. Garvine (1974) examined the effect of the tide on the Connecticut River plume and showed a much more substantial and unambiguous effect with the plume to the east of the river on the ebb (as shown in Fig. 8.1a), and almost symmetrically to the west during the flood. The transition between these phases has not yet been observed, though O'Donnell (1990) interpreted the results of a numerical model to suggest that the plume mixed vertically with the water of Long Island Sound at the turn of the tide.

The most significant difference between the Connecticut plume and that of the Mississippi is at the larger scale. Figure 8.1h shows the salinity 3 m below the surface in the northern Gulf of Mexico presented by Sahl *et al.* (1997). Though there are a few minor rivers along the coastline in this area, the dominant source of freshwater is the Mississippi River and it is therefore quite clear that the influence of the Mississippi extends on the order of 1000 km to the west. Cochrane and Kelly (1986), Vastano *et al.* (1995), and Cho *et al.* (1998) showed that this shelf water was strongly influenced by winds, and Sahl *et al.* (1997) showed that it also interacted with mesoscale eddies at the shelf edge. However, the general westward transport pattern is consistent with the offshore-directed baroclinic pressure gradient.

This large-scale behavior is quite similar to that found in the plume of the Columbia River (Hickey *et al.*, 1998), the Delaware coastal current (Munchow and Garvine, 1993a,b), the Chesapeake Bay outflow plume (Rennie *et al.*, 1999), and the Gulf of Maine coastal current (Fong *et al.*, 1997), among others. This example of an overlap in the characteristics of very small-scale rivers and very large plumes is the central challenge that must be faced in the development of a comprehensive theory for the dynamics of coastal plumes.

8.3. Classification schemes

The examples in Fig. 8.1 demonstrate that river plumes are found with a wide range of sizes and scales. When this structural complexity is combined with the fact that plumes are often subject to unsteady forcing by winds, tides and river discharge, it should not be surprising, therefore, that a general theory of plumes remains to be formulated. An approach to simplifying complex problems that has proven profitable in fluid mechanics is to apply "scaling analysis" (see, e.g., Kundu, 1990) to estimate the relative importance of the physical processes in a particular situation and then neglect the less important effects to develop a mathematically tractable theory. In the problem of interest here, the processes that control the spreading and mixing of the buoyant river effluent appear to depend sensitively on space and time scales and a wide range of plume morphologies can result. Garvine (1995) used scaling analysis to develop a systematic classification system for buoyant coastal flows based on the magnitude of the terms in the time-averaged, vertically integrated momentum balance (convective

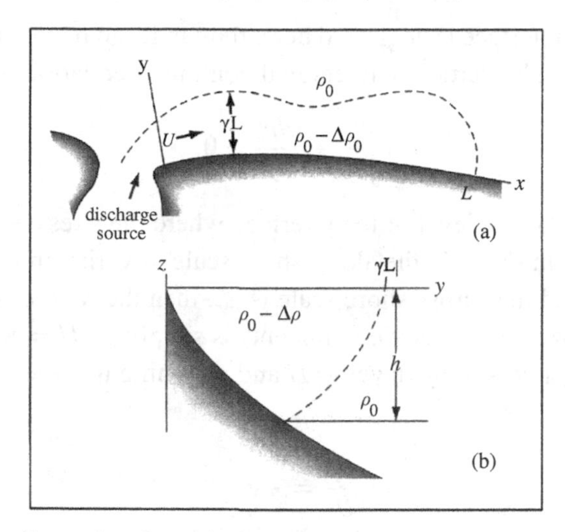

Figure 8.2. A schematic of an idealized buoyant plume showing (a) the alongshore, x, and across-shore, y, coordinates and corresponding length scales, L and γL. The dashed line illustrates the location of the bounding isohaline, or front. The ambient shelf water has density ρ_o, and the plume has a characteristic density $\rho - \Delta\rho$. The vertical structure of the plume is illustrated in (b), which shows the plane of the vertical, z, and offshore coordinates, y, and the maximum buoyant layer depth h. Reproduced from Garvine (1995).

and Coriolis accelerations, wind and bottom stress) relative to the horizontal baroclinic pressure gradient.

For the idealized plume geometry sketched in Fig. 8.2 (from Garvine, 1995) in which a buoyant layer is oriented with its long axis almost parallel to the coast, the vertically averaged momentum equations in the upper layer can be written as

$$\frac{\partial u^2}{\partial x} + \frac{\partial uv}{\partial y} - fv = -\frac{1}{\rho_0}\frac{\partial p}{\partial x} + \frac{\tau_{wx}}{h\rho_0} - \frac{\tau_{bx}}{h\rho_0} \tag{8.1}$$

and

$$\frac{\partial uv}{\partial x} + \frac{\partial v^2}{\partial y} + fu = -\frac{1}{\rho_0}\frac{\partial p}{\partial y} + \frac{\tau_{wy}}{h\rho_0} - \frac{\tau_{by}}{h\rho_0}, \tag{8.2}$$

where u and v are the velocity components in the x- and y-directions, p is the pressure, and the surface and interfacial stresses are τ_w and τ_b, which have components in the x- and y-directions. It is important to note that these equations are approximations and neglect terms that are associated with the vertical integral of the advective acceleration (see Dean and Dalrymple, 1991). The coefficients f and ρ_0 are the Coriolis parameter and a reference density. Note that time mean entrainment of fluid across the interface is neglected here, though it can be included as shown by O'Donnell (1988). The vertically integrated continuity equation is

$$\frac{\partial hu}{\partial x} + \frac{\partial hv}{\partial y} = 0, \tag{8.3}$$

where $z = -h(x, y)$ is the level of the interface where the stress acts.

If, as illustrated in Fig. 8.2, the along-shore scale of variation in the velocity and layer depth is L and the across-shore scale is γL, then the ratio of the scales for the layer average horizontal velocity components is simply $V/U = \gamma$. Taking the scale for the vertical thickness of the layer as H and assuming that the pressure field is in hydrostatic balance, i.e.,

$$\frac{\partial p}{\partial z} = -g\rho, \tag{8.4}$$

then the magnitude of the pressure anomaly, P, can be estimated as

$$P = g\Delta\rho H. \tag{8.5}$$

If the horizontal scale of the layer depth variations in the along-shore direction (x) is L, the wind stress magnitude is τ_W, and the velocity magnitude is U, then the x momentum equation (8.1) can be rewritten with the magnitude of each term written below it using these scales as:

$$\frac{\partial u^2}{\partial x} + \frac{\partial uv}{\partial y} - fv = -\frac{1}{\rho_0}\frac{\partial p}{\partial x} + \frac{\tau_{wx}}{h\rho_0} - \frac{\tau_{bx}}{h\rho_0},$$

$$\frac{U^2}{L}, \quad \frac{\gamma U^2}{L}, \quad f\gamma U, \quad g\frac{\Delta\rho}{\rho_0}\frac{H}{L}, \quad \frac{\tau_W}{H\rho_0}, \quad \frac{rU}{H\rho_0}, \tag{8.6}$$

where a linear drag law with coefficient r has been adopted to simplify the representation of the friction at the bottom of the buoyant layer. The six quantities in equation (8.6) can be made dimensionless by dividing by the magnitude of the pressure gradient term to yield estimates of the relative magnitude of the terms as:

$$\frac{\partial u^2}{\partial x} + \frac{\partial uv}{\partial y} - fv = -\frac{1}{\rho_0}\frac{\partial p}{\partial x} + \frac{\tau_{wx}}{h\rho_0} - \frac{\tau_{bx}}{h\rho_0},$$

$$\frac{U^2}{C^2}, \quad \frac{\gamma U^2}{C^2}, \quad \frac{fL\gamma U}{C^2}, \quad 1, \quad \frac{\tau_W L}{C^2 H\rho_0}, \quad \frac{rUL}{HC^2\rho_0}, \tag{8.7}$$

where the shallow water internal wave speed $C^2 = gH(\Delta\rho/\rho_0)$ has been introduced. The *Froude number*, $F = U/C$, squared is clearly a measure of the relative importance of the rate of advection of momentum. The relative magnitude of the

Coriolis acceleration can then be written as FK, where $K = fL\gamma/C$ is the *Kelvin number*. This is simply the ratio of the width of the plume, γL, to the internal Rossby deformation scale, C/f.

Estimating the scale for the wind stress influence is not as straightforward. Pedlosky (1987, chapter 4), for example, details the development of well-established Ekman layer theory. It predicts that a steady wind over an unbounded and unstratified ocean will drive a steady current with a magnitude that diminishes with distance from the surface in a direction that rotates to the right (in the northern hemisphere) of the wind stress direction. The scale depth for the *Ekman layer* is $\delta_E = \sqrt{2A_v/f}$, where A_v is the vertical eddy viscosity coefficient. The integral of the velocity over the Ekman layer is the volume transport due to the wind, which can be shown to be zero in the direction of the stress and $V_E = \tau_W/\rho_0 f$ to the right of the stress (in the northern hemisphere). Using $V_E\rho_0 f$ as the scale for the wind stress, the relative magnitude of the wind stress in the momentum equations can be written as

$$\frac{V_E fL}{C^2 H} = \frac{V_E}{\gamma UH}\frac{fL\gamma}{C}\frac{U}{C} = \frac{V_E}{UH}\gamma^{-1}FK.$$

Note that the first factor is the ratio of the offshore Ekman flux divided by the alongshore flux. Similarly, the scale for the stress at the bottom of the buoyant layer can be written as

$$\frac{rUL}{HC^2\rho_0} = \frac{r}{fH}\gamma^{-1}FK.$$

Following Bowden (1967), we assume the linear drag law represents the bottom stress, $ru = A_v(\partial u/\partial z)$, which implies $O(rU) = O(A_V U/H)$, and allows us to estimate the eddy viscosity magnitude as $A_V = rH$. The relative magnitude of the bottom stress can then be rewritten as $\frac{1}{2}E\gamma^{-1}FK$, where $E = 2A_V/fH^2$ is the Ekman number. Note that $E = \delta_E^2/H^2$, the square of the ratio of the Ekman layer thickness to the plume layer thickness.

There are now five dimensionless parameters that are important to the steady-plume dynamics: the aspect ratio, γ; the Kelvin number, K; the Froude number, F; the Ekman number, E; and the wind stress magnitude, V_E/UH. Using these, the relative importance of the terms in the x-momentum equation is:

$$\frac{\partial u^2}{\partial x} + \frac{\partial uv}{\partial x} - fv = -\frac{1}{\rho_0}\frac{\partial p}{\partial x} + \frac{\tau_{wx}}{h\rho_0} - \frac{\tau_{bx}}{h\rho_0},$$

$$F^2, \quad F^2, \quad FK, \quad 1, \quad \gamma^{-1}\frac{V_E}{UH}FK, \quad \gamma^{-1}EFK. \tag{8.8}$$

It is a useful exercise to demonstrate that a similar approach applied to the across-shore (y) momentum equation yields

$$\frac{\partial uv}{\partial y}+\frac{\partial v^2}{\partial y}+fu=-\frac{1}{\rho_0}\frac{\partial p}{\partial y}+\frac{\tau_{wy}}{h\rho_0}-\frac{\tau by}{h\rho_0},$$

$$\gamma F^2,\quad \gamma^2F^2,\quad Fk,\quad 1,\quad \frac{V_E}{UH}FK,\quad \gamma EFK. \tag{8.9}$$

Comparison of the magnitudes of the wind stress terms in equations (8.8) and (8.9) leads to the conclusion that for plumes that are elongated along the coastline, i.e., $\gamma \ll 1$, the wind stress is much more significant in the along-shore momentum budget (relative to the horizontal pressure gradient) than in the across-shore budget. Note that if $\gamma^{-1}(V_E/UH)FK \gg 1$, then the layer should not be thought of as buoyancy forced since the wind stress is then larger than the pressure gradient. With this definition, the upper limit to the magnitude of the wind stress in a buoyancy-driven plume is:

$$\frac{V_E}{UH}=\frac{\gamma}{FK} \tag{8.10}$$

which, for $O(K)=1$ and $O(\gamma)=O(F)=0.1$, restricts the across-shore Ekman flux to be less than the along-shore flux. A similar comparison of the terms representing friction at the base of the layer and the pressure gradient shows that bottom friction is also more important in the along-shore momentum budget. Further, comparison to the components of the Coriolis acceleration shows that friction at the bottom will always be relatively small, $O(\gamma E)$, in the across-shelf equation but may be significant, $O(\gamma^{-1}E)$, in the along-shelf budget.

Garvine's (1995) estimates of the values of the classification parameters for a variety of plumes described in the literature are presented in Table 8.1. Plumes with $K \ll 1$ are termed "small-scale", since the width of the plume is narrow compared to the internal Rossby radius. The dynamics of this class must then be non-linear, since the convective acceleration must balance the pressure gradient and $F^2=O(1)$. Garvine (1982, 1984, 1987) and O'Donnell (1988, 1990) have developed models of such plumes and qualitatively compared the results to observations of the Connecticut River (cf. Garvine, 1974, 1975, 1977) and demonstrated reasonable consistency.

The opposite limit, $K \gg 1$, is found in "large-scale" plumes. A dynamic balance in the along-shore direction then requires that the magnitude of the Coriolis acceleration component can be no larger than the imposed pressure gradient in equation (8.8), i.e., $FK = O(1)$, so then $F = O(K^{-1})$ and $F^2 = O(K^{-2})$. The velocity must then be very slow compared to the internal wave speed, and the convective acceleration can be neglected in both momentum equations without introducing a large error.

When the plume is also elongated along the coast, as is often the case for the large-scale plumes listed in Table 8.1, we can exploit the fact that $\gamma \ll 1$ to further simplify the dynamics. When the wind stress magnitude is limited so that buoyancy forcing dominates and the plume is large-scale, $FK = O(1)$, then since $VE/UH = \gamma/FK = O(\gamma) \ll 1$, the relative magnitude of the wind stress in the across-shore momentum equation is γ and the lowest-order approximation to the momentum budget, equations (8.8) and (8.9), can be written as:

$$-fv = -\frac{1}{\rho_0}\frac{\partial p}{\partial x} + \frac{\tau_{wx}}{h\rho_0} - \frac{\tau_{bx}}{h\rho_0} \tag{8.11}$$

and

$$fu = -\frac{1}{\rho_0}\frac{\partial p}{\partial y}. \tag{8.12}$$

Since the frictional effects remain as leading-order terms in the along-shore balance, this regime is termed *semigeostrophic*. Note that when the horizontal scales are large compared to the vertical scale, the vertical accelerations can be neglected in the vertical momentum equation. The pressure field is said to be in hydrostatic balance and can be written as:

$$\frac{\partial p}{\partial z} = -g\rho(x, y, z, t), \tag{8.13}$$

where the density is generally a function of all three space coordinates and time. Taking the vertical derivative of equation (8.12) and substituting equation (8.13) yields

$$\frac{\partial u}{\partial z} = \frac{g}{f\rho_0}\frac{\partial \rho}{\partial y}, \tag{8.14}$$

which is often termed the thermal wind balance. This is an attractive result since linear approximations of the momentum equations are quite accurate.

This classification scheme is attractive in its simplicity and certainly successful in discriminating between the highly non-linear Connecticut River plume and those of large estuaries like the Delaware Bay. However, a seemingly paradoxical result was that the Earth's largest river, the Amazon, and the largest in North America, the Mississippi, were two of Garvine's examples of "small-scale" plumes. This conclusion arose because the Amazon enters the ocean at low latitude and has a small value of the Coriolis parameter and the length scale chosen for the Mississippi was characteristic of the width of a narrow distributary of the delta. Yet both rivers influence large regions of the adjacent ocean (see Lentz, 1995; Lentz and Limeburner, 1995; Sahl *et al.*, 1997), suggesting that the Earth's rotation should have an important influence. It is clear, therefore, that though helpful, this classification scheme is not comprehensive. Simplicity is achieved by the neglect of the effects of high-frequency processes and the consideration of only two spatial scales, L and γL. As discussed above, it has been demonstrated in several observation campaigns that, outside the immediate vicinity of the source, larger-scale plumes are sensitive to wind and we must consider the effect of intermittent forcing as well.

There is also recent evidence, see Marmorino and Trump (2000), that very small-scale (tens of meters) structures characteristic of non-linear processes appear in the outflow plume of the Chesapeake Bay, a "large-scale" plume in the Garvine (1995) classification system. It seems likely that the requirement that a single scale is chosen to represent the plume is an important limitation in Garvine's (1995) argument. He noted that the use of the observed plume length scales was a weakness in the classification scheme and pointed out that a more complete understanding would allow the characteristics of the plume to be determined from the source geometry and discharge characteristics. The use of a single length scale and classification may not be appropriate, especially when non-linear dynamics are predicted. Perhaps it is better to think of plumes as containing subdomains with different dynamics. This approach would have the additional advantage that the estuary could then be included in the paradigm, adjacent perhaps to an inertial near field, which links the estuary and the far field, or coastal current, with a frictional coastal layer and semigeostrophic stratified region.

Since there is, as yet, no single coherent theory for the dynamics of buoyancy-influenced shelf flows, we now summarize the results of a variety of process studies. Each of these is relevant to some plumes at least some of the time. We hope that these simple ideas will guide the development of more complete theories and well-designed observation campaigns.

8.4. Evaluation of the semigeostrophic balance

The first quantitative evidence of the dynamic balance represented by equation (8.14) in a buoyant coastal plume was reported by Blanton (1981) in the coastal current created by the many rivers that drain South Carolina and Georgia, USA. Using two ships with lowered CTDs and current meters, he mapped the salinity distribution along a section approximately normal to the coast, recording the vertical structure of the currents at two stations at which the ships anchored at hourly intervals for two and a half days during a period of weak winds. The geography of the study area is shown in Fig. 8.3a and an example of the vertical and across-

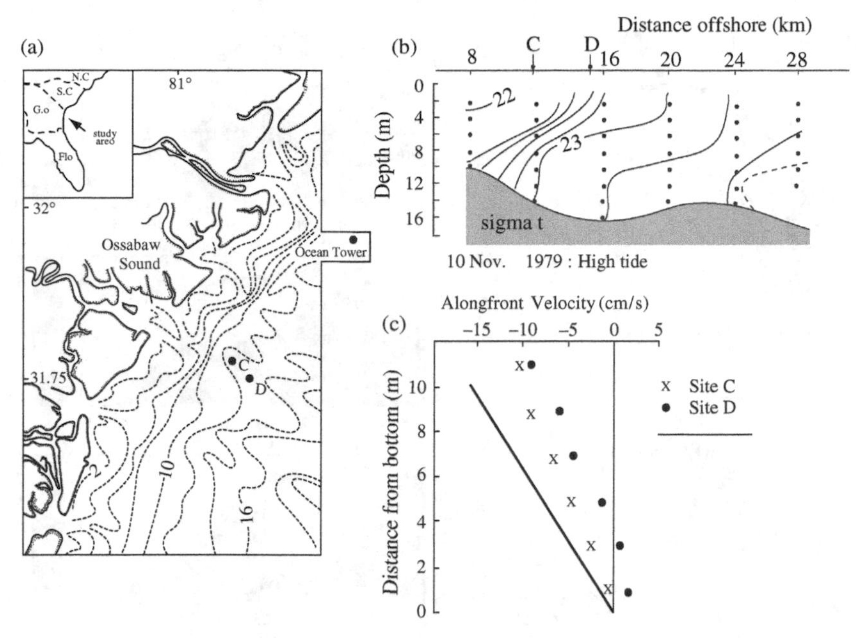

Figure 8.3. (a) The inset shows a map of the coastline and state boundaries of the southeastern United States with an arrow indicating the location of Ossabaw Sound. The more detailed map shows the coastal geometry and bathymetry of Blanton's (1981) study area, with the location of anchor stations C and D. Part (b) shows a vertical cross-section of the density (σ) along a line through C and D in (a). The vertical structure of the mean current component in the direction along-shore at stations C ($\times$) and D ($\bullet$) is shown in (c). The vertical shear predicted by the thermal wind balance is shown by the solid line. From Blanton (1981).

shore density structure is shown in Fig. 8.3b. Note that at both the inshore and offshore boundaries of the domain the vertical stratification is relatively weak and that higher values of vertical and horizontal density variations are restricted to the vicinity of stations C and D. The mean current profiles at C and D are shown in Fig. 8.3c together with the vertical shear predicted by the thermal wind balance, equation (8.14), using the mean horizontal density gradient between the stations. Clearly, the trend in the data almost parallels the theoretical prediction.

A more extensive description of the dynamics of the coastal plume created by the Hudson River in Middle Atlantic Bight was reported by Garvine (2004). Figure 8.4a shows the coastline of the north-eastern seaboard of the United States and the bathymetry of the adjacent continental shelf. The Hudson River forms a plume where it enters the ocean near New York City (see Bowman, 1988) and during periods of weak or northerly winds, the brackish water spreads south along the coast of New Jersey. In the study area defined by the box in Fig. 8.4a and shown in a higher-resolution map in Fig. 8.4b, Garvine conducted surveys of the salinity,

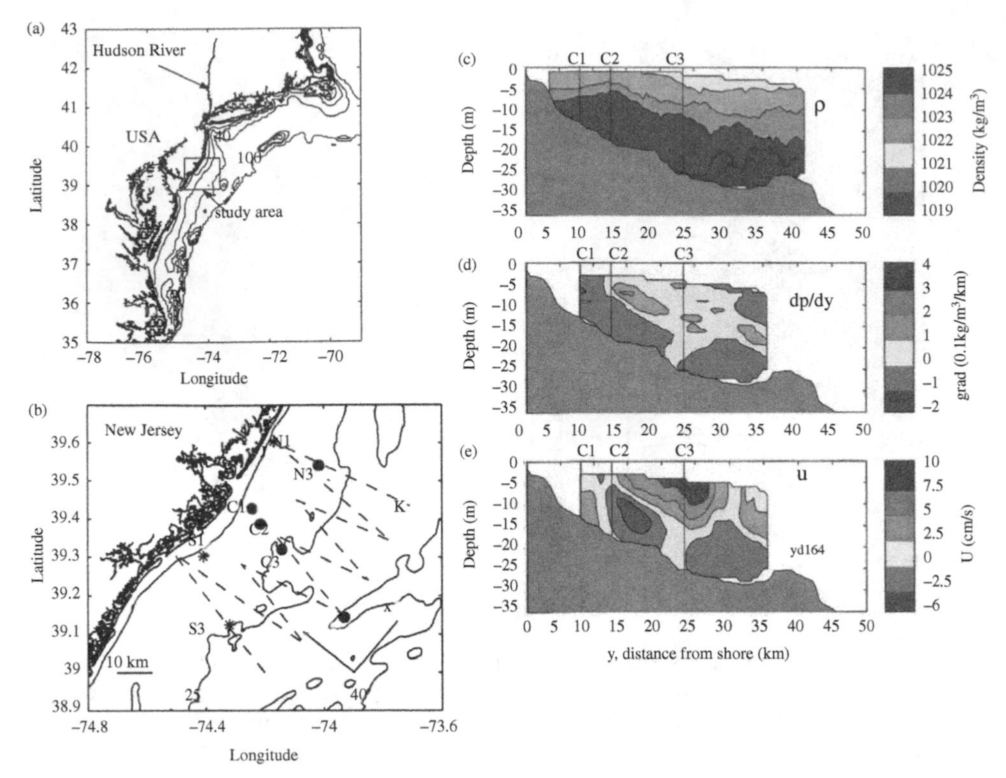

Figure 8.4. (a) A map of the coastline and bathymetry of the Middle Atlantic Bight. The box defines the area of the study which is shown in higher resolution in (b). The coast of New Jersey and the 10, 25, and 40 m isobaths are shown by the solid lines in (b) and the ship tracks are represented by the dashed lines. The locations of moored ADCP and conventional single-level current meters are indicated by the • and * symbols, respectively. (c), (d) and (e) show across-shelf sections of density ρ, density gradient, $\partial\rho/\partial y$, and the along-shelf current component estimated from the thermal wind equation assuming no flow at the bottom. From Garvine (2004).

temperature, and current with towed and profiling CTDs and vessel-mounted ADCPs. Two ships were employed concurrently to limit the effects of aliasing of the tidal motions. The dashed line in Fig. 8.4b shows the ship tracks. Moored acoustic profiling current meters were also deployed at the five sites shown by the filled black circles, and single-level current meters were located at the points shown by the three asterisk symbols.

Figure 8.4c–e shows the vertical and across-shelf variation along line C1–C4 of: (c) density, ρ; (d) horizontal density gradient, $\partial\rho/\partial y$; and (e) the along-shelf velocity component computed using equation (8.14) with the additional assumption of no motion at the bed, $u(z = h(x)) = 0$, on day 164 of 1996. The low-pass filtered wind stress was small (~0.02 Pa) and upwelling favorable (from the south) at this

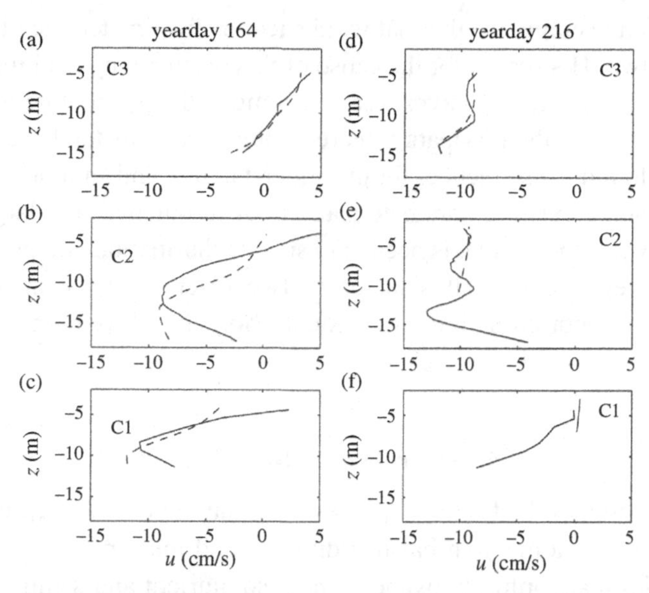

Figure 8.5. Comparison of the vertical structure of the low-pass filtered along-shelf velocity component (shown by solid lines) measured at C3 (a and d), C2 (b and e), and C1 (c and f) to the estimates obtained by integration of the thermal wind equation (shown by the dashed lines) using the observed horizontal density gradients. The observations used for (a–c) and (d–f) were obtained on days 164 and 216 of 1996, respectively, when the winds were light. From Garvine (2004).

time, and the velocity components inferred from the hydrography were to the southwest, as is clear in Fig. 8.4e. Note that the vertical shear is largely negative, consistent with the sign of the density gradient in Fig. 8.4d.

Evaluation of the thermal wind balance hypothesis requires a more direct comparison of the horizontal density gradient and the velocity shear. In Fig. 8.5, therefore, we reproduce Garvine's (2004) presentation of the vertical structure of the low-pass filtered currents observed by the moored ADCPs at C3, C2, and C1 on days 164 and 216 (solid lines) and that predicted by the thermal wind balance using the measured across-shelf density gradients (dashed lines). Note that the depth of the frictional layer was estimated and measurements within that region were excluded. Winds were light on both days. These comparisons show quite good agreement, especially at mooring C3. Using all 15 available comparisons of the thermal wind shear at the moorings in the buoyant layer (C3 and N3) to the low-pass filtered mooring estimates, Garvine found a significant correlation of 0.7 and concluded that the semigeostrophic approximation is useful in light winds even in quite shallow water. Yankovsky (2006) performed a more detailed examination of the same data set and compared the vertical shear observed by the shipmounted ADCPs to the thermal wind shear. This provided many more independent comparisons and revealed that, on average, the thermal wind shear overestimated the observed shear by a factor of two. His search for the cause of this discrepancy led him to speculate that large-amplitude internal waves were breaking in the pycnocline of the buoyant plume and enhancing the dissipation there. If this idea withstands additional scrutiny, it has profound consequences for the understanding and simulation of stratified coastal flows since existing parameterizations of turbulence rely largely on local shear to supply the energy that is needed to sustain dissipation and potential energy increases. However, despite this factor of two discrepancy, it appears that the semigeostrophic approximation is adequate for the development of scaling relationships.

8.5. The role of bottom friction

It is helpful to think of the frictional processes as having two conceptually different effects: one on the momentum balance directly, and the other on the transport of scalar quantities that control buoyancy (and also nutrient and sediment concentrations) through the vertical mixing associated with bottom-generated turbulence. There are obviously important dynamic interactions. However, the history of the development of physical oceanography demonstrates the utility of thinking, at least initially, about the important processes in isolation. For example, the effect of boundary friction on the dynamics of a geophysical-scale flow of homogeneous fluid results in Ekman layers (see, e.g., Pedlosky, 1987) and it is clear in many locations (Brink, 1991) that these play a central role in the response of the shelf circulation to winds. In contrast, the control of the surface temperature and thermal stratification of mid-latitude shelf seas by the horizontal distribution of tidal currents was explained in the seminal work of Simpson and Hunter (1974). This simple theory used the heat budget and ignored the effects of wind and the buoyancy-driven circulation in causing mixing and transporting material. Numerical process studies allow the interaction of Ekman dynamics and the associated modification of the baroclinic pressure gradients through the spatially variable mixing to be explored. We focus first on the effect of bottom friction in the momentum balance and then in Section 8.6 address the effect of vertical mixing by bottom-generated turbulence. We then turn to the effect of wind in the momentum balance of estuary outflow plumes in Section 8.7 and follow with an overview of the role of transient wind events on the vertical mixing of coastal currents in Section 8.8.

Chapman and Lentz (1994) pointed out that some buoyant coastal currents had isopycnals that were almost vertical (e.g., the South Atlantic Bight coastal current; Blanton, 1981), while in others they are almost horizontal (e.g., the Connecticut River plume; Garvine, 1974). They argued that the effect of the vertical stratification in the latter group inhibited the vertical turbulent flux of momentum and restricted

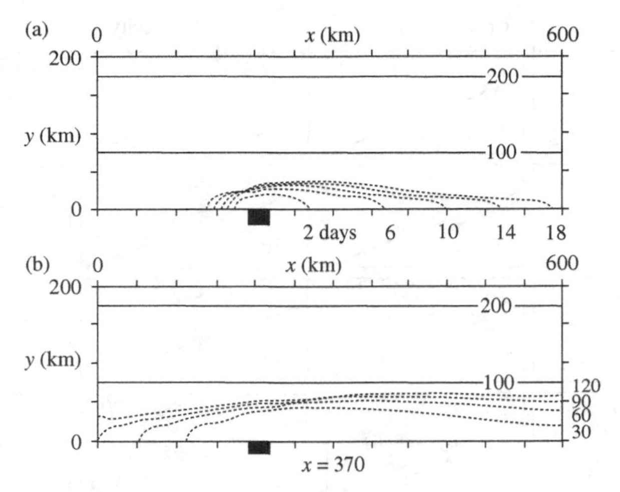

Figure 8.6. Plan view of the model domain of Chapman and Lentz (1994). The coast is at $y = 0$ and the solid lines show the 100 m and 200 m depth contours. The brackish water begins to enter the shelf at $x = 200$ km at $t = 0$. The dashed lines show the extent of the freshened water at times (a) $t = 2$, 6, 10, 14, 18, and (b) $t = 30$, 60 90, 120 days later. The cross-sections shown in Fig. 8.7 were obtained at $x = 370$ km, which is indicated in the diagrams by the black rectangle. From Chapman and Lentz (1994).

the influence of bottom friction to the near-bottom waters. They then used a three-dimensional model to examine the dynamics of the buoyancy forced flow along a straight shelf with linearly increasing bottom depth in the offshore direction when bottom friction was important throughout the water column, $E = O(1)$, and the density gradient was predominantly horizontal. Figure 8.6 shows a plan view of their domain. The coastline, $y = 0$, is at the bottom of each frame and the bathymetric contours (solid lines) show that the depth increases offshore. Using a density advecting numerical model, Chapman and Lentz (1994) demonstrated that brackish water discharged into the shelf domain at $x = 200$ km beginning at $t = 0$ spreads rapidly from the source both along-shore and across-shore. The dashed lines in Fig. 8.6 show the extent of the freshened water at times (a) $t = 2$, 6, 10, 14, 18 and (b) $t = 30$, 60, 90, 120 days later. The buoyant layer evidently spreads across-shore quickly and then slows. The along-shelf propagation is asymmetric, with a much more rapid motion to the right of the source. Analysis of the numerical solution revealed that the along-shelf flow was driven by the baroclinic pressure gradient and the offshore flux of buoyancy was largely established in the bottom Ekman boundary layer. The first column of frames in Fig. 8.7 shows the across-shore distribution of the density anomaly along the line $x = 370$ km that is indicated in Fig. 8.6b. The basic structure of the distribution appears to be established by $t = 20$ days and it then propagates slowly offshore. The second column in Fig. 8.7 shows contours of the

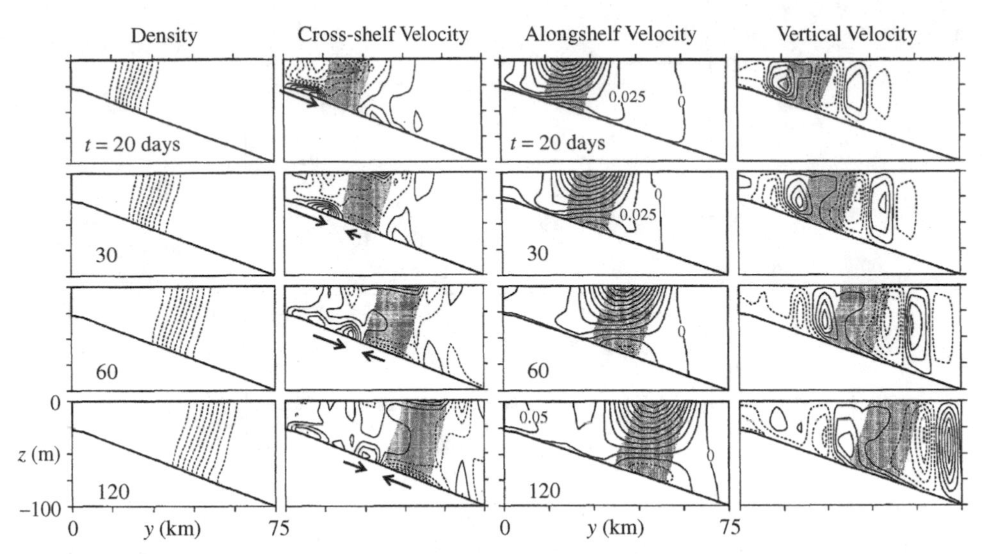

Figure 8.7. Vertical and across-shelf sections of the numerical solution for density, across-shelf velocity, along-shelf velocity, and vertical velocity at $x = 370$ km, 20, 30, 60, and 120 days after the initial discharge of the buoyant fluid at the coast. The contours of the density anomaly in the left column are at 0.1 intervals between -0.9 and -0.1. This vertically stratified frontal region is shaded gray in the other frames to facilitate comparison. In all of the velocity component graphs, values ≥ 0 are shown by the solid lines and negative contour values are dashed. The contour intervals for the across- and along-front velocity components are 0.005 and 0.025 m/s, respectively. For the vertical component the contour interval is 10^{-5}m/s. Note that the frontal zone moves off-shore with time and that the across-shelf transport in the bottom boundary layer is positive to the left of the front and negative within the front, as indicated by arrows in the second column. This Ekman layer convergence is associated with the upwelling on the shallow (left) side of the front and downwelling on the deeper side, as indicated by the arrows on the right column. Adapted from Chapman and Lentz (1994).

across-shelf velocity corresponding to the density field. Note that the location of the density front is shown by the shaded region to facilitate interpretation and that zero and positive values are shown by solid contour lines and negative values are shown by dashed lines. It is important to recognize the offshore-directed transport in the bottom Ekman layer on the shallow side of the frontal zone. As the frontal zone moves offshore, the offshore transport in the bottom Ekman layer reduces and a zone of onshore flow develops. Arrows are drawn below these regions for emphasis. The offshore-directed bottom Ekman layer flux shoreward of the frontal zone tends to bring buoyant water beneath the denser shelf water where it destabilizes the water column, causing it to be unstratified. The frontal zone (the vertically stratified region of high horizontal density gradient) is then moved offshore into deeper water.

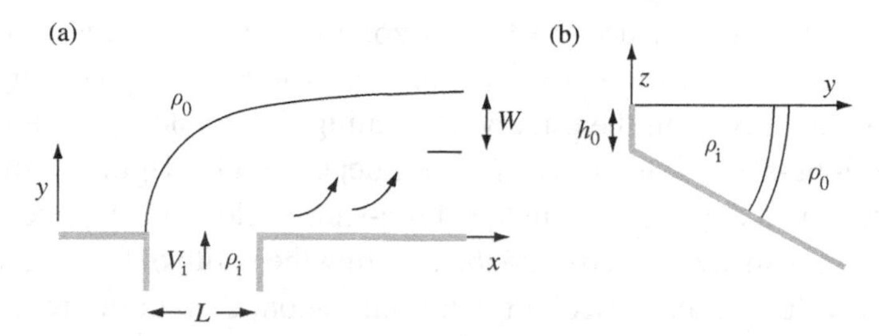

Figure 8.8. Geometry of the bottom advected plume model of Yankovsky and Chapman (1997). (a) A plan view of the geometry of the source region. The buoyant fluid enters the shelf at the coast from a channel of width L and depth h_o. The effluent has density ρ_i and discharges at velocity v_i into an ambient fluid of density ρ_o. The black curve illustrates the outer edge of the plume front, which has an asymptotic width W. The curved arrows illustrate the premise that the role of the bottom Ekman layer is to transport the fluid across-shelf into the frontal zone. (b) A vertical cross-section at large x showing the frontal zone. h_b is the depth at the inshore edge of the stratified region.

The along-shore velocity distribution is shown in the third column of Fig. 8.7. This component is positive everywhere at $t = 20$ days but begins to reverse near the bottom by 60 days when the front has moved further from shore. Within the frontal zone the dynamics evolve toward the steady thermal wind balance, with the vertical shear in the along-shelf velocity component proportional to the horizontal density gradient. As the front moves offshore, the horizontal density gradient remains steady and the along-shelf velocity at the bottom, and its associated offshore Ekman flux, fall to zero, bringing the across-shelf migration of the frontal region to a halt.

The fourth column in Fig. 8.7 shows the vertical velocity distribution predicted by the model of Chapman and Lentz (1994). It is evident that the convergence in the bottom Ekman layer flow on the shoreward side of the front leads to an upwelling circulation there. A downwelling flow is found further from shore. Recent observational evidence with tracers supports the existence of this convergence pattern (Barth *et al.*, 1998; Houghton and Visbeck, 1998) in the Middle Atlantic Bight shelf break front, but its importance on the inner shelf is still in question.

The conditions that result in vertical or horizontal stratification in buoyant coastal outflows were developed in an insightful paper by Yankovsky and Chapman (1997), which exploited the results of Chapman and Lentz (1994) and proposed criteria for the formation of "bottom-advected" plumes, the regime simulated by Chapman and Lentz (1994). The characteristics of bottom-advected plumes are sketched in Fig. 8.8. The fundamental idea is that the bottom Ekman layer transports the buoyant fluid discharged at the coast with density ρ_i and velocity v_i from an estuary of width L and depth h_0 across-shelf to the frontal zone, which is indicated in Fig. 8.8a,b by the curve. Once the frontal zone has moved sufficiently far from shore, the along-shelf transport due to the thermal wind-induced velocity can carry all of the volume flux from the estuary. Assuming that in steady state the vertical shear in the along-shore velocity, $\partial u/\partial z$, is independent of depth and the front has moved offshore to a depth h_b, where the along-shore velocity component at the top of the bottom Ekman layer is zero, and then writing the width of the frontal region as W, allows the volume flux carried in the frontal region, T, to be approximated as:

$$T = \int_{-h_b}^{0} \int_{y}^{(y'+W)} u \, dz \, dy = \frac{1}{2} h_b{}^2 W \frac{\partial u}{\partial z}. \tag{8.15}$$

The thermal wind balance, equation (8.14), allows this to be written in terms of the density field as:

$$T = \frac{1}{2} h_b{}^2 W \frac{g}{\rho_o f} \frac{\partial \rho}{\partial y}. \tag{8.16}$$

Approximating the derivative by the density difference between the ocean and the coastal current divided by the front width, equation (8.16) can be written as:

$$T = \frac{1}{2} h_b{}^2 \frac{g'}{f}, \tag{8.17}$$

where $g' = g(\rho_0 - \rho')/\rho_0$. If, in the steady state, all of the buoyancy flux from the estuary ($v_i h_o L$) is transported along-shore in the frontal zone, then

$$h_b{}^2 = \frac{2Tf}{g'} = \frac{2fLv_i h_o}{g'}. \tag{8.18}$$

Representing the topography by $h(y) = h_o + sy$, where s is the bottom slope, as is illustrated in Fig. 8.8b, the width of the coastal current can be written as:

$$y_b = \frac{h_o}{s} \left\{ \sqrt{\frac{2fLv_i}{g' h_o}} - 1 \right\}. \tag{8.19}$$

Note that this result is based on the premise that $h_b > h_o$. If that is not the case for a particular plume, then the bottom Ekman layer cannot transport the effluent across-shelf and another mechanism must be dominant. They proposed a cyclostrophic dynamic balance to develop a scale for surface-advected plumes, however, we will not discuss it here. Further insight to the dynamics of the turning region of a large estuary outflow can be found in Valle-Levinson *et al.* (1996).

Yankovsky and Chapman (1997) investigated the veracity of their simple model by comparing the predictions to the solutions obtained from three-dimensional

Table 8.2. *Plume parameters of Yankovsky and Chapman (1997)*

Example	L	v_i	h_o	ρ_i	ρ_0	$f \times 10^{-4}$	y_s	h_b
Niagara 1	1.5	0.4	10	0.5	1000	0.99	12.2	15.6
Niagara 2	1.5	0.4	10	0.9	1000	0.99	14.8	11.6
Connecticut	1.5	0.34	2	9.0	1020	0.93	20.1	1.5
Hudson River	7	0.1	10	2.0	1021	0.93	20.1	8.2
Delaware 1	10	0.1	15	0.5	1025	0.90	12.8	23.8
Delaware 2	10	0.1	15	1.0	1025	0.90	18.0	16.8
Delaware 3	10	0.15	15	3.5	1025	0.90	33.5	11.0
Chesapeake	15	0.1	10	3.0	1020	0.87	26.3	9.5

numerical simulations and with a few examples from the available literature. Table 8.2 lists some examples of the scales observed in plumes and the corresponding h_b values. Only the Connecticut River is predicted to have $h_o > h_b$, though for the Hudson and Chesapeake the values are almost equal. This work has provoked many further advances in modeling and laboratory experiments (e.g., Avicola and Huq, 2002; Lentz and Helfrich, 2002). Though the effects of tides and wind were omitted, and there is no evidence from observations that the cyclostrophic balance is an appropriate model for river plumes, this work has yielded valuable insight.

8.6. The role of vertical mixing

Another important role of bottom friction is the production of turbulent kinetic energy that can result in a vertical buoyancy flux through mixing. In the conceptual model discussed in the preceding section, the vertical mixing in the coastal current resulted solely from the gravitational instability of the buoyant water being directed offshore in the bottom Ekman layer under saltier, less buoyant, water. In much of the inner shelf, however, tidal currents are much larger than buoyancy-driven currents and strongly influence the structure of the water column. A series of models have followed from that of Simpson and Hunter (1974), who considered the potential energy budget of water heated by radiation at the surface and stirred by turbulence generated by tidal currents at the bottom. They considered the energy budget of a column of water of thickness h that is subject to a stratifying buoyancy flux at the surface, H (Watts m^{-2}). If the fluid is continually mixed by bottom turbulence at a rate sufficient to keep it vertically homogeneous, then the rate of change of heat per unit surface area in the layer is

$$h\rho_0 C_p \frac{dT}{dt} = H. \qquad (8.20)$$

Here, C_p is the heat capacity at constant pressure and ρ_0 is the density of the water. Changes in density associated with heating have been neglected here. The rate of change of temperature in the layer can then be expressed as:

$$\frac{dT}{dt} = \frac{H}{h\rho_0 C_p}. \tag{8.21}$$

Adopting a linear approximation for the equation of state in a limited range of temperatures around T_0, and neglecting variations in salinity, we can write

$$\rho = \rho_o(1 - \alpha(T - T_o)), \tag{8.22}$$

where ρ_0 is the density at T_0 and the thermal expansion coefficient is $\alpha\ (S, T) \sim 10^{-4}$. For water at the surface of the ocean at 10°C and practical salinity 30, α (30, 10) = 1.562×10^{-4} °C^{-1} (see McDougall, 1987). The rate of change of density due to heating is then

$$\frac{d\rho}{dt} = -\rho_0 \alpha \frac{dT}{dt} = -\frac{\alpha H}{hC_p}. \tag{8.23}$$

Since $\alpha > 0$, heating decreases the density of the water, which tends to increase its buoyancy. The buoyancy force per unit volume can be expressed as $B = g(\rho_0 - \rho)$, where g is the gravitational acceleration. The change in the potential energy per unit surface area in the layer of thickness h as a consequence of heating is then the integral of the product of the buoyancy force and distance from the reference level, i.e.,

$$\phi = \int_{-h}^{o} g(\rho - \rho_0) z dz. \tag{8.24}$$

With this definition, ϕ is equivalent to the work that is required to homogenize the water column. If we define the reference density as the average in the water column,

$$\rho_o = \frac{1}{h} \int_{-h}^{o} \rho(z) dz, \tag{8.25}$$

then a homogeneous layer has zero potential energy, and the rate of change of potential energy per unit surface area is

$$\frac{d\phi}{dt} = \int_{-h}^{o} gz \frac{d\rho}{dt} dz. \tag{8.26}$$

When stirring is just sufficient to keep the water column uniform, the integral is simple since the time derivative is independent of z. Using equation (8.23), the rate of change of potential energy is

$$\frac{d\phi}{dt} = -\frac{d\rho}{dt} g \frac{h^2}{2} = \frac{\alpha g H h}{2C_p}. \tag{8.27}$$

The work done by the currents at the seabed is the product of the velocity and the stress at the top of the viscous boundary layer, i.e., $\tau_b u_b$. A somewhat uncertain fraction of this work, γ, must be used to mix the water column. If the water column is to remain homogeneous then the rate of work done at the bed to mix the fluid must be equal to the rate at which potential energy is added, i.e.,

$$\gamma \tau_b u_b = \frac{\alpha g H h}{2C_p}. \tag{8.28}$$

Using the quadratic drag law, $\tau_b = \rho_0 C_D {u_b}^2$, this can be restated as

$$\gamma \rho_0 C_D {u_b}^3 = \frac{g\alpha H h}{2C_p}, \tag{8.29}$$

which implies

$$\frac{{u_b}^3}{h} = \frac{g}{2} \frac{\alpha H}{C_p} \frac{1}{\gamma \rho_0 C_D}. \tag{8.30}$$

The quantity $S_{HS} = {u_b}^3/h$ has become known as the Simpson–Hunter parameter. In regions that are unstratified, we would expect that ${u_b}^2/h \geq S_{SH}$, and conversely, if ${u_b}^2/h \leq S_{SH}$ the water column would be stratified. Though often criticized because it is a dimensional quantity and values depend on the units, it has frequently been cited.

S_{SH} has proven useful because, though the values of γ, C_d, and H are difficult to measure and uncertain, they can often be assumed to be horizontally uniform on the scale of the coastal topography and geometry. Then the distribution of S_{SH} is easily computed since the water depth, h, can be obtained from navigation charts and the current speed, u_0, whether interpreted as the amplitude of the tidal current or the mean current during a tidal period, can be obtained from tidal current almanacs or from numerical models.

Contours of $\log_{10} S_{HS} \approx 2$ (using meters and meters/second) have been shown in many areas to separate areas of vertically mixed waters from stratified areas and these boundaries are often clearly distinguishable in infrared images of the sea surface from satellites. A recent demonstration of this is provided by Mavor and Bisagni (2001), who analyzed a decade of sea surface temperature (SST) maps of the Georges Bank, an important fishing ground separating the northwest Atlantic from the Gulf of Maine. The geography of this area is shown in Fig. 8.9a. Mavor and Bisagni (2001) computed the distribution of a dimensionless variant of the Simpson–Hunter parameter, $\log_{10} S'_{SH} = \log_{10} \frac{h}{\rho_0 C_D |U_T|^3}$, contoured in Fig. 8.9b

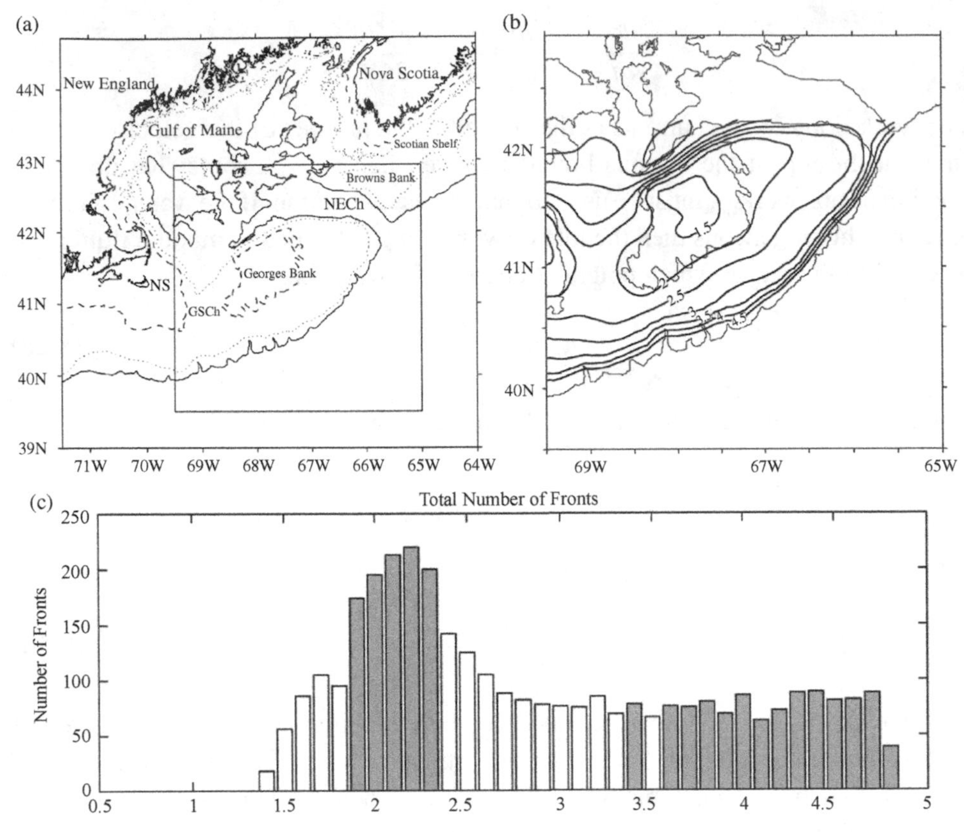

Figure 8.9. (a) Map of the coastline of the Gulf of Maine showing the location of Georges Bank. The bathymetry of the Bank is shown by the thin black contour lines in (b). The contours of $\log_{10} h/\rho C_D |U_T|^3$ shown by the thick lines in (b) were computed from a numerical model (Loder and Greenberg, 1986). (c) The number of times a frontal zone was detected in intervals of $\log_{10} h/\rho C_D |U_T|^3$. From Mavor and Bisagni (2001).

using a numerical model to obtain U_T, the M_2 tidal current amplitude. Note that in m.k.s. units, $\rho_0 C_D \approx 2$, and therefore $\log_{10} S'_{SH} = \log_{10} S_{SH} + 0.3$. Mavor and Bisagni (2001) applied an objective method to determine where temperature fronts, regions of high gradient in SST, occur and correlated the frequency of occurrence of fronts with the magnitude of $\log_{10} S'_{SH}$. Figure 8.9c shows an example of the frequency of the occurrence of fronts in their study region in intervals of $\log_{10} S'_{SH}$. This clearly demonstrates that fronts were most frequently found in regions in which $1.9 < \log_{10} S'_{SH} < 2$, as predicted by the argument of Simpson and Hunter (1974).

In buoyancy-driven coastal currents, vertical stratification can also be changed by the horizontal advection of water from a region of higher buoyancy over denser water by a vertically sheared flow. This mechanism has been termed "straining" by Simpson *et al.* (1990). This restratification mechanism can compete with vertical mixing in the same manner as surface heating, and Nunes Vaz *et al.* (1989), Simpson *et al.* (1990), Simpson and Sharples (1991), and Monismith and Fong (1996) have generalized the approach of Simpson and Hunter (1974) to include the effects of straining by both tides and the buoyancy-driven motion. These simple one-dimensional (vertical) models have been quite successful in simulating mooring time series, but are inconsistent with observations of frontal boundaries in the neighborhood of estuarine outflows (see, e.g., Marmorino and Trump, 2000).

Many large-estuary outflow plumes have been found to flow through regions of strong, tidally driven mixing. The recognition that the competition between mixing and straining could be a major influence on the character of the estuarine outflows has led to considerable recent work on the mechanisms and rates of vertical mixing in these regimes. The crucial importance of the model of vertical mixing adopted in numerical models has been clearly demonstrated by Garvine (1999), who simulated the flow resulting from the discharge from an estuary onto a straight, linearly sloping shelf with the three-dimensional numerical model of Blumberg and Mellor (1987). The model included forcing by tides. Vertical mixing was described by the model (level 2.5) of Mellor and Yamada (1982), as it is in many coastal circulation models. Garvine (1999) explored the across- and along-shelf scales of the plume resulting from many combinations of estuarine buoyancy flux, discharge width, latitude, and tidal amplitude. The effect of the vertical mixing model was made clear in a set of numerical experiments that varied the tidal amplitude as measured by the ratio of the imposed sea level variation to the depth at the edge of the model shelf. With no tidal forcing, the tendency for the outflow plume to turn anticyclonically after exiting the estuary, as illustrated by the Chapman and Lentz (1994) simulations shown in Fig. 8.6, created a vertical stratification and shear that was sufficient to shut down vertical mixing by turbulence. Setting the vertical eddy viscosity and diffusivity coefficients to constant values, another common model of turbulent mixing, approximately halved the along- and across-shelf scales of the plume. Varying the amplitude of the tidal forcing from zero to 0.4 (very strong forcing) also decreased the plume scales by a factor of two. This model demonstrated that getting the vertical mixing right can be as important to the overall structure of the plume as simulating the tides accurately. That Yankovsky's (2006) analysis of shear and stratification observations in the Hudson plume led to the conjecture that large-amplitude internal waves, or some other mechanism, must enhance the vertical shear enough to drop the Richardson number below the critical threshold for the onset of instabilities and mixing, emphasizes this point.

Evaluating models of vertical mixing in the coastal ocean has been handicapped by the difficulty in obtaining measurements of the evolution of the turbulence properties predicted by the models. One potentially important development has been the application of the purposefully introduced tracer techniques of Ledwell *et al.* (1998) to the coastal ocean by Sundermeyer and Ledwell (2001) and Houghton and Ho (2001). Another is the more common use of microstructure profilers in the shallow, salt stratified coastal ocean by, for example, Simpson *et al.* (1996) and Peters (1997). These types of observations now allow the evaluation of models of vertical mixing directly by comparing measured values of the rate of dissipation of turbulent kinetic energy to those predicted by models. The simulation of the Simpson *et al.* (1996) data by Burchard *et al.* (1998) has already revealed that the dissipation rate was underpredicted by the Mellor and Yamada (1982) model and a competing $\kappa - \varepsilon$ model. They concluded, in agreement with the suggestion of Garvine (1999) and Yankovsky (2006), that the mixing associated with internal waves unresolved by the horizontal circulation model must be parameterized and used as a source of turbulent kinetic energy production. This type of direct evaluation of turbulence models in complex flows should have high priority.

8.7. The role of wind

The hydrographic surveys of Bowman (1978) first described the structure and variability of the Hudson River plume. He noted that it could be directed to either the left or the right of the river mouth, and suggested that this was controlled by the wind direction. Munchow and Garvine (1993a, b) reported on a more extensive program at the mouth of the Delaware Bay, and their combination of current measurements and hydrographic surveys allowed a deeper dynamic interpretation of the observations. They found that the brackish water was frequently to the south (right) of the estuary and could be traced up to 100 km along-shore from the source. However, during upwelling-favorable wind events they observed that vertical stratification near shore was increased and the brackish layer thinned. Drifter trajectories suggested that this was the result of a surface Ekman layer response which requires surface waters to be transported offshore and to be replaced by the onshore transport of deeper water. Masse and Murty (1992) documented a similar response in the outflow plume of the Niagara River in Lake Ontario. The description of the Gulf of Maine Coastal Current, essentially the combined plumes from Maine's many rivers, by Fong *et al.* (1997), is also consistent with this interpretation. Recent studies of the Hudson plume by Yankovsky and Garvine (1998) and Yankovsky *et al.* (2000), and the Chesapeake Bay plume by Rennie *et al.* (1999), showed that downwelling-favorable winds were associated with along-shore propagation of narrow plumes far from the estuary mouth. At the mouth of the Columbia River in the northwest of North America, Hickey *et al.* (1998) have also documented a similar asymmetric response of a river plume to wind direction.

The fundamental effects of the wind on estuary plumes were studied theoretically by Chao (1988) using a three-dimensional numerical model. He simulated the discharge from a simple rectangular estuary onto a straight, flat, or linearly sloping shelf in the absence of tides and described the character of the asymmetric response to wind. In four separate experiments he allowed a buoyant estuarine discharge to evolve in the presence of winds directed both on and offshore, and along shore in both the upwelling and downwelling-favorable directions. Downwelling winds cooperate with the tendency for the outflow to turn anti-cyclonically and be confined and accelerated in the along-shore current. Upwelling winds compete with the buoyancy-driven motion. In this model the result was rapid offshore transport and separation of the plume from shore. The principal effects of the across-shore winds were to modify the transport of water from the estuary, with offshore winds increasing the flux and onshore winds decreasing it.

The crude spatial resolution of the Chao (1988) model and the simplistic model of vertical mixing (constant eddy coefficients) tended to emphasize the effect of friction. It is not surprising, therefore, that the Ekman layer behavior dominated the model's response. Almost a decade later, Kourafalou *et al.* (1996) extended this approach by using higher resolution and a more sophisticated closure model (Mellor–Yamada 2.5) to revisit the problem of Chao (1988) with a much broader exploration of discharge, shelf depth gradient, and vertical mixing effects. The complications associated with tide were again omitted. A more complex geometry, the South Atlantic Bight of the east coast of North America, was employed in the examination of the effects of wind. The simulations of the growth of a plume in steady up and downwelling winds were qualitatively similar to those depicted by Chao (1988). The difference in geometry clouded the effect of the influence of stratification on vertical mixing.

Further experiments were conducted to explore the effect of wind on a buoyant plume that had formed with no wind forcing. Again, modest upwelling-favorable winds of 5 m/s caused rapid offshore transport of the buoyant surface water and the tilting of isopycnals toward the horizontal. In contrast, the onset of downwelling winds accelerated the buoyant water along shore in the downwind direction and tilted the isopycnals toward the vertical. Kourafalou *et al.* (1996) argued that this response was consistent with the field observations in the area by Blanton *et al.* (1989, 1994).

Recently, Lentz and Largier (2006) presented a concise integration of the previous work on the response of coastal currents to wind based largely on the theory and laboratory experiments of Lentz and Helfrich (2002). Figure 8.10a shows their

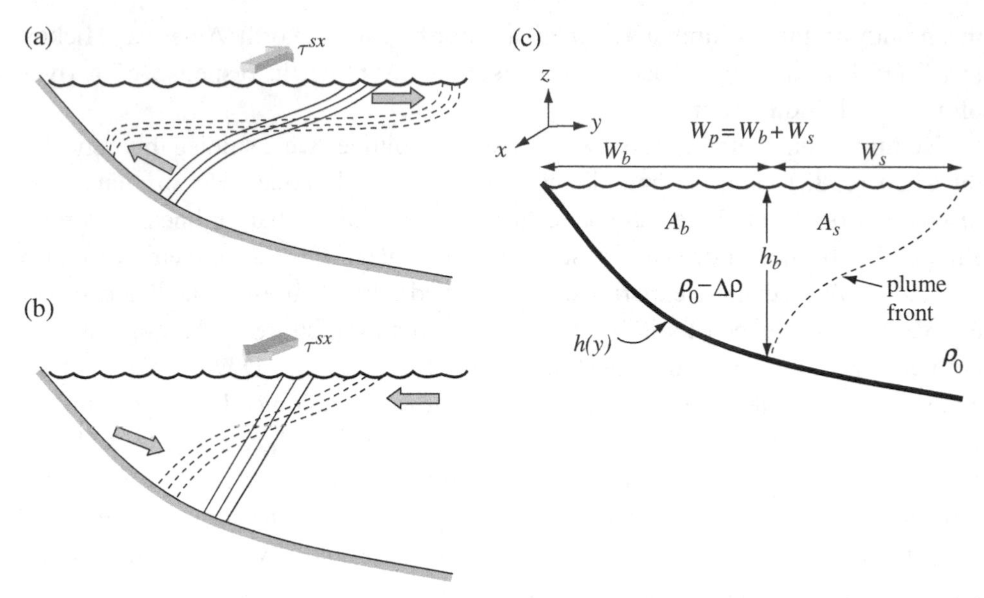

Figure 8.10. (a) A conceptual illustration of the response of a coastal current to an upwelling wind. The solid lines represent the pycnocline in a coastal current that is in steady state. The arrows show the currents induced by an upwelling-favorable along-shelf wind stress. The motion modifies the density field and moves the isopycnals to the locations shown by the dashed lines. (b) The response to a downwelling wind. (c) Definition of the coastal current geometry in the model described by Lentz and Largier (2006). The width of the stratified region is W_s and the width of the buoyant but unstratified region is W_b. From Lentz and Largier (2006).

schematic across-shelf section of a coastal current with the location of the pycnocline in the presence of weak ($|\tau_w| < 0.02$ Pa) winds illustrated by a dashed line. An upwelling-favorable wind is shown to drive an offshore surface Ekman layer flux and a compensating onshore flow in the lower layer. This tends to stretch the buoyant layer from the original position (solid lines) across the shelf, making it thinner and wider (dashed lines). Figure 8.10b illustrates the response to a modest ($0.02 < \tau_w < 0.07$ Pa) downwelling wind stress. In this case the stress drives an onshore surface Ekman layer flux which tends to steepen the pycnocline compared to the unforced slope (dashed lines) and reduce the coastal current width. Strong downwelling-directed wind stress causes vigorous vertical mixing and reduces both the vertical and horizontal density gradients.

Following the Yankovsky and Chapman (1997) idea that bottom friction in shallow water causes transport of the estuary effluent offshore in a bottom Ekman layer, Lentz and Largier (2006) proposed that once the plume had formed a shore-parallel coastal current, the along-shore transport in the plume was carried in the vertically stratified region where the thermal wind balance pertains. This argument

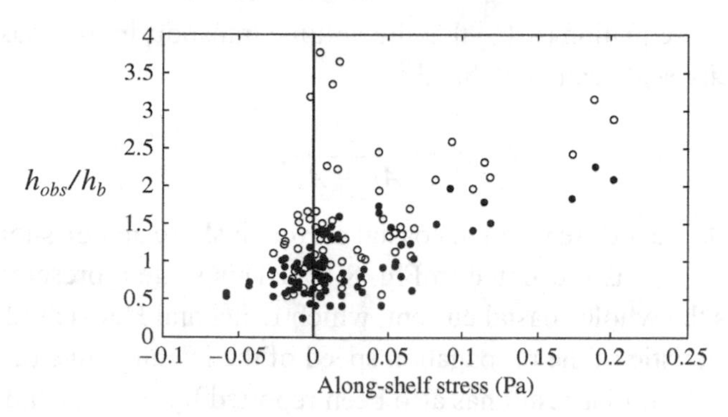

Fig. 8.11. The dependence on along-shelf wind stress of the observed maximum thickness of the Chesapeake coastal current (h_{obs} in Fig. 8.10) normalized by the predicted thickness in the absence of wind forcing, h_b, given by equation (8.18). From Lentz and Largier (2006).

predicts that the depth of the inshore edge of the frontal zone, h_b, is given by equation (8.18). Figure 8.10c is a sketch defining the quantities used by Lentz and Largier (2006) to represent a coastal current in the absence of wind. The dashed line labeled "plume front" shows the offshore extent of the buoyant water, and this intersects the bottom at depth h_b. Figure 8.11 shows Lentz and Largier's (2006) values of h_{obs}/h_b, the ratio of values of the maximum coastal current depth estimated from hydrographic sections across the Chesapeake coastal current to that predicted based on the coastal current transport, as a function of along-shelf wind stress. Note that the transport was estimated two different ways, resulting in the open and filled symbols in Fig. 8.11. It is clear that both estimates agree that at low wind stress, the ratio h_{obs}/h_b clusters around unity, supporting the basic idea that Ekman dynamics control the coastal current width. However, for downwelling wind stress in excess of 0.05 Pa, the layer thickness increases beyond the model predictions.

Lentz and Largier (2006) developed a scaling for the effect of wind by considering its effect on the volume transport of the current and the propagation speed of the leading edge of the coastal current as it advances along-shore during downwelling wind events. Since the frontal zone is in thermal wind balance, the width, defined as W_s in Fig. 8.10c, can be estimated as the internal Rossby radius based on h_b, i.e.,

$$W_s = \frac{c_b}{f} = \frac{\sqrt{g' h_b}}{f}, \tag{8.31}$$

where c_b is the interfacial wave phase speed. The total width of the plume is then

$$W = W_s + W_b \tag{8.32}$$

where, as illustrated in Fig. 8.10c, W_b is the location at which $h(y = W_b) = h_b$.

Recalling from equation (8.15) that the volume transport in the coastal current is T, then a velocity scale can be defined as

$$c_p = \frac{T}{A_b + A_s}, \tag{8.33}$$

where A_b and A_s are the areas of the coastal current inshore and offshore of the base of the front, $y = W_b$, as illustrated in Fig. 8.10c. This scale represents the average velocity across the whole coastal current, which Lentz and Helfrich (2002) demonstrate is characteristic of the propagation speed of the leading edge of a developing coastal current. A similar result has also been reported by Avicola and Huq (2002). Substitution of equation (8.31) in equation (8.17) allows the transport to be expressed as $T = \frac{1}{2} h_b f W_s^2$, then estimating the area of the stratified region as $A_s = \frac{1}{2} W_s h_b$, equation (8.33) can be written as

$$c_p = \frac{c_b}{1 + \dfrac{A_b}{A_s}}. \tag{8.34}$$

Note that the area ratio is controlled by the bottom slope. When the bottom slope is steep, the area ratio is small and the transport velocity scale is $c_p \approx c_b$. Conversely, when the slope is shallow, the area ratio is larger and the velocity scale is much smaller, with $c_p \approx T/A_b$. When there is wind forcing, the scale for the wind-driven velocity, c_w, can be obtained by assuming that the bottom friction balances the wind stress, which can be expressed as

$$c_w = \frac{\tau_w}{\rho_0 r}, \tag{8.35}$$

where the conventional linear bottom drag law has been adopted. Note that this scale is different from that adopted by Garvine (1995). Assuming that the wind does not modify the dynamics in the stratified area by diapycnal mixing, plausible only during light winds, then the wind-driven transport adds to the along-shore transport and the velocity scale, c_w, times the total area can be expressed as

$$c_w(A_b + A_s) = \frac{\tau_w}{\rho_0 r}(A_b + A_s) + T \tag{8.36}$$

which, using equation (8.33), yields

$$c_w = \frac{\tau_w}{\rho_0} + c_p. \tag{8.37}$$

This result provides a scale estimate for the relative importance of a downwelling wind stress and, as Lentz and Largier (2006) point out, it is similar to that developed by Whitney and Garvine (2006). Figure 8.12 shows a comparison of the observed

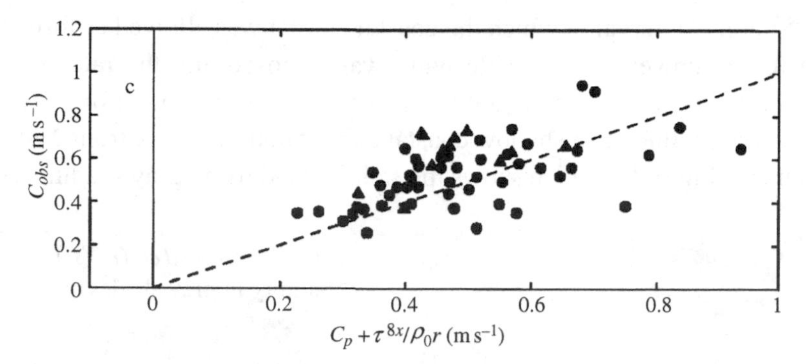

Figure 8.12. The dependence of the along-shore propagation speed of a buoyant coastal current C_{obs}, on the scale derived in equation (8.37). From Lentz and Largier (2006).

along-shore propagation velocity of the Chesapeake coastal current to that predicted from equation (8.37). The dashed line indicates the trend if the model was perfect. Note that transport is difficult to estimate since, as shown by Valle-Levinson *et al.* (2001), the wind can modulate the outflow from the Chesapeake Bay. This may contribute to the scatter in Fig. 8.12, however, the trend in the measurements lends support for the basic assumptions of this very simple theory, in which the effect of downwelling wind is to move the stratified region offshore steepening the pycnocline, and to augment the geostrophically adjusted transport in the frontal zone with a wind-driven current that is balanced by bottom friction on the inner shelf. Of course, at high winds the flow in the frontal zone would also be modified by the wind by enhancing cross-isopycnal mixing.

There are few observations of wind-induced mixing in buoyant coastal currents. Souza and Simpson (1997) augmented an analysis of mooring and survey observations at the mouth of the Rhine with simulations using a one-dimensional model based on that of Simpson and Sharples (1991). They inferred that the wind-driven mixing was comparable to that due to bottom friction when averaged over a year. Since the wind effect varied by orders of magnitude, there were many periods when it dominated. It seems likely that this conclusion has broader applicability. Fong and Geyer (2001) recognized that the interplay between the wind-induced circulation and the stratification in the plume water was likely to be central to the understanding of the response to upwelling-favorable wind. They developed a simple model of the response of the plume to upwelling-favorable winds and then simulated a buoyancy-driven coastal current along a straight coast and almost linearly sloping shelf using a variant of the model of Blumberg and Mellor (1987) that employed the Mellor–Yamada 2.5 turbulence closure scheme (Mellor and Yamada, 1982), which has been demonstrated to be effective in representing the influence of strong winds on stratified shear flows. Tides were neglected in order to illuminate the wind effects.

After a "spin-up" interval in which the coastal current was allowed to form without wind, a 0.1 Pa upwelling-favorable wind was imposed and the rates of vertical mixing of salt were then computed from the solutions. At the onset of the wind, the buoyancy was confined to a shallow coastal current bounded by a frontal zone. This is illustrated in Fig. 8.13, which shows the salinity field using gray shading. After 24

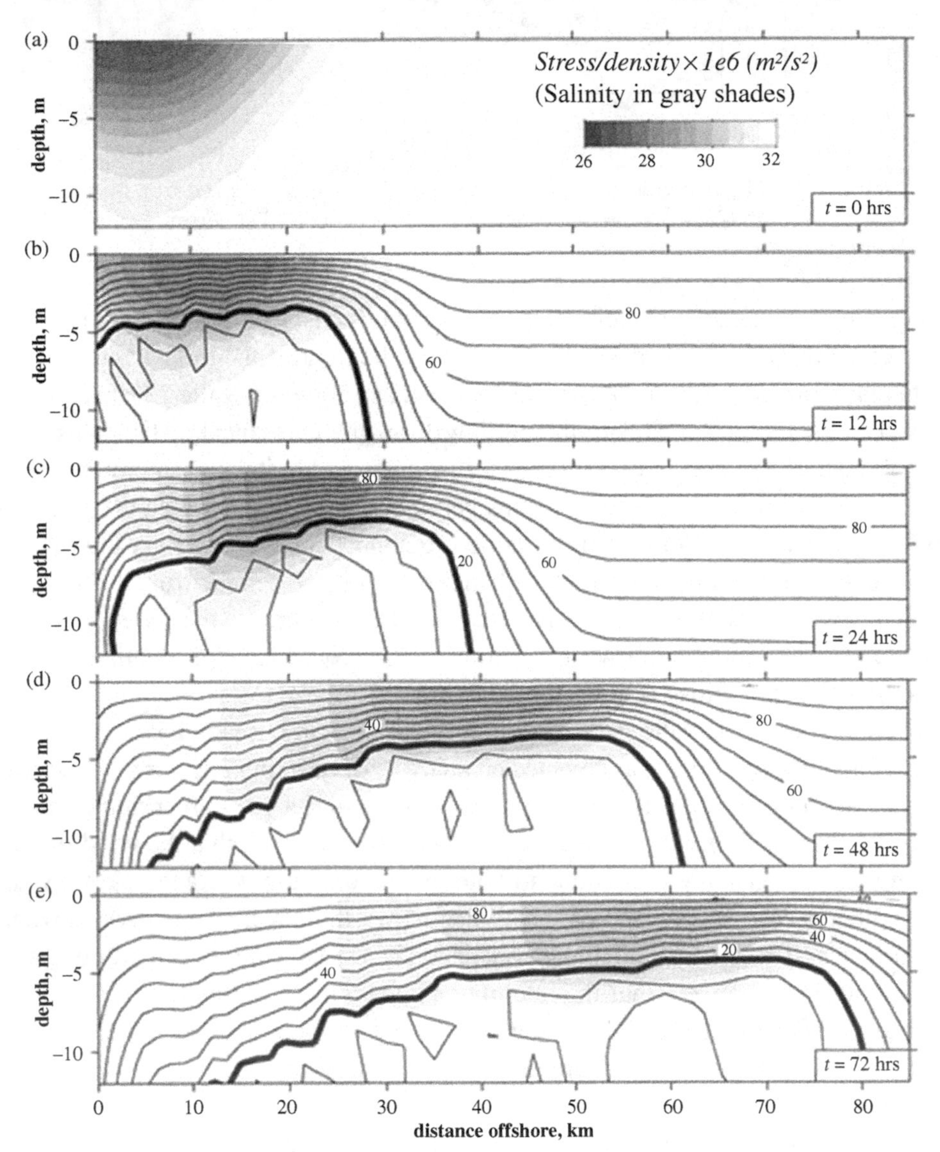

Figure 8.13. Contours of stress and salinity for $t = 0$, 12, 24, 48, and 72 hours. Shaded areas are used to indicate salinity at intervals of 0.5. Superimposed on the shading are contours of magnitude of the total stress in solid contour lines at 0.01 Pa intervals. The thick contour line is the stress contour that is 10% of the applied surface stress. From Fong and Geyer (2001).

hours of upwelling wind (Fig. 8.13b), the Ekman response causes the surface waters to move offshore. Contours of stress are superimposed on the evolution of the salinity field and demonstrate that the stratification in the coastal current confines the stress near the surface. As the stratified region moves offshore and the inner shelf becomes well mixed, a brackish water lens detaches from the coast (see Fig. 8.13c–e).

Figure 8.14 shows the evolution of the vertical turbulent salt flux during this transition. In the initial phase of the adjustment, the shear generated in the coastal current leads to enhanced vertical mixing throughout the buoyant layer. This is clearly seen by comparing Fig. 8.14a,b. After 24 hours, Fig. 8.14c shows that the mixing persists throughout the layer, but a clear maximum emerges at the seaward side of the frontal zone where the surface buoyancy is low and the layer shallow. Note that the low-salinity water is detached from the coast at this time and the maximum vertical salt flux is at the seaward side of the plume. This maximum then moves offshore at the leading edge of the brackish layer. Fong and Geyer's (2001) careful examination and comparison of the vertical profiles of stress in the plume and frontal zones (region of maximum vertical salt flux) revealed that the vertical structure was similar in both regions. They concluded that the high vertical salt flux in the frontal zone was, therefore, due to the shoaling of the level of maximum stratification toward the surface where the stress was largest.

For stratification to persist in the frontal zone the vertical turbulent flux of salt there must be countered by an across-front flux of freshwater. Fong (1999) and Fong and Geyer (2001) showed that the offshore Ekman layer transport provided this flux, and they referred to it as "Ekman straining" by analogy to the mechanism that Simpson *et al.* (1990) described in an estuary where the ebb tide reinforces vertical stratification by advecting fresher water from the landward end of the estuary.

Observations of vertical salt flux and turbulent mixing on the shelf are very difficult to obtain, and since the theoretical results discussed above suggest that the spatial structure of the vertical salt (and presumably other dissolved and suspended materials) distributions is quite variable and sensitive to wind direction, it is not surprising that there have been few direct observations of mixing rates in coastal currents and plumes. Recently, Houghton *et al.* (2004) employed a Lagrangian approach which circumvented the need to resolve the details of the spatial structure by tracking a patch of judiciously deployed Rhodamine-WT fluorescent dye to trace the motion of salt in the coastal current formed by the Delaware Bay outflow during upwelling-favorable winds. Figure 8.15 shows the coastal geometry and the distribution of salinity at the mouth of the Delaware Bay, together with the location of the dye discharge. Near the beginning of a two-day upwelling period, Houghton *et al.* (2004) injected dye into the halocline in the outer part of the coastal current between 5 and 6 m below the surface. They then continuously surveyed the region with a towed, undulating instrument package.

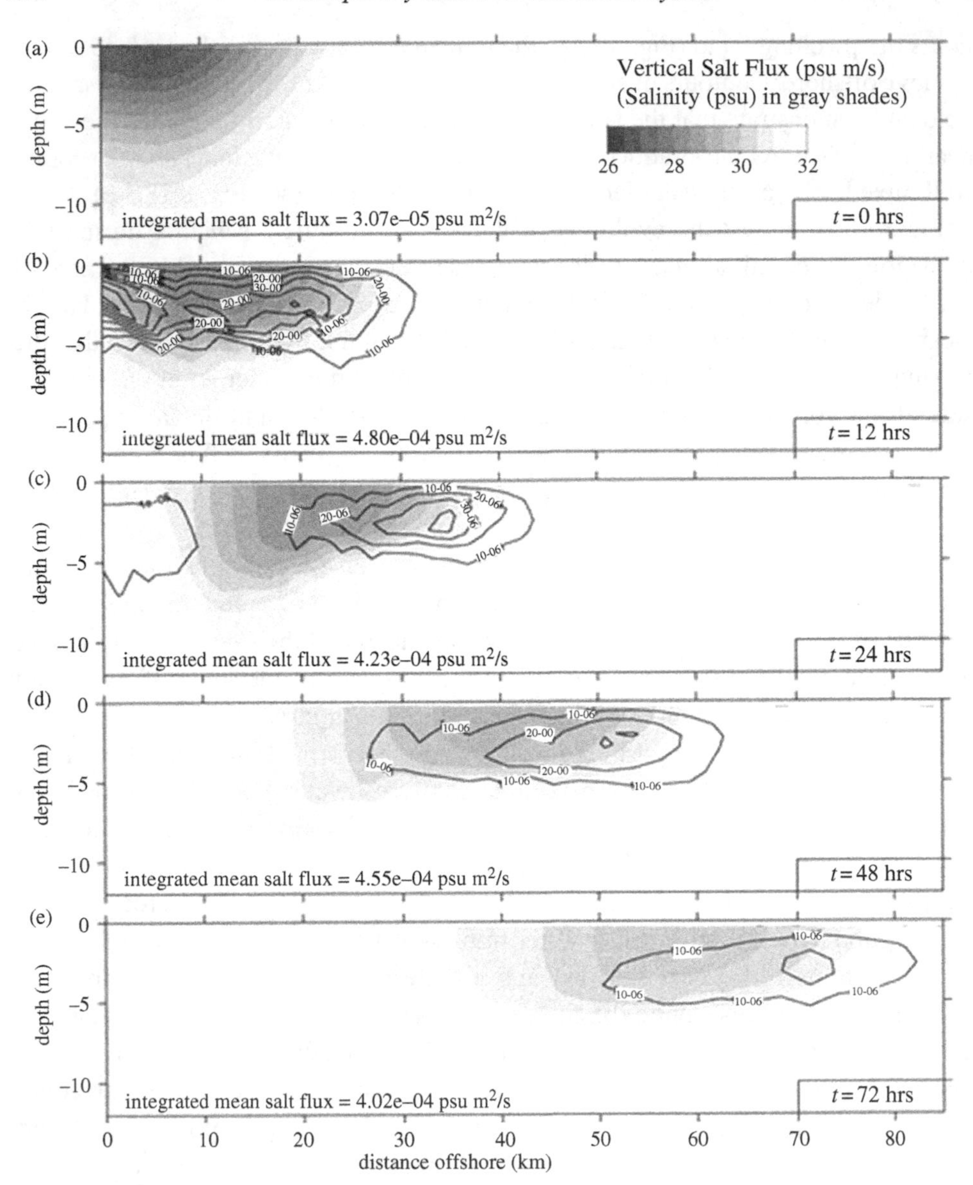

Figure 8.14. Contours of vertical salt flux and salinity for $t = 0$, 12, 24, 48, and 72 hours. Shaded contours are used to indicate practical salinity contours in intervals of 0.5. Superimposed on the shading are contours of vertical salt flux in solid contour lines at 10^{-5}m/s intervals. The cross-shore integrated mean vertical salt flux is indicated in the lower left corner of each panel. From Fong and Geyer (2001).

Figure 8.16a,c shows maps of the distribution of the vertically integrated dye concentration shortly after the deployment and then half a day later. Clearly, two patches of higher concentration are formed. Figure 8.17a,c presents the observed vertical distribution of both the dye concentration and salinity along the straight line

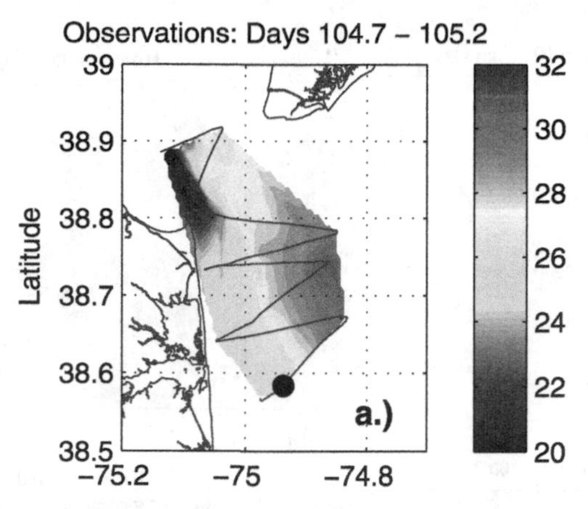

Figure 8.15. Map of the surface salinity distribution at the mouth of the Delaware Bay in 2003, days 104.7–105.2. The black circle indicates location of the dye release. From Tilburg *et al.* (2007).

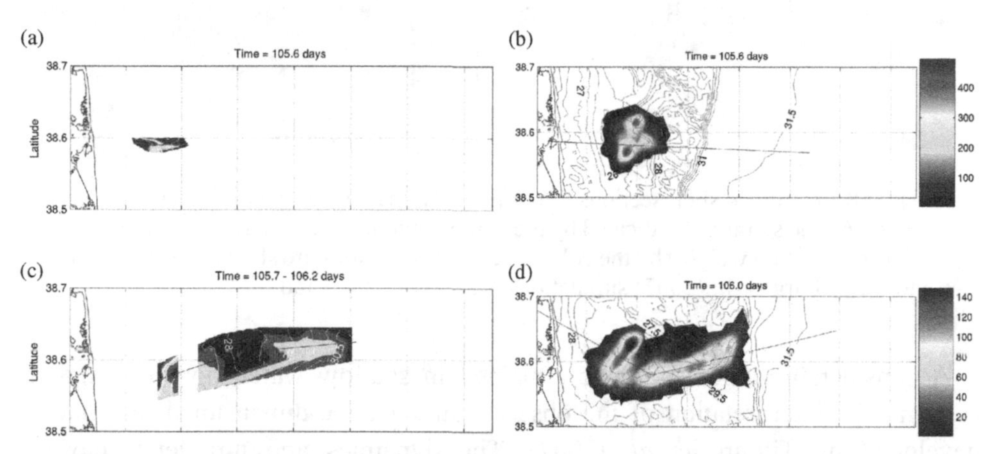

Figure 8.16. Map of the dye distribution in the Delaware coastal current showing the evolution of the integrated dye concentrations (parts per 10^{11} by weight × m) for the observed [(a) and (c)] and simulated [(b) and (d)] dye injections on days 105.6 and 106. The simulated surface salinity is shown by the contour lines in (b) and (d). The straight lines show the ship transects used to construct the cross-sections shown in Fig. 8.17. From Tilburg *et al.* (2007).

shown in Fig. 8.16. These distributions show that the plume has responded to the wind forcing largely in accord with the predictions of the Ekman layer idea, and that the eastern dye patch has moved almost 28 km offshore. Houghton *et al.* (2004) performed a quantitative analysis of the rates of mixing necessary to account for the distribution of the dye and salt in the offshore patch and concluded that they were consistent with the Fong and Geyer (2001) estimates.

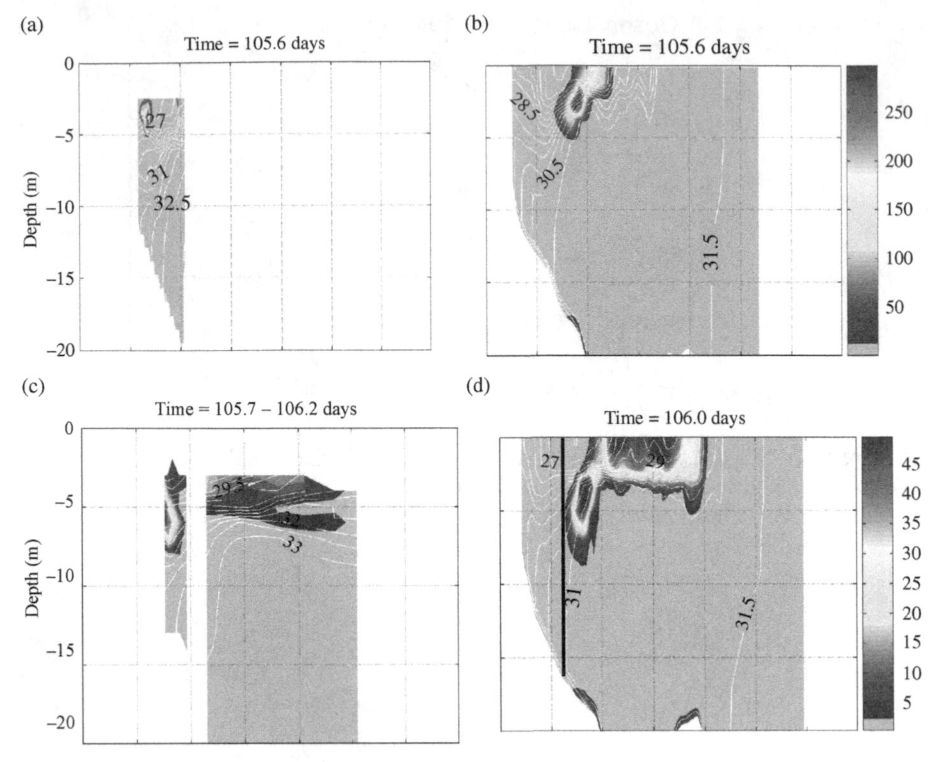

Figure 8.17. Across-shelf sections through the dye patch along the tracks shown in Fig. 8.16. The salinity is indicated by the white contour lines and dye concentration (parts per 10^{11} by weight) by the color levels. Observations are shown in (a) and (c) and the Tilburg *et al.* (2007) simulations are shown in (b) and (d).

A considerable amount of dye remained in shallow water, however, and to assist in the interpretation of the observations, a three-dimensional model was developed by Tilburg *et al.* (2007). The dynamics and turbulence closure scheme included in this model were similar to those of Fong and Geyer (2001), but also included tides and used realistic bathymetry and observed winds and river discharge conditions. Whitney and Garvine (2006) discuss the performance of the model. Figure 8.16b,d shows the simulated distributions of the vertically integrated dye concentrations, while Fig. 8.17b,d shows the corresponding vertical structure. Analysis of these simulations demonstrated that the horizontal and vertical shear in the inner shelf dictated the division of the dye into two separate patches. Though the simple theory that emphasized the enhanced mixing at the frontal zone maintained by Ekman straining occurs in nature and in sophisticated three-dimensional models, much more needs to be understood about across-shelf transport of terrestrially derived materials and the fate of coastal currents.

8.8. Small-scale front classification

The existence of frontal boundaries within estuaries and estuarine outflow plumes or coastal currents has been well established for several decades. Bowman (1988) and O'Donnell (1993) reviewed the early literature in these areas. In this section we summarize the mechanisms that can lead to the development of fronts and then outline an early model for the steady-state maintenance of a frontal zone and present some recent comparison of the model predictions to observations. However, it is clear that this model is not applicable to all situations. For example, recent observations of the plume from the Columbia River of Nash and Moum (2005) have highlighted that the buoyant effluent from a river may, on each ebb tide, advance into the remnants of a previous outflow plume that has partially mixed with ocean water. They showed that the leading edge of the Columbia River plume (the front) generates large-amplitude internal waves on the pycnocline formed by the remnants of earlier plumes. The geometry sketched in Fig. 8.2 may, therefore, be too simplistic for all circumstances and the boundary zone of estuaries and the ocean may be better characterized as a region in which high-frequency and high-wavenumber processes associated with the distortion of the density field by tidal currents, bathymetry, and internal hydraulics combine in complicated ways to determine the character of the coastal currents classified in Section 8.3.

To begin the discussion of small-scale structures, we define a "front" to be a region of high spatial gradient relative to surrounding areas. Such regions tend to draw attention to themselves in aerial photographs and satellite images of surface ocean temperature and chlorophyll distributions. In the coastal ocean the term "front" has been applied to features with scales ranging from 1–10^4 m. In situ sampling of such features is made difficult by their scale and, in many circumstances, by their transient nature. Here we mainly focus only on more recent work, after a brief introduction to the variety of fronts that have been observed.

O'Donnell (1993) introduced a classification scheme for estuarine fronts that is based upon the mechanism leading to the intensification of horizontal density gradients. These can be understood conceptually by consideration of the spatial derivative of the equation expressing the conservation of density, i.e.,

$$\frac{\partial}{\partial y}\frac{D\rho}{Dt} = \frac{\partial}{\partial y}\left\{\frac{\partial \rho}{\partial t} + u\frac{\partial \rho}{\partial x} + v\frac{\partial \rho}{\partial y} + w\frac{\partial \rho}{\partial z}\right\} = -\frac{\partial}{\partial y}\left\{\frac{\partial F_z}{\partial z} - S\right\}. \tag{8.38}$$

Here we have assumed that the density is changed by a source term, S (representing, for example, heating) and that only vertical turbulent transport, F_z, is significant. We only consider the rate of change in the y-direction here, but the following arguments are unchanged if the x-direction is considered instead. Equation (8.38) can be rewritten as

$$
\begin{aligned}
&\frac{\partial}{\partial t}\frac{\partial \rho}{\partial y} + \left\{ u\frac{\partial}{\partial x} + v\frac{\partial}{\partial y} + w\frac{\partial}{\partial z} \right\} \frac{\partial \rho}{\partial y} \\
&= -\frac{\partial}{\partial y}\left\{ \frac{\partial F_z}{\partial z} - s \right\} \\
&-\frac{\partial u}{\partial y}\frac{\partial \rho}{\partial x} \\
&-\frac{\partial v}{\partial y}\frac{\partial \rho}{\partial y} \\
&-\frac{\partial w}{\partial y}\frac{\partial \rho}{\partial z},
\end{aligned} \tag{8.39}
$$

where the terms on the left constitute the material derivative of the spatial gradient. The local rate of change of the gradient at a mooring is, therefore, clearly modulated by the advection of water with a different gradient past it. The remaining terms that modulate the material rate of change of $\partial\rho/\partial y$ can be divided into four groups, and these are written on separate lines on the right of equation (8.39). The terms on the second line are significant when there is a spatial gradient in the difference between the rate of buoyancy addition and the rate of vertical transport by turbulence. These terms lead to "tidal mixing fronts" of the type discussed in Section 8.6 and described by Simpson and Hunter (1974).

The schematic in Fig. 8.18 illustrates a simple channel flow in which the term on the third line, $\frac{\partial u}{\partial y}\frac{\partial \rho}{\partial x}$, leads to an increase in the magnitude of the gradient $\partial\rho/\partial y$.

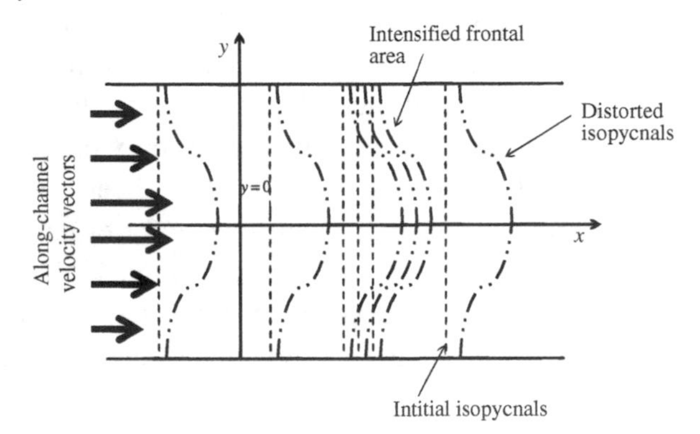

Figure 8.18. A schematic channel flow that illustrates the formation of lateral density fronts by differential advection of a longitudinal gradient by lateral shear. The thick black arrows illustrate the velocity field, which is maximum in the center of the channel. The dashed lines show the density contours at $t = 0$. Note that the along-channel gradient is larger where the lines are close together. The dash–dot lines show the distribution a short time later when the velocity field has deformed the distribution. They show lateral gradients with the same sign as the shear and with larger magnitude in areas where the along-channel gradients are larger.

The black arrows in the figure represent velocity vectors in the x-direction with shear in the y-direction. The thin dashed lines show isopycnals at an initial time, $t = 0$. Note that the gradient is not constant in the schematic. The dashed-dot lines show the position of the isopycnals a short time later, after they have been translated and distorted by the velocity field. Since the larger velocities in the center of the channel advect the isopycnals further than at the edges, regions of positive and negative $\partial\rho/\partial y$ have been created. In the figure, the lateral (y) gradients are shown to have larger magnitude where $\partial\rho/\partial x$ is larger. Fronts created in this way may reasonably be termed "shear fronts". Examples of these have been documented by Simpson and Turrell (1986), Huzzey and Brubaker (1988), Valle-Levinson *et al.* (2003), and Neill *et al.* (2004).

The black arrows in Fig. 8.19 illustrate a converging flow in the x–y plane with $\partial v/\partial y < 0$ and $\partial u/\partial x > 0$. Note that the x-coordinate is directed downward and that the y-coordinate is to the right. The dashed lines again represent isopycnals at an initial time t_0 and the solid lines show the isopycnals after they have been distorted by the flow. In this regime the convergence of the v-component in the presence of y-direction gradients leads to an intensification of the gradient, as predicted by the third term on the right of equation (8.39). Fronts generated by this term may be termed "convergence fronts".

The final term on the right of equation (8.39) is non-zero when there is vertical stratification and a y-direction gradient in the vertical velocity component, as may occur when flow moves over topography or winds along the coast drive an offshore Ekman flux and coastal upwelling. Fronts arising in this way may be classed as "upwelling fronts".

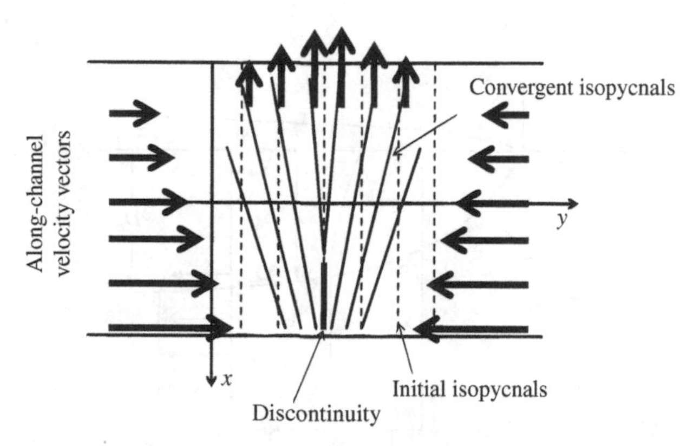

Figure 8.19. A schematic convergent flow illustrating the formation of convergence fronts. As in Fig. 8.18, the black arrows show the velocity vectors and the dashed lines the distribution of density at $t = 0$. The solid lines show the density field after it has been distorted by the velocity field. The region of highest gradient is shown by the thick solid line where contours merge, indicating a jump discontinuity.

8.9. Structure and dynamics of small-scale fronts

It is important to recognize that the classification introduced in Section 8.8 was simply a kinematic analysis of gradient intensification. However, it is clear that density variations may also drive circulation and mixing which combine to accelerate the rate of convergence, as was demonstrated by the laboratory experiments of Linden and Simpson (1986). Garvine (1974) developed an early diagnostic model of the steady structure of two-dimensional, small-scale density fronts with this character. Here, "diagnostic" implies that the model defines a hypothesis about what constitutes the essential dynamics that can be tested by prescribing a density field and comparing the predictions of the velocity field to observations.

Garvine's (1974) model assumed that the frontal zone bounds a larger-scale buoyant layer that overlies a uniform layer of denser water. The lower layer is assumed to move toward the buoyant layer, with velocity u_∞ in a frame of reference moving with the front, and to be sufficiently deep that the flow is not appreciably affected by the presence of the upper layer. A schematic of this geometry is shown in Fig. 8.20. Within the frontal zone the flow was assumed to be in hydrostatic balance. The horizontal dynamic balance was assumed to be between the sum of the horizontal baroclinic pressure gradient, the flow inertia, and the divergence of the turbulent stress, dominantly the vertical variation of the vertical flux of horizontal momentum. Using the across-front coordinate, x, which is taken to be normal to, and move with, the surface expression of the front and to be directed toward low density as is sketched in Fig. 8.20, the momentum equation, incorporating the Boussineq approximation, can be expressed as

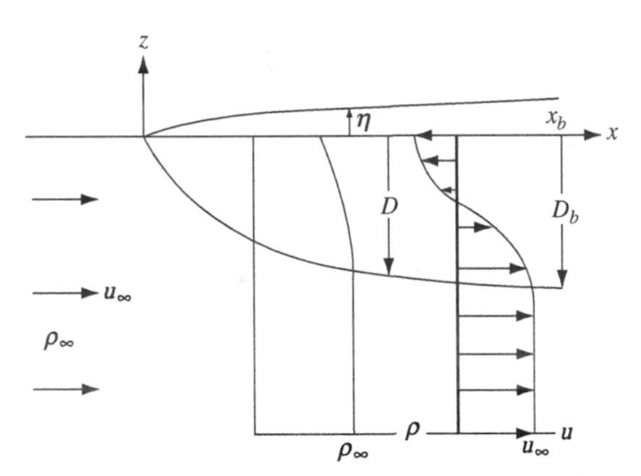

Figure 8.20. The geometry adopted by Garvine (1974) for his steady, diagnostic model of the circulation in a small-scale river plume front. The frame of reference is taken to move with the speed of the leading edge of the buoyant layer so that the ambient fluid moves to the right (x direction) at u_∞. The pycnocline is taken as $z = -D(x)$ and the sea surface is $z = \eta(x)$. From Garvine (1974).

$$u\frac{\partial u}{\partial x}+w\frac{\partial u}{\partial z}=-\frac{1}{\rho_\infty}\frac{\partial p}{\partial x}+\frac{1}{\rho_\infty}\frac{\partial \tau_{xz}}{\partial z}, \tag{8.40}$$

where ρ_∞ is the density in the lower layer. Defining the level of the pycnocline as $z = -D(x)$, where $D(x=0)=0$ at the front and $D(x=x_b)=D_b$ where the frontal zone joins the buoyant layer, then the density field was prescribed in terms of $D(x)$ and the function $r(x)$ such that

$$\frac{\rho(x,z)}{\rho_\infty}=\left\{\begin{matrix} 1-2\gamma r(x)\left\{1+\frac{z}{D(x)}\right\}, & 0\geq z\geq -D(x) \\ 1, & z\leq -D(x)\end{matrix}\right\}, \tag{8.41}$$

where the constant $\gamma=\dfrac{(\rho_\infty-\rho(x_b))}{2\rho_\infty}$ is the magnitude of the buoyancy at x_b, the boundary of the frontal zone and the plume. Note that $D(x_b)=D_b$, and both γ and $r(x)$ must be specified and that the density increases linearly from the surface to the pycnocline. The depth of the pycnocline, $D(x)$, is the only characteristic of the density field that is predicted by the model. Neglecting terms of order γ, and η/D, the ratio of the sea level displacement to the thickness of the buoyant layer, the horizontal pressure gradient can be written as

$$\frac{1}{\rho_\infty}\frac{\partial p}{\partial x}=\left\{\begin{matrix} \dfrac{d\eta}{dx}+2\gamma z\dfrac{dr}{dx}\left\{1+\dfrac{z}{2D}\right\}-\dfrac{\gamma r}{D^2}\dfrac{dD}{dx}z^2, & 0\geq z\geq -D(x) \\ \dfrac{d\eta}{dx}-\gamma\dfrac{d}{dx}(rD), & z\leq -D(x)\end{matrix}\right\}. \tag{8.42}$$

Note that if the pressure field in the lower layer is to be unmodified by the upper layer and $\eta(0)=D(0)=0$, then $\eta=\gamma rD$, i.e., the sea level slope offsets the influence of density variations on the lower layer. This is indicated schematically in Fig. 8.20.

Garvine (1974) further prescribed the velocity field in the upper layer to have the structure

$$u(x,z)=\left\{\begin{matrix} u_s(x)+\{u_\infty-u_s(x)\}\left\{\frac{z}{D(x)}\right\}^2, & 0\geq z\geq -D(x) \\ u_\infty, & z\leq -D(x)\end{matrix}\right\} \tag{8.43}$$

and sought the functions $u_s(x)$ and $D(x)$ for which the forms assumed in equations (8.41) and (8.43) are solutions to equation (8.40) and the continuity equation.

Garvine (1974) integrated the continuity equation across the upper layer to obtain

$$\frac{d}{dx}\int_{-D}^{0}udz=\frac{d}{dx}\left\{\frac{D(x)}{3}\{2u_s(x)+u_\infty\}\right\}=q_e, \tag{8.44}$$

where q_e is defined as the mean velocity associated with the turbulent entrainment of fluid across the pycnocline in the frontal zone, which he parameterized as $q_e = -u_\infty E(x)$. The function $E(x)>0$ describes the distribution of the volume transport out of the upper layer. Garvine (1974) also parameterized the magnitude of the interfacial stress as $\tau_{xz} = \rho_\infty u_\infty^2 C_f(x)$, so that the spatial structure and magnitude of momentum and mass exchange between the layers are defined by C_f and E, which are central components of the hypothetical dynamics.

Using the continuity principle, we can write the vertical velocity within the frontal zone as

$$w(x, z) = \int_z^0 \frac{\partial u}{\partial x} dz \tag{8.45}$$

and vertical integration of the momentum balance, equation (8.40), across the upper layer then yields

$$\begin{aligned} \frac{d}{dx}\int_{-D}^{0} u^2 dz &= \frac{d}{dx}\left\{\frac{D}{15}\{8u_s^2 + 4u_\infty u_s + 3u_\infty^2\}\right\} \\ &= -\frac{g\gamma}{3}\frac{d}{dx}\{rD^2\} - u_\infty^2(E - C_f). \end{aligned} \tag{8.46}$$

The divergence of the horizontal momentum flux, on the left of equation (8.46), is balanced by the pressure gradient and the across-interfacial transport by entrainment and turbulent stress on the right.

Equations (8.44) and (8.46) form a pair of non-linear, first-order ordinary differential equations for $u_s(x)$ and $D(x)$. The solution is sensitive to the choice of $E(x)$ and $C_f(x)$, and Garvine (1974) explored several alternatives. However, in order for the frontal zone variables to merge smoothly with the parent plume, he prescribed them to decay exponentially with distance from the surface front as

$$E(x) = a\exp\left\{-\frac{ax}{\beta D_b}\right\} \tag{8.47}$$

and

$$C_f(x) = fa\exp\left\{-\frac{ax}{\beta D_b}\right\}. \tag{8.48}$$

Here, a is the magnitude of the entrainment of momentum and f is the ratio of the magnitude of the entrainment and friction. Note also that this formulation implies that $x_b = D_b/a$ is the scale width of the frontal zone, or $a = D_b/x_b$ is the scale for the frontal pycnocline slope, and β dictates the decay scale for mixing and friction within the frontal zone.

Calculation of the model solutions for $u_s(x)$ and $D(x)$ then requires the numerical integration of two simultaneous differential equations. This is most easily accomplished by introducing the dimensionless variables, $\zeta = ax/D_b$, $q = u_s/u_\infty$ and $\delta = D/D_b$, so that by integrated continuity, equation (8.44) can be written as

$$\frac{dq}{d\zeta} = -\frac{(2q+1)\delta' + 3e^{-\zeta/\beta}}{2\delta} \tag{8.49}$$

and the momentum equation as

$$\frac{d\delta}{d\zeta} = \frac{-5\delta^2 Ri_b r' + (15f - 9 + 24q)e^{-\zeta/\beta}}{1 - 8q - 8q^2 + 10Ri_b r\delta}, \tag{8.50}$$

where the primes indicate differentiation with respect to ζ and $Ri_b = g\gamma D_b/u_\infty^2$. Note that the constant a does not appear explicitly in these equations since it sets the scale for the frontal zone. To initiate the integration in ζ it is helpful to know $q(0)$ and $\delta'(0)$. This is most easily accomplished by introducing Taylor series expansions for small ζ and exploiting $\delta(0) = 0$. This allows equation (8.49) to be expanded and rewritten as

$$\delta'(0) = -\frac{3}{2q_0 + 1} \tag{8.51}$$

and, after a little more algebra, equation (8.50) yields

$$q(0) = \frac{1}{8}\left\{3 - 5f + 5\sqrt{f^2 - 2.8f + 1}\right\}. \tag{8.52}$$

It is clear from equations (8.51) and (8.52) that the conditions at the front depend only on the friction and entrainment parameter f. Further, since we expect $\delta > 0$, then equation (8.51) requires $q_0 < -\frac{1}{2}$, which is assured if $f \sim 2.38$.

Additional constraints on the parameter choices must be imposed to assure that the pycnocline depth matches that of the plume, i.e., $\delta(\zeta \to \infty) = 1$. These can be developed by integration of the continuity and momentum equations across the frontal zone. Using the dimensionless variables and the parameterization of the entrainment flux, equation (8.47), then the integral of the continuity principle, equation (8.44), can be written as

$$\int_0^\infty \frac{d}{d\zeta}\left\{\frac{\delta}{3(2q+1)}\right\} d\zeta = \frac{2q+1}{3} = -\int_0^\infty e^{-\zeta/\beta} d\zeta = -\beta. \tag{8.53}$$

Using $\delta(0) = 0$ and $\delta(\zeta \to \infty) \to 1$,

$$q_b = -\frac{3\beta + 1}{2}. \tag{8.54}$$

Applying the same procedure to the momentum balance, equation (8.46) becomes

$$\int_0^\infty \frac{d}{d\zeta}\left\{\frac{\delta}{15}\{8q^2+4q+3\}\right\}d\zeta$$
$$= -\frac{Ri_b}{3}\int_0^\infty \frac{d}{dx}\{r\delta^2\}d\zeta - (1-f)\int_0^\infty e^{-\zeta/\beta}d\zeta, \tag{8.55}$$

which simplifies to

$$\frac{1}{15}\{8q_b{}^2+4q_b+3\} = -\frac{Ri_b}{3} - (1-f)\beta \tag{8.56}$$

or

$$Ri_b = 3\beta f - \frac{18\beta^2 - 21\beta + 4}{5}. \tag{8.57}$$

To evaluate the solutions, values of a, f, and β are required. In addition, the across-front variation of the surface density, $\gamma r(x)$, and the pycnocline depth at the plume boundary, D_b, must be specified and equation (8.57) must be satisfied for the parameter choices to be consistent.

Garvine and Monk (1974) reported high-resolution observations of the density and velocity fields at the front of the Connecticut River plume and Garvine (1974) used these to motivate the choice of parameters for his model. An example distribution of $\rho(x,z)$ from Garvine and Monk (1974) is shown in Fig. 8.21a. These density measurements were obtained by pumping water from the locations of the dots in the figure through a shipboard conductivity–temperature sensor and the positioning was accomplished using floats and lines. Garvine's (1974) representation of this section is shown in Fig. 8.21c, which is based on equation (8.41) with $\gamma = 7.5\times10^{-2}$, $r(\zeta) = 1-\exp\left(\frac{-\zeta}{0.055}\right)$, $D_b = 2\,\text{m}$, and $a = 0.017$. Note that the origin of the horizontal axis is offset by 25 m in Fig. 8.21a since the field observations used a color change as the reference location. This parameter set implies that the length scale of the frontal zone, where $\zeta = ax/D_b \sim 1$, is $x_b = D_b/a = 118$ m. The second critical length scale is determined by β which, through equations (8.47) and (8.48), sets how rapidly entrainment and friction decay within the frontal zone, and $\beta = 0.1$ sets this scale as 11.8 m.

With these values and $u_\infty = 0.5\,\text{m/s}$, then equation (8.57) requires $f = 5.52$ and integration of equations (8.49) and (8.50) yields $q(x)$ and $\delta(x)$ and the velocity field shown in Fig. 8.21d. For comparison with these predictions, Garvine and Monk's (1974) observations of the across-front velocity component relative to the front measured by lowered current meter from a ship are shown in Fig. 8.21c.

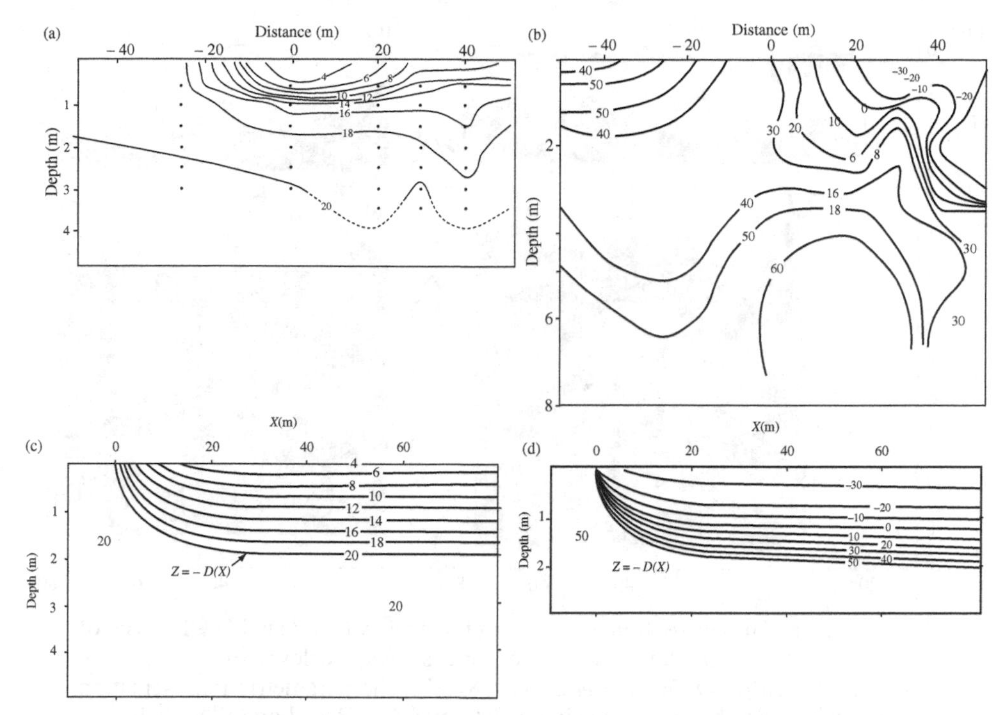

Figure 8.21. (a) The across-front structure of the density (σ_T) field at the front of the Connecticut River plume. The dots show the locations of measurements. From Garvine and Monk (1974). (b) The across-front variation of the across-front velocity component at the front of the Connecticut River plume accompanying the density observations shown in (a). From Garvine and Monk (1974). (c) The parametric representation of the density field shown in (a) using equation (8.41). From Garvine (1974). (d) The model-predicted across-front velocity component corresponding to the density field shown in (c). From Garvine (1974).

Measurement locations are again shown by the dots. The sparseness of the data make it difficult to assess whether the structure of the flow is in agreement with the model predictions or not. However, the observations do reveal convergence at the front and that the magnitude of the velocity is consistent with predictions.

A more critical evaluation of this model was made possible by improved navigation and instrument technology. O'Donnell (1997) constructed a rigid array of electromagnetic current meters and conductivity–temperature sensors and mounted it on a small boat equipped with motion sensors and a GPS navigation system. He then employed this surface current and density (SCUD) array to observe the small-scale structure and motion field in the upper 3 m of the water column at the Connecticut River plume front in the same area as Garvine and Monk's (1974) observations. Subsequently, O'Donnell *et al.* (1998) employed the SCUD array in

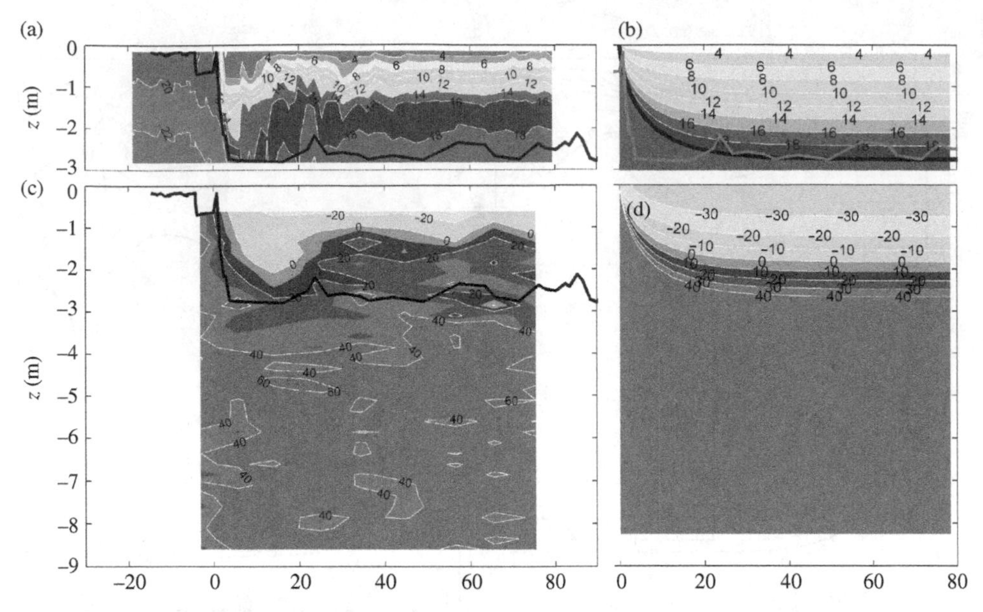

Figure 8.22. (a) The across-front structure of the density ($\sigma \downarrow T$) field at the front of the Connecticut River plume. The black line shows the level of the $\sigma_T = 19$ isopycnal. Based on O'Donnell *et al.* (1998). (b) The parametric representation of the density field shown in (a) using equation (8.41). Based on O'Donnell *et al* (1998). (c) The across-front variation of the across-front velocity component at the front of the Connecticut River plume accompanying the density observations shown in (a). Based on O'Donnell *et al.* (1998). (d) The model-predicted across-front velocity component corresponding to the density field shown in (c). Based on O'Donnell *et al.* (1998).

conjunction with a towed acoustic Doppler current profiler (TOAD; see Trump *et al.*, 1995) to improve the accuracy and vertical extent of the current measurements. Figure 8.22a shows the density field in the upper 3 m of the water column as reported by O'Donnell *et al.* (1998), with the thick black line showing the level of the $\sigma_T = 19$ contour. Note that in the interval $5 < x < 17$, the $\sigma_T = 19$ contour drops below the level of the lowest sensor on the SCUD array (2.8 m) so that the level represented in the figure is only an upper bound in that interval. Comparison of Fig. 8.22a and Fig. 8.21a shows some significant differences. It is clear in the higher-resolution observations that the pycnocline depth does not increase monotonically in depth with distance from the front, but overshoots its asymptotic level and then rises before leveling out at a large distance from the front. This feature cannot be represented in Garvine's model. The later measurements also reveal that the frontal zone is much narrower than the Garvine and Monk (1974) measurements suggested, and for comparison with their measurements, O'Donnell *et al.* (1998) set $a = D_b/x_b = 0.055$, which requires that the pycnocline slope in the frontal zone be three times larger than that used by Garvine (1974). They also specified the surface density structure to vary more rapidly as $r(\zeta) = 1 - \exp\left(\frac{-\zeta}{0.005}\right)$ and chose similar values for the other parameters of the density field: $D_b = 2.75$, $\gamma = 0.0085$. These choices lead to the density field shown in Fig. 8.22b, in which the thick black line represents the parameterized $\sigma_T = 19$ contour and the thick line shows the observed level transferred from Fig. 8.21a. The general structure of these two fields is very similar, though the change in sign of the x-direction gradients of the isopycnal depths is not represented in the parameterization.

Figure 8.22c displays the velocity observations obtained from the SCUD array and TOAD profiler. The average velocity below the $\sigma_T = 19$ level is taken as the ambient flow magnitude $u_\infty = 0.45\,\text{m/s}$, so $Ri_b = 1.13$. Choosing $\beta = 0.13$ so that mixing and friction decays in approximately the same fraction of the frontal zone width as suggested by Garvine (1974), then equation (8.57) requires $f = 6$. Integration of equations (8.49) and (8.50) then produces the velocity field shown in Fig. 8.22d. Comparison of Fig. 8.22c,d illustrates that though there are qualitative similarities, the model does not adequately describe some significant features of the observations. Note that the highest observations are at 0.75 m and that the model predictions extend to the surface. This makes direct comparison a little more difficult. It is clear, however, that the near-surface velocity in the buoyant fluid layer is slightly less negative than in the model. The level of the 0 isotach is significantly shallower in the observations than is predicted, indicating that the shear layer is significantly thicker than predicted. The most distinct difference is in the interval $0 < x < 20$, where the observations show that water of negative velocity drops down to −2 m. This is consistent with the density field variations in the same area, which indicate buoyant surface water being transported downward to the level of the shear layer at the base of the plume. This overshoot in the pycnocline depth cannot be explained in the frictional momentum balance of Garvine's (1974) model and is more evocative of the inviscid inertia–baroclinic pressure gradient dynamics of a gravity current described theoretically by Benjamin (1968) and empirically in the monograph of Simpson (1987).

Laboratory experiments on the structure of, and circulation within, gravity currents have direct relevance to that of small-scale fronts and there is an extensive literature on this subject. Britter and Simpson (1978) pioneered laboratory studies of two-dimensional gravity currents. A schematic of the structure of a turbulent gravity current in which mixing and friction are active is shown in Fig. 8.23a. In their terminology, gravity currents are typically composed of two distinct regions: a small-scale "head-wave" region in which they propose a Kelvin–Helmholtz instability grows rapidly and the horizontal density gradients are large, and the

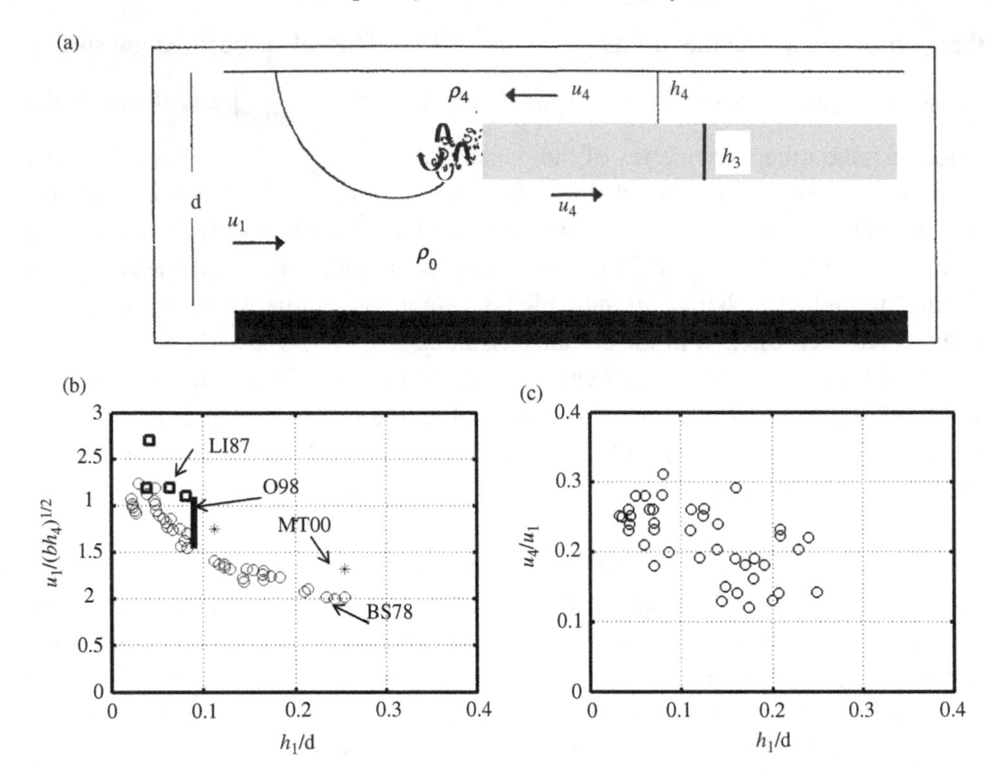

Figure 8.23. (a) Schematic of the density field in a gravity current in a channel of depth *d* in a frame of reference moving with the leading edge of the buoyant layer. The water of density ρ_o in the lower layer is advancing to the right at speed u_1. The upper layer forms a large overturning roller, which breaks up into three-dimensional turbulence which is advected into the shear layer (gray region) of thickness h_2. The uppermost layer of density ρ_4 and thickness h_4 moves to the left at u_4. (b) Observation of $u_1/\sqrt{b_4 h_4}$ in the gravity current experiments of Britter and Simpson (1978), shown by circles, together with the plume observations of Luketina and Imberger (1987), squares; Marmorino and Trump (2000), asterisk; and O'Donnell *et al.* (1998), black line. (c) Observation of u_4/u_1 in the gravity current experiments of Britter and Simpson (1978).

much larger, slowly varying "tail" which is stably stratified. In Fig. 8.23a the stable region is divided into a vertically uniform region of thickness h_4, velocity u_4, and density ρ_4, which overlies a stratified shear layer of thickness h_2. In river plumes, the head region is often identified as the front. When a gravity current moves into a stationary, inviscid, deep, ambient fluid without mixing, von Karman (1940) and Benjamin (1968) predicted the translation speed of the front relative to the deeper water to be $u_1 = \sqrt{2c_0}$ where, in the notation defined in Fig. 8.23a, $c_0 = \sqrt{bh_0}$ is the phase speed of long interfacial waves in the buoyant layer that follow the head region with thickness $h_0 = h_2 + h_4$ and buoyancy $b = g(\rho_0 - \rho_4)/\rho_0$. Benjamin (1968) also established that this proportionality should be valid when the ambient layer depth, d, is finite but with a proportionality coefficient that depends on h_0/d. Many experiments on the motion of gravity currents in channels have been performed to empirically establish the influence of mixing and friction on these results, which have been summarized by Simpson (1987) and are shown in Fig. 8.23b. The Britter and Simpson (1978) experiments were performed in a tank, and the frontal velocity scaled by the phase speed $\sqrt{bh_4}$ is shown by the open circles for a range of relative layer depths. Note that the frontal velocity gets smaller as the layer depth ratio increases. There have been several studies that have estimated the propagation speed of estuarine fronts. Luketina and Imberger's (1987) measurements were obtained in a shallow thermal plume in deeper water and these are shown in Fig. 8.23b by the open squares. O'Donnell *et al.* (1998) estimated a range of values for the rate of propagation of the front of the Connecticut River plume and these are shown by the black bar. Marmorino and Trump's (2000) observations from the Chesapeake Bay plume are shown by the * symbols. Measurements at sea are much more difficult to obtain, and uncertainties are larger than in the laboratory experiments. However, the laboratory and field estimates and the trend with h_4/d are largely consistent, which supports the utility of the gravity current model for plume front propagation.

Britter and Simpson (1978) also measured the velocity with which the buoyant water approached the front, u_4. Figure 8.23c shows the variation of their results with the relative layer thickness h_4/d. There is strong evidence that the convergence is larger in deeper water where u_4/u_1 approaches 0.25. Though O'Donnell *et al.* (1998) made measurements close to the surface, the uncertainties in these measurements were too large to estimate u_4 and this important prediction remains untested.

Real river plumes and plume fronts are seldom two-dimensional as assumed in the aforementioned laboratory experiments. The front propagation velocity may have a component in the direction normal to the ambient fluid velocity, as observed in the plume of the Connecticut River. The effects of cross-flowing ambient fluid on the spread of a positively buoyant plume from a radially symmetric source have been investigated in the laboratory by Huq (1983). These experiments showed that the shape of the plume front and the distribution of mass in the plume were both very sensitive to the cross-flow velocity. However, the scale of the experiments and the resolution of the instruments were inadequate to resolve any variation in the structure of the front in the along-front direction.

The monograph of Thorpe (2005) and Chapter 7 of this volume provide an introduction to the field of turbulence and mixing in the ocean. It has been well established by theoretical and experimental studies (e.g., Rohr *et al.*, 1988; Smyth and Moum, 2000) that the rate of growth or decay of turbulence in stratified shear flows is controlled by the gradient Richardson number, $Ri = N^2/S^2$ where $N = \left(-\frac{g}{\rho_0}\frac{\partial \rho}{\partial z}\right)^{\frac{1}{2}}$ is the buoyancy frequency and $S = \partial u/\partial z$. A fundamental quantity that measures turbulence in a fluid is the rate of dissipation of mechanical energy per unit volume by viscosity at small scales. Representing the velocity vector as $\vec{u} = [u_1, u_2, u_3]^T$, then the dissipation rate can be shown (see Kundu, 1990) to be

$$\varepsilon = \frac{\nu}{2}\sum_{i=1}^{3}\sum_{j=1}^{3}\left\{\frac{\partial u_i}{\partial x_j} + \frac{\partial u_j}{\partial x_i}\right\}\left\{\frac{\partial u_i}{\partial x_j} + \frac{\partial u_j}{\partial x_i}\right\}. \tag{8.58}$$

This is a very complicated quantity to measure and evaluate since it involves derivatives of all the velocity components in all directions. However, when the statistics of the derivatives and their products are isotropic, then

$$\varepsilon = \frac{15}{2}\nu\left\langle\frac{\partial u}{\partial z}\right\rangle^2, \tag{8.59}$$

where the $\langle..\rangle$ notation indicates the mean over the time and space scales of turbulent eddies. The molecular viscosity of seawater, ν, is approximately $10^{-6}\,\mathrm{m}^2/s$ and varies with temperature, salinity, and pressure. The units of ε are therefore $\mathrm{m}^2/\mathrm{s}^{-3}$, or equivalently, W/kg. It is not uncommon in the literature for the dissipation rate to be reported as $\mathrm{W/m}^3$ of seawater, which is approximately 1000W/kg but it is important to keep the units straight. In a quiescent part of the deep ocean, values as small as 10^{-10} W/kg have been measured, and in the coastal ocean in a vigorous bottom boundary layer 10^{-3} W/kg has been observed.

Techniques have been developed to measure this rate in the ocean using small metal "shear" probes whose resistance varies with the torque applied by small-scale relative velocities. An indirect approach is based on the argument of Ozmidov (1965), who suggested that the length scale of the largest turbulent eddies in a stably stratified flow should be determined only by the magnitude of the stratification as measured by N (s^{-1}) and the dissipation rate ε ($\mathrm{m}^2/\mathrm{s}^3$). Dimensional considerations then require

$$L_o = \left(\frac{\varepsilon}{N^3}\right)^{\frac{1}{2}}. \tag{8.60}$$

At the smallest scales of the motion the viscosity must be important and then the dissipation rate and molecular viscosity imply that the Kolmogorov scale is

$$L_k = \left(\frac{\nu^3}{\varepsilon}\right)^{\frac{1}{4}}. \tag{8.61}$$

In a weakly stratified and energetic tidal boundary layer with $\varepsilon \sim 10^{-4} W/kg$ and $N \sim 3 \times 10^{-2}\ \mathrm{s}^{-1}$, we find $L_o \sim 2$ m and $L_k \sim 0.3$ mm.

Vertical profiles of salinity, temperature, and density in the ocean obtained with very high resolution (say 1 mm) often exhibit regions within which $\partial\rho/\partial z > 0$, i.e., they are statically unstable. These have been interpreted as a manifestation of active overturning eddies temporarily disrupting the stable structure. Thorpe (1977) proposed that such profiles could be reordered vertically so that the $\partial\rho/\partial z \leq 0$ and the distance a sample had to be displaced from its original level, *d*, could be squared and averaged in vertical bins to yield an estimate of the mean size of the eddies disrupting the stratification, L_T. This has become known as the Thorpe or overturning scale. He then proposed that L_T was proportional to L_o, so that the dissipation rate could be estimated from profile observations and equation (8.60) as

$$\varepsilon = cL_T^2 N^3, \tag{8.62}$$

where the proportionality constant is $O(1)$. The observations of Peters *et al.* (1988) suggest $c = (L_0/L_T)^2 = 0.9$.

The Ozmidov scale has also been used to estimate the dissipation rate in the laboratory. If a time series of measurements of salinity or temperature is available at a point and the mean (average over several eddy scales) vertical gradient is known, then the vertical displacements required to account for deviation from the mean in the time series (ρ') can be estimated as

$$l' = \rho' \left(\frac{\partial \langle\rho\rangle}{\partial z}\right)^{-1}. \tag{8.63}$$

Using this assumption, Ellison (1957) proposed the length scale for energetic turbulent eddies as

$$L_E = \langle l'^2 \rangle^{\frac{1}{2}} \tag{8.64}$$

and using laboratory observations, Caldwell (1983) and Itsweire (1984) showed $L_T = L_E$. Later, direct numerical simulations of Itsweire *et al.* (1993) and Smyth and Moum (2000) supported this conclusion.

Rohr *et al.* (1988) and Itsweire *et al.* (1993) performed careful and extensive laboratory experiments on the growth and decay of grid-generated turbulence in stratified shear flows. They used a water tunnel in which the vertical density gradient and the velocity structure could be specified independently and then created three-dimensional eddies within the flow by vibrating a metal grid and measured the variation of the spectra of velocity and density fluctuations with distance downstream of the grid. One of the most important results was the demonstration that in experiments with gradient Richardson number, $Ri > Ri_{cr} = 0.25$, vertical mixing was suppressed rapidly as water left the stirring grid. Further, for values of the dissipation rate, $\varepsilon < 16\nu N^2$, the vertical turbulent buoyancy flux was effectively zero. In experiments with a mean flow in the shear layer U and $Ri > Ri_{cr}$, Itsweire *et al.* (1993) concluded that the vertical buoyancy flux went to zero within five buoyancy periods of passing through the grid, or equivalently after a distance $L_I = 5U/N$ from the generation location.

The gravity current and stratified shear layer experiments led O'Donnell *et al.* (2008) to propose a scale for the width of the zone of active mixing in the frontal zone of a river plume. Based on the flow visualizations of gravity currents of Britter and Simpson (1978), which revealed that the coherent overturning structures at the leading edge of the front broke up into three-dimensional turbulence within a few buoyant layer depths from the leading edge, O'Donnell *et al.* (2008) argued that the scale L_I should provide a guide to the width of the active mixing zone in plume fronts with $Ri > Ri_{cr}$. They tested this idea at the front of the Connecticut River in the area studied by Garvine and Monk (1974) and O'Donnell *et al.* (1998). They used two salinity sensors, one near the surface and a second towed a few meters below the ship, to estimate the mean vertical gradients and then used the high-frequency fluctuation at the lower level to estimate ρ'. To measure the current and shear they employed a ship-mounted ADCP. To assess the validity of the Ellison scale approach they also deployed an autonomous vehicle to measure the distribution of the dissipation rate using shear probes. Figure 8.24a shows the ship track as the curved line with six pairs of arrows indicating the local across- and along-front coordinate system at the surface expression of the frontal boundary between 09:33 and 10:17. A line between the origins of these axes shows the approximate location of the frontal line at maximum ebb tide in Long Island Sound. The short straight line to the west of the frontal line shows the track of the AUV with shear probes which sampled between 11:04 and 11:09 when the plume had expanded to the west.

Figure 8.24b displays the across-front variation of the salinity measured along section 1 by the near-surface and towed instruments averaged in 2 s intervals to remove the high-frequency fluctuations. The rapid drop in salinity at the front, $x = 0$, is clear in both sensor records. The across-front and along-front velocity components are shown in Fig. 8.24c,d. Note that the depth of the ADCP and the bin size do not allow the near-surface velocity to be resolved, however, there is evidence of negative across-front flow within 10 m of the front. The along-front velocity is shown with a different color range and is small. The relative acoustic backscatter intensity is shown in Fig. 8.24e, with the black dashed line indicating the depth of the towed CTD relative to the plume structure. The bin-averaged salinity and the

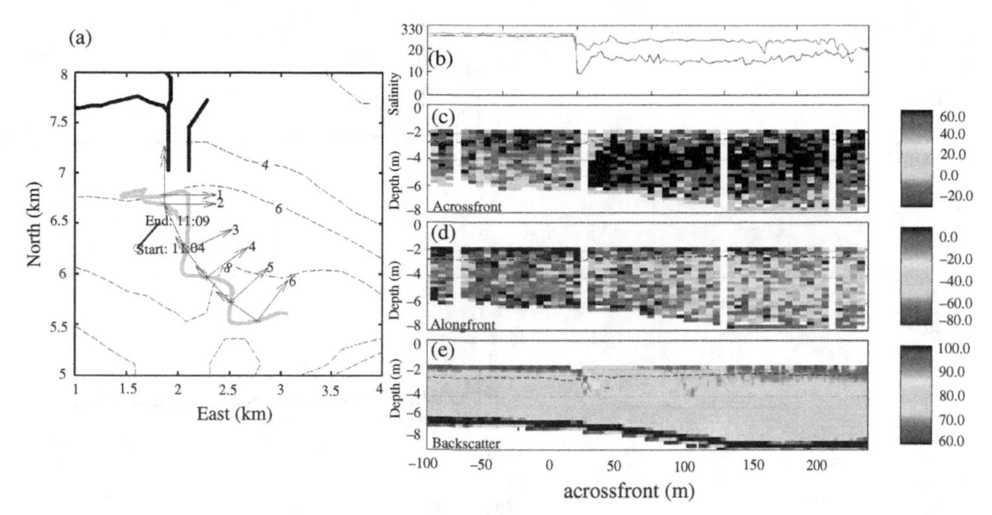

Figure 8.24. (a) The coastline, breakwaters (solid black lines), and bathymetry (dashed lines) at the mouth of the Connecticut River. The gray line shows the ship track of the O'Donnell *et al.* (2008) survey. The "+" symbols show the location of front crossings and the arrows indicate the along- and across-front coordinate system for the observations shown in (b)–(e) and Fig. 8.25 and Fig. 8.26. (b) The across-front variation of the near-surface salinity (darkest shading) and the low-pass filtered salinity at approximately 3 m. The sensor level is shown by the black dashed line in (e). (c) The vertical and across-front structure of the across-front velocity component observed on section 2. (d) The along-front velocity component (positive toward the river mouth) observed on section 2. (e) The across-front structure of the relative acoustic backscatter (dB) on section 2 with the level of the towed salinity sensor shown by the dashed black line. Adapted from O'Donnell *et al.* (2008).

variance at the lower sensor within each bin for all six sections are shown in Fig. 8.25. Observations from both sensors were not obtained on sections 2 and 6, so vertical gradients could not be estimated. Figure 8.25e shows the measurements from section 5. It is clearly different from the others in that the salinity at the lower sensor does not decrease appreciably at the front and the variance does not increase. Examination of the acoustic backscatter and the sensor depth shows that the plume layer was shallower than the instrument at this section and, therefore, the instrument didn't sample the stratified shear layer.

For the four sections, 1, 3, 4, and 5, from which data from two levels was available, O'Donnell *et al.* (2008) showed that $Ri > 0.25$ for $x > 0$ and computed the Ellison scale and the dissipation rate in 5 m intervals in the across-front direction. The across-front structure is shown in Fig. 8.26a. The three solid lines show the best fit logarithmic regressions for sections 1, 3, and 4 during which the lower sensor was within the shear layer. Note that the estimates for section 5

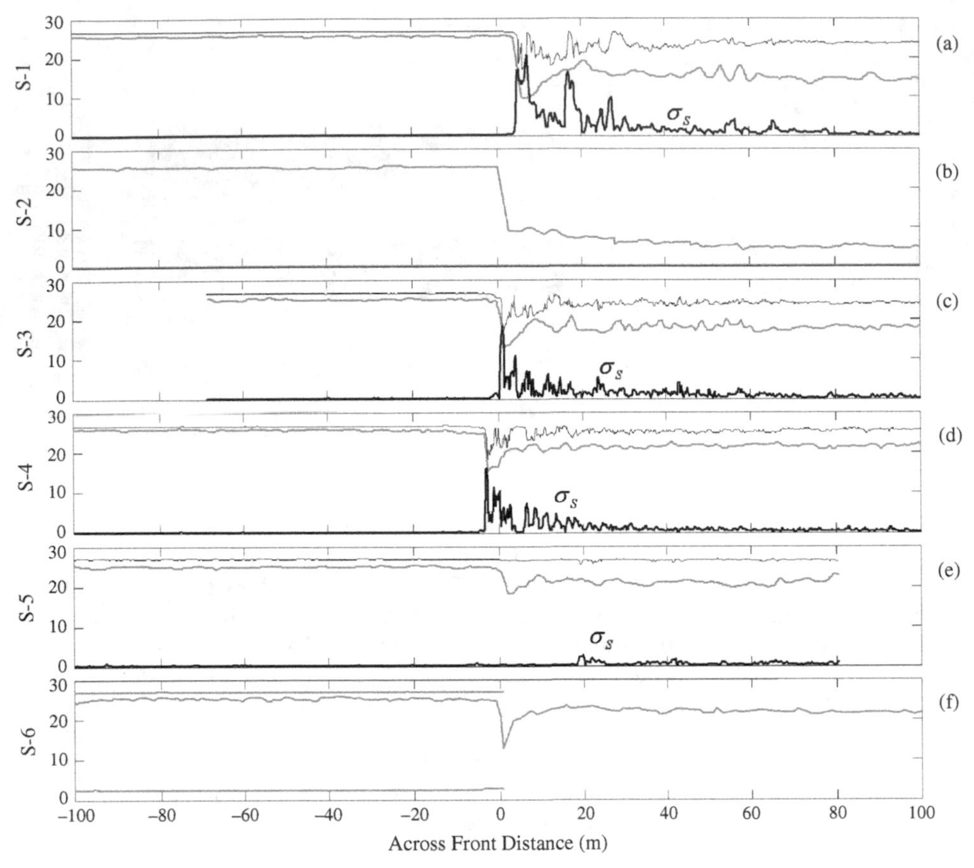

Figure 8.25. (a) The across-front structure of the near-surface (~ 1 m) salinity, thin black line, and the low-pass filtered salinity at approximately 3 m, thick black line. The variance of the lower-level salinity sensor in 2 s bins is shown by the lower line labeled σ_s in the graph. (b–f) show the same quantities for sections 2–6. From O'Donnell *et al.* (2008).

are shown by the * symbols and are all less than 5×10^{-6} W/kg. The thin dashed lines show $\varepsilon_m = 16\nu N^2$, which Itsweire (1984), Itsweire *et al.* (1986), and Rohr *et al.* (1988) have argued is the dissipation rate required to sustain active overturning in a stratified shear flow with $Ri > 0.25$. Close to the front, dissipation rates reach $10^{-2}\mathrm{W/kg}$ and decrease exponentially to $10^{-6}\mathrm{W/kg}$, 100 m from the front. This is approximately equivalent to $\varepsilon(x) = \varepsilon_0 \exp\{-x/L_G\}$ with $L_G =$ 15 m, which is illustrated in Fig. 8.26a by the thick dashed line. The length scales derived from regression through the data obtained on each section are listed in Table 8.3 and the trends are shown by the solid lines in Fig. 8.26. Though dissipation rate estimates suggest that the maximum decreases with

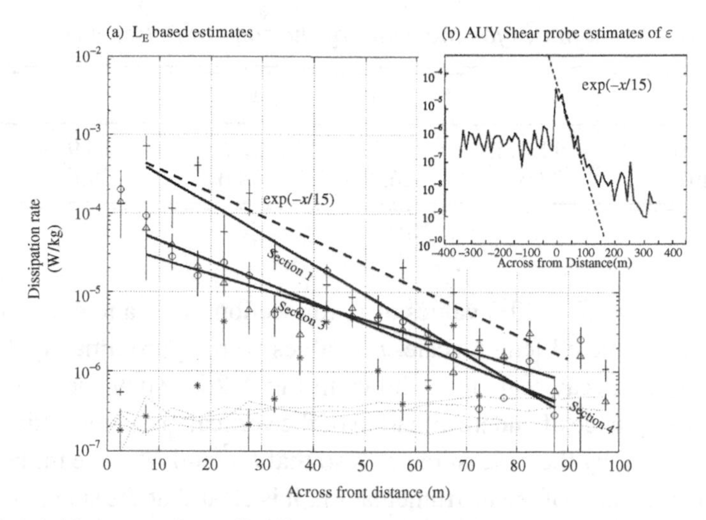

Figure 8.26. (a) Across-front distribution of the turbulent kinetic energy dissipation rate for sections 1, "+" symbol; 3, "O" symbol; 4, "Δ" symbol; and 5, "*" symbol. The thick solid lines illustrate the exponential decay of the dissipation rate in the across-front direction. The dashed lines show the Rohr *et al.* (1988) estimate of the level of dissipation at which mixing ceases. From O'Donnell *et al.* (2008). (b) Across-front distribution of the turbulent kinetic energy dissipation rate computed from the shear probe measurements by the REMUS AUV.

distance from the source and that L_G increases, these differences are not statistically significant and additional observations are required to address these issues.

Though the shear probe measurements of ε were acquired more than an hour after the ship measurements and, as is shown in Fig. 8.24, approximately 300 m west of the frontal crossing on section 3, they provide an independent estimate using well-established technology of the dissipation rate distribution. Figure 8.26b displays the dissipation rate estimate assuming isotropy at viscous dissipation scales. These were computed from microscale velocity gradients measured by the shear probes following the processing approach detailed by Goodman *et al.* (2006) and averaged in 10 m across-front bins. As in Fig. 8.26a, the maximum value of the dissipation rate, $\varepsilon = 7 \times 10^{-5}$W/kg, occurs near the front and decreases with distance from the front. Though the maximum is smaller than that obtained by the towed instrument, it is still very large. The discrepancy could be due to the larger spatial averaging window employed in the shear probe estimates, or the time difference between the observations. Figure 8.26a,b shows the trend $\exp\{-x/L_G\}$ with $L_G = 15$ by solid dashed lines to aid in comparing the rate of decline in the estimates of ε with distance from the front shown in the two graphs. It is clear that the across-front decay rates in the shear probe and towed measurements are consistent.

Table 8.3. *Estimates for the width of the zone of active mixing*

Section	1	3	4	Mean
L_G (m)	11.4	16.7	22.6	16.9
L_I (m)	23.5	26.7	38.6	29.6

The O'Donnell *et al.* (2008) values of L_I for sections 1, 3, and 4 are presented in Table 8.3. The predicted length scale, L_I, values were approximately 1.5–2 times larger than L_G, the decay scale for ε shown in Fig. 8.26a. They speculated that this could be a consequence of underestimation of the vertical gradients which were only minimally resolved by their measurements so that L_E^2 and N^2 were underestimated. Despite the inadequacy of the instrumentation, it is clear that the spatial decay scale, L_G, is the same order of magnitude as L_I. More importantly, these measurements provide strong evidence that Garvine's (1974) intuitive choice of an exponential decrease in the magnitude of mixing and friction across the frontal zone is in fact a reasonable representation of nature.

The role of small-scale frontal structures on the larger-scale dynamics of river plumes remains unresolved at the moment. Interpretation of the model results in which fronts are explicitly parameterized (e.g., Garvine, 1982, 1984; O'Donnell, 1988, 1990) suggests that there is an important effect of the frontal friction and mixing on the large-scale shape of the plume layer thickness. Other, more sophisticated numerical model studies have not provided sufficient spatial resolution to allow strong fronts to develop and so diagnosis of their effects awaits further study. Recent developments in both computing facilities and numerical techniques will soon allow more relevant calculations to be performed.

8.10. Conclusion

Two fundamental difficulties have handicapped the development of an understanding of coastal plumes. The first is the non-linearity of the dynamics. That the density distribution strongly modifies the pressure field that drives the motion and the velocity field modifies the density structure through mixing and advection means only very simple problems are tractable analytically. Though much has been learned from the use of numerical models that couple mixing and transport processes in estuarine outflow plumes, it is critical to appreciate that numerical solutions must be interpreted carefully and skeptically. Not only are the advective non-linearities in salt and momentum transport equations numerically troublesome, with spurious spatial and temporal oscillations often appearing in regions of high gradient, the parameterizations of turbulent transport are not well tested. Several numerical techniques have been developed and employed to deal with these problems and produce smooth density and velocity fields. However, as Hyatt and Signell (2000) have demonstrated, the solutions can be substantially different depending upon the approximation adopted for the advection terms, and it is not yet clear how we should select among the various approximate solutions. A hierarchy of tests for numerical schemes needs to be developed so that the interpretation of numerical process studies is not confused by the approximation techniques.

The second major handicap is lack of instrumentation for simultaneously observing the small and large scales to link cause and effect. As pointed out by Yankovsky (2006), for example, the role of internal waves in transporting energy that results in mixing on the inner shelf has not been evaluated with measurements. And the interaction of mixing and advection in coastal currents at higher wind velocities may dictate the dispersion of coastal currents, yet we are unable to sustain measurements in the right places for very long in rough seas. However, the recent development of low-cost autonomous vehicles with sophisticated instrumentation may, in combination with moored arrays, lead to new insights in the near future.

References

Avicola, G. and P. Huq (2002) Scaling analysis for the interaction between a buoyant coastal current and the continental shelf: experiments and observations. *J. Phys. Oceanogr.* **32**, 3233–3248.

Barnes, C. A., A. C. Duxbury and B.-A. Morse (1972) Circulation and selected properties of the Columbia River effluent at sea. In A. T. Pruter and D. L. Alverson (eds), *The Columbia River Estuary and Adjacent Ocean Waters*. University of Washington Press, Washington, DC.

Barth, J. A., D. Bogucki, S. D. Pierce and P. M. Kosro (1998) Secondary circulation associated with a shelfbreak front. *Geophys. Res. Lett.* **25**, 2761–2764.

Benjamin, T. B. (1968) Gravity currents and related phenomena. *J. Fluid. Mech.* **31**, 209–248.

Blanton, J. O. (1981) Ocean currents along a nearshore frontal zone on the continental shelf of the southeastern United States. *J. Phys. Oceanogr.* **11**, 1627–1637.

Blanton, J. O., L.-Y. Oey, J. Amft and T. Lee (1989) Advection of momentum and buoyancy in a coastal frontal zone. *J. Phys. Oceanogr.* **19**, 98–115.

Blanton, J. O., F. Werner, C. Kim, L. Atkinson, T. Lee and D. Savidge (1994) Transport and fate of low-density water in a coastal frontal zone. *Cont. Shelf Res.* **14**, 401–427.

Blumberg, A. F. and G. L. Mellor (1987) A description of a three dimensional coastal ocean circulation model. In N. Heaps (ed.), *Three Dimensional Coastal Ocean Models*. American Geophysical Union, Washington, DC.

Bowden, K. F. (1967) Circulation and diffusion. In G. H. Lauff (ed.), *Estuaries*. AAAS Publications, New York, pp. 17–36.

Bowman, M. J. (1978) Spreading and mixing of the Hudson River effluent into the New York Bight. In J. C. J. Nihoul (ed.), *Hydrodynamics of Estuaries and Fjords*. Elsevier, Amsterdam, pp. 373–386.

Bowman, M. J. (1988) Estuarine fronts. In B. J. Kjerfve (ed.), *Hydrodynamics of Estuaries*. CRC Press, Boca Raton, FL, pp. 85–132.

Brink, K. (1991) Coastal-trapped waves and wind driven currents over the continental shelf. *Ann. Rev. Fluid Mech.* **23**, 389–412.

Britter, R. E. and J. E. Simpson (1978) Experiments on the dynamics of a gravity current head. *J. Fluid. Mech.* **88**, 223–240.

Burchard, H., O. Petersen and T. P. Rippeth (1998) Comparing the performance of the Mellor–Yamada and k–epsilon, two equation turbulence models. *J. Geophys. Res.* **103**, 10,543–10,554.

Caldwell, D. (1983) Oceanic turbulence: big bangs or continuous creation? *J. Geophys. Res.* **88**, 7543–7550.

Chao, S.-Y. (1988) Wind driven motions of estuarine plumes. *J. Phys. Oceanogr.* **18**, 1144–1166.

Chapman, D. C. (2000) Boundary layer control of buoyant coastal currents and the establishment of a shelfbreak front. *J. Phys. Oceanogr.* **30**, 2941–2955.

Chapman, D. C. and S. J. Lentz (1994) Trapping of a coastal density front by the bottom boundary layer. *J. Phys. Oceanogr.* **24**, 1464–1479.

Cho, K. W., R. O. Reid and W. D. Nowlin (1998) Objectively mapped stream function fields on the Texas–Louisiana shelf based on 32 months of moored current meter data. *J. Geophys. Res.* **103**(C5), 10,377–10,390.

Cochrane, J. D. and F. J. Kelly (1986) Low-frequency circulation on the Texas–Louisiana continental shelf. *J. Geophys. Res.* **91**, 10,645–10,659.

Dean, R. and R. Dalrymple (1991) *Water Wave Mechanics for Engineers and Scientists*. World Scientific, Singapore.

Ellison, T. H. (1957) Turbulent transport of heat and momentum from an infinite rough plane. *J. Fluid Mech.* **2**, 456–466.

Fong, D. A. (1999) Dynamics of freshwater plumes: observations and numerical modeling of the wind forced response and along-shore freshwater transport. PhD dissertaion, MIT-WHOI Joint Program in Oceanography, Woods Hole, 1998.

Fong, D. A. and W. R. Geyer (2001) The response of a river plume during an upwelling favorable wind event. *J. Geophys. Res.* **106**, 1067–1084.

Fong, D. A., W. R. Geyer and R. P. Signell (1997) The wind-forced response of a buoyant coastal current: observations of the western Gulf of Maine plume. *J. Mar. Syst.* **12**, 69–81.

Garvine, R. W. (1974) Physical features of the Connecticut river outflow during high discharge. *J. Geophys. Res.* **79**, 831–846.

Garvine, R. W. (1975) The distribution of salinity and temperature in the Connecticut River estuary. *J. Geophys. Res.* **80**, 1176–1183.

Garvine, R. W. (1977) Observations of the motion field of the Connecticut River plume. *J. Geophys. Res.* **82**, 441–454.

Garvine, R. W. (1981) Frontal jump conditions for models of shallow, buoyant surface layer dynamics. *Tellus* **33**, 301–312.

Garvine, R. W. (1982) A steady state model for buoyant surface plumes in coastal waters. *Tellus* **34**, 293–306.

Garvine, R. W. (1984) Radial spreading of buoyant surface plumes in coastal water. *J. Geophys. Res.* **89**, 1989–1996.

Garvine, R. W. (1987) Estuary plumes and fronts in shelf waters: a layer model. *J. Phys. Oceanogr.* **17**, 1877–1896.

Garvine, R. W. (1995) A dynamical system for classifying buoyant coastal discharges. *Cont. Shelf Res.* **15**, 1585–1596.

Garvine, R. W. (1999) Penetration of buoyant coastal discharge on to the continental shelf: a numerical model. *J. Phys. Oceanogr.* **29**, 1892–1909.
Garvine, R. W. (2004) The vertical structure of the subtidal dynamics of the inner shelf off New Jersey. *J. Mar. Res.* **62**, 337–371.
Garvine, R. W. and J. D. Monk (1974) Frontal structure of a river plume. *J. Geophys. Res.* **79**, 2251–2259.
Goodman, L., E. R. Levine and R. Lueck (2006) On measuring the terms of the turbulent kinetic energy budget from an AUV. *J. Atmos. Ocean. Technol.* **23**(7), 977–990.
Hickey, B. M., L. J. Pietrafesa, D. A. Jay and W. C. Boicourt (1998) The Columbia River Plume Study: subtidal variability in the velocity and salinity fields. *J. Geophys. Res.* **103**, 10,339–10,368.
Hickey, B., S. Geier, N. Kachel and A. MacFadyen (2005) A bi-directional river plume: the Columbia in summer. *Cont. Shelf Res.* **25**, 1631–1656.
Houghton, R. W. and C. Ho (2001) Diapycnal flow through the Georges Bank tidal front: a dye tracer study. *Geophys. Res. Lett.* **28**, 33–36.
Houghton, R. W. and M. Visbeck (1998) Upwelling and convergence in the Middle Atlantic Bight shelfbreak front. *Geophys. Res. Lett.* **25**, 2765–2768.
Houghton, R. W., C. E. Tilburg, R. W. Garvine and D. A. Fong (2004) Delaware River plume response to a strong upwelling-favorable wind event. *Geophys. Res. Lett.* **31**, L07302, doi:10.1029/2003GL018988.
Huq, P. (1983) Experimental investigations of three-dimensional density currents in stratified environments. M. S. Thesis, Cornell University, Ithaca, NY.
Huzzey, L. M. and J. M. Brubaker (1988) The formation of longitudinal fronts in a coastal plain estuary. *J. Geophys. Res.* **93**, 1329–1334.
Hyatt, J. and R. P. Signell (2000) Modeling surface trapped river plumes: a sensitivity study. In M. Spaulding and H. L. Butler (eds), *Estuarine and Coastal Modeling*, Proceedings of the 6th International Conference. ASCE Press, New York, pp. 452–465.
Itsweire, E. C. (1984) Measurements of vertical overturns in a stably stratified turbulent flow. *Phys. Fluids* **27**, 764–766.
Itsweire, E. C., K. Helland and C. Van Atta (1986) The evolution of grid-generated turbulence in a stably-stratified fluid. *J. Fluid Mech.* **162**, 299.
Itsweire, E. C., J. R. Koseff, D. D. Briggs and J. H. Ferziger (1993) Turbulence in stratified shear flows: implications for interpreting shear-induced mixing in the ocean. *J. Phys. Oceanogr.* **23**(7), 1508–1522.
Kourafalou, V. H., L. Oey, J. Wang and T. N. Lee (1996) The fate of river discharge on the continental shelf. 1. Modeling the plume and inner shelf coastal current. *J. Geophys. Res.* **101**, 3415–3434.
Kundu, P. K. (1990) *Fluid Mechanics*. Academic Press, New York.
Ledwell, J. R., A. J. Watson and C. S. Law (1998) Mixing of a tracer in the pycnocline. *J. Geophys. Res.* **103**(C10), 21,499–21,529.
Lentz, S. J. (1995) The Amazon river plume during AMASSEDS: subtidal current variability and the importance of wind forcing. *J. Geophys. Res.* **100**, 2355–2376.
Lentz, S. J. and K. R. Helfrich (2002) Buoyant gravity currents along a sloping bottom in a rotating fluid. *J. Fluid Mech.* **464**, 251–278.
Lentz, S. J. and J. Largier (2006) The influence of wind forcing on the Chesapeake Bay buoyant coastal current. *J. Phys. Oceanogr.* **36**, 1305–1316.
Lentz, S. J. and R. Limeburner (1995) The Amazon River plume during AMASSEDS: spatial characteristics and salinity variability. *J. Geophys. Res.* **100**, 2355–2376.
Linden, P. F. and J. E. Simpson (1986) Gravity-driven flows in a turbulent fluid. *J. Fluid Mech.* **172**, 481–497.

Loder, J. W. and D. A. Greenberg (1986) Predicted positions of tidal fronts in the Gulf of Maine. *Cont. Shelf Res.* **6**, 397–414.
Luketina, D. A. and J. Imberger (1987) Characteristics of a surface buoyant jet. *J. Geophys. Res.* **92**(C5), 5435–5447.
Marmorino, G. O. and C. L. Trump (2000) Gravity current structure of the Chesapeake Bay outflow plume. *J. Geophys. Res.* **105**, 28,847–28,861.
Masse, A. K. and C. R. Murty (1992) Analysis of the Niagara River plume dynamics. *J. Geophys. Res.* **97**, 2403–2420.
Mavor, T. P. and J. J. Bisagni (2001) Seasonal variability of sea-surface temperature fronts on Georges Bank. *Deep-Sea Res.* **II**(48), 215–243.
McDougall, T. J. (1987) Neutral surfaces. *J. Phys. Oceanogr.* **V**(17), 1950–1964.
Mellor, G. L. and T. Yamada (1974) A hierarchy of turbulence closure models for planetary boundary layers. *J. Atmos. Sci.* **31**, 1791–1806.
Mellor, G. L. and T. Yamada (1982) Development of a turbulence closure model for geophysical fluid problems. *Rev. Geophys. Space Phys.* **20**, 851–875.
Monismith, S. G. and D. A. Fong (1996) A simple model of mixing in stratified tidal flows. *J. Geophys. Res.* **101**, 28,583–28,595.
Munchow, A. and R. W. Garvine (1993a) Buoyancy and wind forcing of a coastal current. *J. Mar. Res.* **51**, 293–322.
Munchow, A. and R. W. Garvine (1993b) Dynamical properties of a buoyancy driven coastal current. *J. Geophys. Res.* **98**, 20,063–20,077.
Nash, J. D. and J. N. Moum (2005) River plumes as a source of large-amplitude internal waves in the coastal ocean. *Nature* **437**, 400–403. doi:10.1038/nature03936.
Neill, S. P., G. J. M. Copeland, G. Ferrier and A. M. Folkard (2004) Observations and numerical modelling of a non-buoyant front in the Tay Estuary, Scotland. *Est. Coast. Shelf Sci.* **59**, 173–184.
Nunes Vaz, R. A., G. W. Lennon and J. R. de Silva Samarasinghe (1989) The negative role of turbulence in estuarine mass transport. *Est. Coast. Shelf Sci.* **28**, 361–377.
O'Donnell, J. (1988) A numerical technique to incorporate frontal boundaries in layer models of ocean dynamics. *J. Phys. Oceanogr.* **18**, 1584–1600.
O'Donnell, J. (1990) The formation and fate of a river plume; a numerical model. *J. Phys. Oceanogr.* **20**, 551–569.
O'Donnell, J. (1993) Surface fronts in estuaries: a review. *Estuaries* **16**(1), 12–39.
O'Donnell, J. (1997) Observations of near surface currents and hydrography in the Connecticut River plume with the SCUD array. *J. Geophys. Res.* **102**, 25,021–25,033.
O'Donnell, J., G. O. Marmorino and C. L. Trump (1998) Convergence and downwelling at a river plume front. *J. Phys. Oceanogr.* **28**, 1481–1495.
O'Donnell, J., S. G. Ackleson and E. R. Levine (2008) On the spatial scales of a river Plume. *J. Geophys. Res. Oceans* **113**, C04017, doi:10.1029/2007JC004440.
Orton, P. M. and D. A. Jay (2005) Observations at the tidal plume front of a high volume river outflow. *Geophys. Res. Lett.* **32**, L11605, doi:10.1029/2005GL02237.
Ozmidov, R. V. (1965) On the turbulent exchange in a stably stratified ocean. *Izv. Acad. Sci. USSR Atmos. Ocean. Phys.* (English trans.) **1**, 853.
Pedlosky, J. (1987) *Geophysical Fluid Dynamics*. Springer-Verlag, New York.
Peters, H. (1997) Observations of stratified turbulent mixing in an estuary: neap to spring variations during high river flow. *Est. Coast. Shelf Sci.* **45**, 69–88.
Peters, H., M. C. Gregg and J. M. Toole (1988) On the parameterization of equatorial turbulence. *J. Geophys. Res.* **93**, 1199–1218.

Pullen, J. D. and J. S. Allen (2000) Modeling studies of the coastal circulation off Northern California: shelf respose to a major Eel River flood event. *Cont. Shelf Res.* **20**, 2213–2238.

Rennie, S. E., J. L. Largier and S. J. Lentz (1999) Observations of a pulsed buoyancy current downstream of Chesapeake Bay. *J. Geophys. Res.* **104**, 18,227–18,240.

Rohr, J. J., E. C. Itsweire, K. N. Helland and C. W. Van Atta (1988) Growth and decay of turbulence in a stably stratified shear flow. *J. Fluid Mech.* **195**, 77–111.

Sahl, L. E., D. E. Weisenburg and W. J. Merrel (1997) Interaction of mesoscale features with Texas shelf and slope waters. *Cont. Shelf Res.* **17**, 117–136.

Simpson, J. E. (1987) *Gravity Currents in the Environment and the Laboratory.* Ellis Horwood, Chichester.

Simpson, J. H. and J. Hunter (1974) Fronts in the Irish Sea. *Nature* **250**, 404–406.

Simpson, J. H. and J. Sharples (1991) Dynamically-active models in the prediction of estuarine stratification. In D. Prandle (ed.), *Dynamics and Exchanges in Estuaries and the Coastal Zone; Coastal and Estuarine Studies*, Vol. 40. Springer-Verlag, New York, pp. 101–113.

Simpson, J. H. and W. R. Turrell (1986) Convergent fronts in the circulation of tidal estuaries. In D. A. Wolfe (ed.), *Estuarine Variability.* Academic Press, New York, pp. 139–152.

Simpson, J. H., J. Brown, J. Matthews and G. Allen (1990) Tidal straining, density currents, and stirring in the control of stratification. *Estuaries* **13**, 125–132.

Simpson, J. H., W. R. Crawford, T. P. Rippeth, A. R. Campell and J. V. S. Cheok (1996) The vertical structure of turbulent dissipation in shelf seas. *J. Phys. Oceanogr.* **26b**, 1579–1590.

Smyth, W. D. and J. N. Moum (2000) Length scales of turbulence in stably stratified mixing layers. *Phys. Fluids* **12**, 1327–1342.

Souza, A. J. and J. H. Simpson (1997) Controls on stratification in the Rhine ROFI system. *J. Mar. Syst.* **12**, 311–323.

Sundermeyer, M. A. and J. R. Ledwell (2001) Lateral dispersion over the continental shelf: analysis of dye release experiments. *J. Geophys. Res.* **106**, 9603–9621.

Thorpe, S. A. (1977) Turbulence and mixing in a Scottish loch. *Philos. Trans. R. Soc. London, Series A.* **286**, 125–181.

Thorpe, S. A. (2005) *The Turbulent Ocean.* Cambridge University. Press, Cambridge.

Tilburg, C. E., R. W. Houghton and R. W. Garvine (2007) Mixing of a dye tracer in the Delaware plume: comparison of observations and simulations. *J. Geophys. Res.* **112**, C12004, doi:10.1029/2006JC003928.

Trump, C. L., G. O. Marmorino and J. O'Donnell (1995) Broadband ADCP measurements of the Connecticut River plume front. In Proceedings of the IEEE Fifth Working Conference on Current Measurement, St. Petersburg, FL, Institute of Electrical and Electronics Engineering, pp. 73–78.

Valle-Levinson, A., J. M. Klinck and G. H. Wheless (1996) Inflows/outflows at the transition between a coastal plain estuary and the coastal ocean. *Cont. Shelf Res.* **16**, 1819–1847.

Valle-Levinson, A., K.-C. Wong and K. T. Bosley (2001) Observations of the wind-induced exchange at the entrance to Chesapeake Bay. *J. Mar. Res.* **59**, 391–416.

Valle-Levinson, A., W. C. Boicourt and M. Roman (2003) On the linkages among density, flow, and bathymetry gradients at the entrance to the Chesapeake Bay. *Estuaries* **26**(6), 1437–1449.

Vastano, A. D., C. N. Barron, Jr. and E. W. Sharr, Jr. (1995) Satellite observations of the Texas Current. *Cont. Shelf Res.* **15**, 729–754.

von Karman, T. (1940) The engineer grapples with nonlinear problems. *Bull. Amer. Math. Soc.* **46**, 615.

Whitney, M. M. and R. W. Garvine (2006) Simulating a coastal buoyant outflow: comparison to observation. *J. Phys. Oceanogr.* **36**, 3–21.

Wright, L. D. and J. M. Coleman (1971) Effluent expansion and interfacial mixing in the presence of a salt wedge, Mississippi River delta. *J. Geophys. Res.* **76**, 8649–8661.

Yankovsky, A. E. (2006) On the validity of thermal wind balance in alongshelf currents off the New Jersey coast. *Cont. Shelf Res.* **26**, 1171–1183.

Yankovsky, A. E. and D. C. Chapman (1997) A simple theory for the fate of buoyant coastal discharges. *J. Phys. Oceanogr.* **27**, 1386–1401.

Yankovsky, A. E. and R. W. Garvine (1998) Subinertial dynamics on the inner New Jersey shelf during the upwelling season. *J. Phys. Oceanogr.* **28**, 2444–2458.

Yankovsky, A. E., R. W. Garvine and A. Munchow (2000) Mesoscale currents on the inner New Jersey shelf driven by the interaction of buoyancy and wind forcing. *J. Phys. Oceanogr.* **30**, 2214–2230.

9

Low-inflow estuaries: hypersaline, inverse, and thermal scenarios

JOHN LARGIER

9.1. Introduction

Worldwide there are many estuaries that experience low inflow at times, resulting in the absence of classical low-salinity, density-driven estuarine circulation. However, these systems function like classical hypopycnal basins during periods following rain and enhanced river flow (even if brief and infrequent). Further, the morphology and ecology of these low-inflow estuaries (LIEs) are clearly similar to those estuaries with more persistent freshwater inflow. Through an increase in the diversity of estuaries studied in recent decades, extending beyond cool, wet regions (e.g., western Europe, northeastern United States), it has become evident that estuaries with low-inflow periods are as common as estuaries with persistent inflow of significant volumes of freshwater (often called "classical" estuaries, as they are the focus of the older/classical estuarine literature). Wherever evaporation exceeds precipitation for a time longer than the basin residence time (e.g., during the dry season), it is possible to find LIEs associated with watersheds of smaller size and storage volume (where runoff approaches zero during the dry season). Recognizing that evaporation is on the order of 1–10 mm per day, a brief review of seasonal precipitation maps (e.g., http://geography.uoregon.edu/envchange/clim_ animations/gifs/pminuse_web.gif) reveals that LIEs may be found along many coastlines of the world. Further, in regions where coastal lands slope steeply to the sea (e.g., California, Chile, South Africa), watersheds are smaller and faster, hydrographs are short-tailed and LIEs are a common occurrence (e.g., Largier *et al.*, 1997).

The approach in this chapter is at the scale of the basin. While there are implications of reduced or reversed salinity and density gradients on small-scale processes like mixing and features like fronts, these are discussed in previous chapters. Secondly, the focus here is on smaller systems and on basins with shorter time scales (i.e., seasonal and shorter), typically ignoring the effects of rotation (i.e., Coriolis term). From a review of observations to date, it appears that estuaries at these smaller scales are more likely to exhibit low-inflow characteristics. This can be expected as smaller watersheds exhibit shorter-tailed hydrographs, and thus periods with little inflow to no inflow to the estuary. Nevertheless, there are some very large low-inflow "estuaries" (marginal seas) that are well studied and exhibit similar features to these smaller systems, but they are not the focus of this review, e.g., Red Sea (Phillips, 1966; Tragou and Garrett, 1997), Mediterranean Sea (Lacombe and Richez, 1982), Gulf of California (Bray and Robles, 1991; Lavin *et al.*, 1998). There are also several examples of larger bays or gulfs that develop hypersaline or hyperpycnal structure and they are briefly discussed here, e.g., Spencer Gulf (Nunes Vaz *et al.*, 1990). Finally, there are many examples of small estuary basins and salt marshes that have limited exchange with the ocean, and which can develop very high levels of hypersalinity during dry periods (in places high enough to precipitate salt, e.g., Milnerton Lagoon; Day, 1981). These are also not the focus of this discussion, and neither are systems where freshwater loss is due to ice formation or desalination facilities – although there are some common features.

9.2. Terminology

In the limited literature on LIEs there have been differences in the use of the words to describe these basins. While there are relationships between freshwater inflow, salinity structure, density structure, and vertical circulation in an estuary, one needs to recognize that these are different phenomena and unique words are required for specific concepts. Following Largier *et al.* (1996) and others, and with reference to Fig. 9.1, the following terms are used in this chapter:

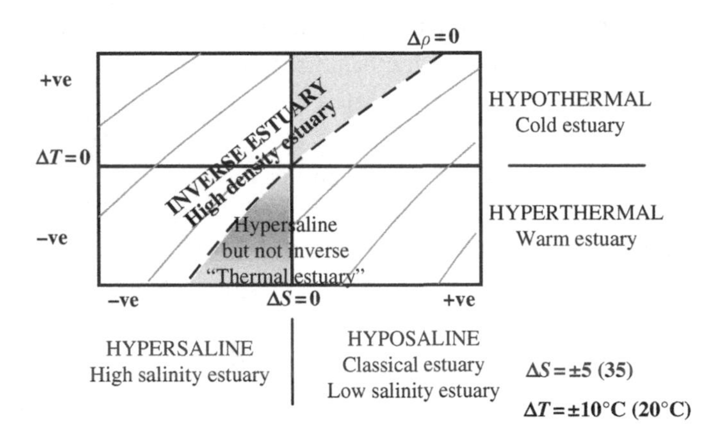

Figure 9.1. An illustration of the relation between changes in temperature (y-axis), salinity (x-axis) and density (contours) for estuary minus ocean salinity varying from –5 to +5 and estuary minus ocean temperature varying from –10 to +10°C. The schematic is drawn for center values 35 and 20°C. Terminology is given for salinity gradients, thermal gradients and density gradients – demonstrating that "hypersaline" is not synonymous wih "inverse".

Negative estuary – an estuary with a negative hydrological balance, i.e., a negative net inflow of freshwater, or, in other words, a net loss of zero-salinity waters due to evaporation (or ice formation or desalination) exceeding the sum of precipitation, river inflow, and groundwater inflow.

Hypersaline estuary – an estuary in which the salinity of water is greater than the salinity of adjacent (ocean) waters – a classical (sic.) estuary may be referred to as hyposaline; for hypersalinity to be statistically significant, we consider times when estuary salinity $S_e - S_o > \sigma$, where σ is the standard deviation of ocean salinity S_o (following Largier *et al.*, 1996, 1997).

Inverse estuary – an estuary in which the density of water is greater than the density of adjacent (ocean) waters – one could also call this a hyperpycnal estuary, and a classical (sic.) estuary could be called a hypopycnal estuary.

Inverse estuarine circulation – the density-driven circulation in an inverse estuary comprised of outflow of more-dense waters at depth and inflow of less-dense ocean waters near surface; inverse circulation may also be observed where the receiving water density suddenly decreases (e.g. de Castro *et al.*, 2004), rather than an increase in estuary water density.

In addition, recognizing that temperature can be important in determining density in estuaries with weak salinity structure, it is worth defining two further words, although these words are only likely to be used when density is dominated (or strongly influenced) by temperature (i.e., when salinity structure is weak):

Hypothermal estuary – an estuary in which the temperature of water is less than the temperature of adjacent (ocean) waters.

Hyperthermal estuary – an estuary in which the temperature of water is greater than the temperature of adjacent (ocean) waters; in a situation where this thermal structure drives a positive estuarine circulation, the term "thermal estuary" has been used (Largier *et al.*, 1996).

The use of the above descriptors recognizes a more cosmopolitan idea of estuaries, with classical (or hyposaline or hypopycnal) estuaries being one type of estuary. Further, one can expect a sequence of estuary types as a given basin develops negative, hypersaline, hyperthermal, and/or inverse structures through the dry season.

9.3. Water, salt and heat budgets

Whether LIEs develop hypersaline and inverse conditions depends on the interaction of fluxes of water, salt, and heat to/from the estuary basin. A negative hydrological balance is a necessary but not sufficient condition for hypersalinity. It is also necessary to clear the estuary basin of excess freshwater from earlier positive hydrological periods, and then to retain waters in the basin long enough for the negative balance to generate a significant salinity excess. In turn, the presence of hypersalinity alone does not imply inverse structures as density also depends on water temperature. One may find inverse systems that are not hypersaline, because

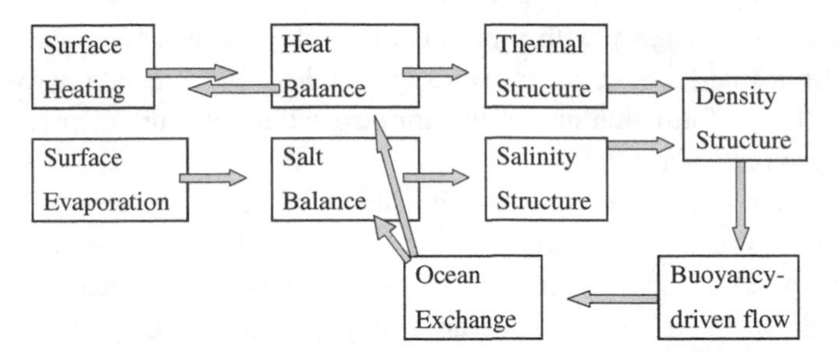

Figure 9.2. The hydrodynamic behavior of low-inflow estuaries is ultimately forced by surface buoyancy fluxes (heating and evaporation) interacting with mechanical stirring due to tides and winds. The complexity of the hydrodynamics is due to the feedback between buoyancy-driven exchange (due to density structure) and the heat and salt balances that control density structure but which are strongly influenced by estuary–ocean exchange.

they are hypothermal (Jervis Bay – Holloway *et al.*, 1992), and one may find hypersaline systems that are not inverse, because they are hyperthermal (Tomales Bay – Largier *et al.*, 1997).

The interest in the physics of LIEs lies in the dynamics of non-gravitational circulation in basins with weak longitudinal density gradients, in the interplay of seasonal cycles in thermal and haline contributions to density, and in the development of inverse (vertically sheared) circulation due to inverse longitudinal density structures in the presence of vertical mixing (Fig. 9.2). To start, one can learn from considering the development of thermal and haline structure through a mass balance approach (i.e., water, salt, and heat budgets). Clearly, there are multiple drivers with river flow, air–water fluxes, and ocean tides delivering or removing water and buoyancy (through heat and salt fluxes).

9.3.1. Water budget

The flow of water to/from the estuary basin must balance. Although changes in volume may balance inflow–outflow differences over short time scales (e.g., tidal rise and fall of water level), over the longer time scales that characterize the low-inflow estuary problem, there is a negligible change in volume. During negative hydrological periods, river plus groundwater inflow is less than the evaporative loss of water, so there must be an inflow of saline ocean water $R_o = ([E - P]\ SA - R_r - R_g)$, where E is evaporation rate (m/s), P is precipitation rate (m/s), R_r is river inflow rate (m^3/s), R_g is groundwater inflow rate (m^3/s), and SA is the surface area (m^2). Typically, $P = 0$ during low-inflow periods and it can be dropped and, in the case of extensive emergent vegetation, E should include evapotranspiration. If this evaporation-driven flow dominates basin flushing, then the basin time scale can be approximated from the water budget by $t_{water} \sim V / E{\cdot}SA \sim H / E$, where V is basin volume and H is basin depth. This is unusual as E is typically less than 0.01 m/day and H is typically more than 1 m, so that flushing times would be many months or more.

9.3.2. Salt budget

The salt budget of the estuary includes advective and diffusive transport of salt across the mouth of the estuary (Fig. 9.3). Diffusive fluxes are primarily due to tidal mixing. When the estuary basin is sufficiently hypopycnal, a density-driven "estuarine circulation" develops that exports low-density (low-salinity) waters – and this loss is matched by the inflow of river water during periods of positive hydrological balance. However, as the ocean–estuary density difference weakens during negative hydrological periods, the density-driven circulation tends to zero and the remaining freshwater content must be exported through a combination of evaporation and non-gravitational exchange with the ocean. In the smaller basins of interest here, this estuary–ocean exchange is dominated by tidal mixing and the time for the residual freshwater to be expelled is approximated by $t_{flush} \sim ½L^2{\cdot}K^{-1}$, where K is an exchange coefficient (with the dimensions of eddy diffusivity). In the case of weak tidal currents and/or long basins, there may not be enough time to mix away the excess freshwater from the wet season and seasonal hypersalinity will not develop. However, if the dry period is prolonged, hypersalinity may still develop in these estuaries (e.g., Cassamance River, Senegal – Pages and Debenay, 1987; Savenije and Pages, 1992; Lake St Lucia, South Africa – Begg, 1978). In many estuaries, however, this time scale t_{flush} is of the order of weeks, shorter than the dry

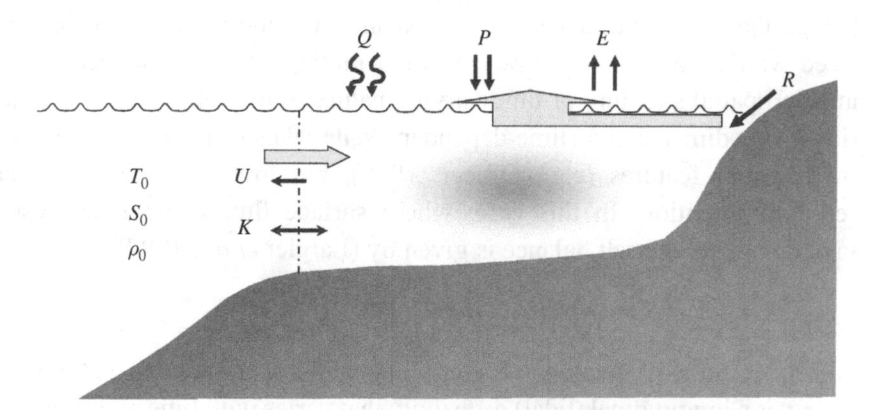

Figure 9.3. The salt and heat balances in an estuarine basin are a function of river inflow R, ocean–estuary mean flow U, ocean–estuary tidal exchange K, surface heating Q, surface evaporation E, surface precipitation P, and ocean water temperature T_0 and salinity S_0. This schematic is further described in Largier *et al.* (1996).

season. After the excess freshwater has been mixed away, the salinity continues to rise owing to the continued loss of freshwater through net evaporation $E_{net} = E - P - (R_r + R_g)/SA$. This salinity increase will continue until the hydrological balance returns to positive, or until a hypersaline steady salt balance is achieved with the advective import of salt R_oS_o balanced by a diffusive export of salt $K{\cdot}\Delta S{\cdot}H{\cdot}B/½L$. This diffusive export of salt is due to tides mixing hypersaline water out of the estuary (Largier *et al.*, 1997), where B is basin width and $\Delta S = S_e - S_o$ represents the difference between average estuary salinity S_e and ocean salinity S_o. Under low-inflow conditions, $E_{net} \sim E$ where $E_{net} = R_o/SA$ and the estuary–ocean salinity gradient can be expressed as follows (Hearn, 1998):

$$\partial_t \Delta S = S_0 E/H - 2K\Delta S/L^2. \tag{9.1}$$

The steady estuary salinity is then $S_e = (1 + EL^2/2HK)\, S_o = (1 + R_oL^2/VK)\, S_o$, which is close to S_o for rapid estuary–ocean exchange (large K/L^2) and higher for weaker exchange (small K/L^2) or large R_o/V due to large evaporative loss from shallow waters, i.e., large E_{net}/H. The time to develop hypersalinity in the basin is approximated by $t \sim (S_e - S_o){\cdot}V \,/\, E_{net}{\cdot}SA{\cdot}S_o$ and this should also be scaled by the basin residence time $L^2/2K$. Seasonal steady state has been observed in several Mediterranean-climate estuaries with moderate estuary–ocean exchange rates (e.g., Tomales Bay and San Diego Bay, USA – Largier *et al.*, 1997). However, seasonal hypersalinity will not be found in small-K basins as it would take too long to export excess freshwater remaining from the wet season (e.g. St. Lucia Estuary; Begg, 1978) and hypersalinity found in large-K basins will be weak (i.e., estuary salinity close to ocean salinity – for example, Saldanha Bay; Monteiro and Largier, 1999).

Although the above introduces the salt balance and the interplay of timescales associated with evaporation, inflow, and tidal mixing, it does not recognize the fundamental spatial structure or time-varying nature of the salt balance. For many estuaries a one-dimensional, time-dependent salt balance adequately represents first-order spatial features (e.g., Officer, 1976), but low-inflow estuaries have received little attention. In this case, where surface fluxes and ocean–estuary exchange dominate, the salt balance is given by (Largier *et al.*, 1997):

$$\partial_t S = \partial_x [K_x \partial_x S - (R_r/BH - Ex/H)S], \tag{9.2}$$

where S is tidal and cross-section average salinity and it is assumed that P and R_g are zero. $K_x(x,t)$ is a longitudinal (tidal) diffusivity that varies with time and location, as does $S(x,t)$, while width $B(x)$ and depth $H(x)$ only vary with distance x from the head (landward end of the basin). In hypersaline periods, $\partial_x S < 0$ in much of the basin, but one finds $\partial_x S > 0$ for small $x < R_r/E{\cdot}B$ as river inflow exceeds evaporative loss near to the river. While this expression can be used numerically to track the time-dependent salt balance, a simple analytical solution can be found for the dry season steady state if one assumes $R_r = 0$ and that K_x scales according to the Prandtl mixing length theory, giving $K_x = k{\cdot}x^2$, where k is a constant (Largier *et al.*, 1997). For the estuary sufficiently seaward of $x = R_r/E{\cdot}B$ (where river influence is negligible), this gives an exponential increase of salinity S with distance from the ocean:

$$S/S_o = (x/L)^{-E/kH}. \tag{9.3}$$

Further, one can obtain values for $K_x(x)$ by fitting this curve to observations of $S(x)$ during the steady-state period. One can also estimate K_x during steady state by requiring a local balance in advective and diffusive salt fluxes for any value of x: $u{\cdot}S = K{\cdot}\partial_x S$. With $u = E{\cdot}x / H$, this gives $K_x(x) = E{\cdot}x{\cdot}S / H{\cdot}\partial_x S$.

Finally, considering a Lagrangian salt balance, one can estimate the age of a water parcel (sensu Monsen *et al.*, 2002) by $t_{age} \sim (S - S_o){\cdot}H/E.S_{av}$, where $S_{av} \sim (S+S_o)/2$ is the Lagrangian average salinity. Thus, salinity indicates age and the increase of salinity with distance from the ocean represents an increase in the age of water.

9.3.3. Heat budget

At times, water temperature may be more important to water density than salinity, thus one must also track the flux of heat to/from the estuary basin during low-inflow periods. Heat budget terms include air–water fluxes, river and groundwater inflows, and ocean–estuary exchanges. Assuming that river and groundwater inflows are weak and water temperatures are similar to the surface water in the estuary, attention turns to the temperature of ocean surface water – whether it is warmer or cooler than saturation temperature for the given location and season, and, thus, whether these waters will be warming or cooling. Ocean waters retained in the shallower estuary basins warm up or cool down faster than the coastal waters that are continuously being advected away. The estuary–ocean temperature difference ΔT changes in response to surface heating and estuary–ocean mixing, where Q is the surface heat flux (W/m^2) and c_p is the heat capacity of water, as given by Hearn (1998):

$$\partial_t \Delta T = Q/\rho H c_p - 2\Delta T \cdot K/L^2. \tag{9.4}$$

Along mid-latitude west coasts, where coastal upwelling maintains cool ocean temperatures, estuarine waters are warmer than ocean waters (hyperthermal). Considering a warm summer steady state, one obtains estuary temperature $T_e = T_o + Q{\cdot}L^2/2K\rho H c_p$, where T_o is ocean temperature. In contrast, along many mid-latitude east coasts, waters are cooling as they are delivered by warm western boundary currents (e.g., Gulf Stream). In these cases, or during cooling in the autumn season, the shallow waters in the estuary cool down faster and the estuary basin is typically hypothermal (e.g., Jervis Bay – Holloway *et al.*, 1992). Along tropical coasts, ocean waters are typically in thermal equilibrium and estuary–ocean thermal gradients are weak. As for the salt budget, the longitudinal structure of the heat budget should also be resolved (but this is left as an exercise for the reader!).

9.3.4. Water density

The density of estuarine waters depends on the salt and heat content of the waters (Fig. 9.1), which are determined by the above salt and heat budgets. More explicitly, following Hearn (1998), the estuary–ocean density difference $\Delta\rho$ changes as follows:

$$\partial_t \Delta\rho = aQ/\rho H c_p - \beta S_o E/H - 2K\Delta\rho/L^2, \tag{9.5}$$

where α is the thermal expansion coefficient and β is the haline contraction coefficient. Clearly, there is more than one possible steady state, with either heating or evaporation balancing the estuary–ocean exchange, or the heating and evaporation terms balancing each other. This leads to the Stommel bistability, which is discussed later (Section 9.4).

Here, let us consider the important difference in how the surface fluxes of heat and mass vary with the age of the water in the estuary – resulting in a spatial separation of thermal and hypersaline effects in many estuaries. For example, in a warming estuary (ocean surface temperature below saturation temperature), the heat flux is greatest for the zero-age cool waters that have just entered the estuary, but this flux decreases as the surface water temperature increases with age (time in the estuary). Ultimately, the net surface heat flux will go to zero when the water is old enough to be in thermal equilibrium. The same is true for heat flux in a cooling estuary. In contrast, the evaporative loss of freshwater from the estuary has negligible dependence on surface water salinity and only weak dependence on surface water temperature. Thus, this flux remains constant with age (assuming constant hydrological conditions through the dry season). Largier *et al.* (1996) discuss this phenomenon, showing that for summer in typical mid-latitude west-coast estuaries one can expect water temperature to increase during the first 20–30 days in the estuary basin, but with slower increase as it approaches an asymptote that is the local saturation temperature (Fig. 9.4). Meanwhile, salinity increases constantly with age. For parameter values typical of this region, one sees an initial decrease in water density (order of 1–2 kg/m^3) due to warming that dominates increasing salinity for the first 20–30 days, but as waters age beyond about 40 days, the persistent evaporative loss of freshwater dominates the buoyancy flux and one sees a similar increase in density. Noting the above discussion that the age of water increases with distance from the mouth, one can expect a region of decreasing density in the outer estuary and increasing density in the inner estuary with a mid-estuary density

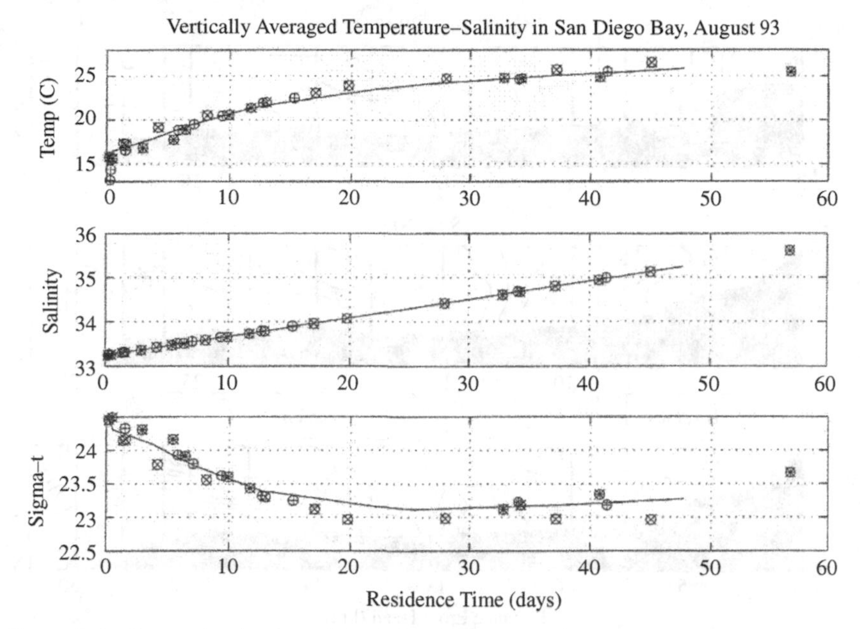

Figure 9.4. Theoretical curves (lines) for the increase in water temperature, salinity and density with the age of water in a west-coast estuary, as calculated by Largier *et al.* (1996). Symbols represent observations on three different days for San Diego Bay (California, USA) in August 1993 (cf. Fig. 9.5). Salinity increases linearly with age, while temperature asymptotes to an equilibrium value.

minimum. This has been observed in several estuaries (e.g., San Diego Bay, Figs 9.4 and 9.5; also Tomales Bay and others – Largier *et al.*, 1997). While the whole basin is hypersaline $\partial_x S < 0$, the outer estuary is also hyperthermal $\partial_x T < 0$ and this explains the observed hypopycnal character $\partial_x \rho > 0$ of the outer estuary. Thus, the outer basin may behave as a thermal estuary, e.g., San Diego Bay (Chadwick *et al.*, 1996; Largier *et al.*, 1996), Hilary Harbor (Schwartz and Imberger, 1988), Knysna Lagoon (Largier *et al.*, 2000), and Saldanha Bay (Monteiro and Largier, 1999). Further landward, density gradients go to zero, but if the basin is long enough to retain water for well over a month, one will see an inverse (hyperpycnal) estuary even further landward (Figs 9.4 and 9.5). Thus, smaller west-coast estuaries (shorter time scales L^2/K) exhibit thermal estuary character while larger west-coast estuaries (longer time scales L^2/K) exhibit hypersaline estuary character.

9.4. Circulation response

Whether hyperpycnal or hypopycnal, low-inflow estuaries tend to have weak longitudinal density gradients and thus weak vertically sheared circulation. Longitudinal mixing is typically dominated by tidal motions, as represented in the

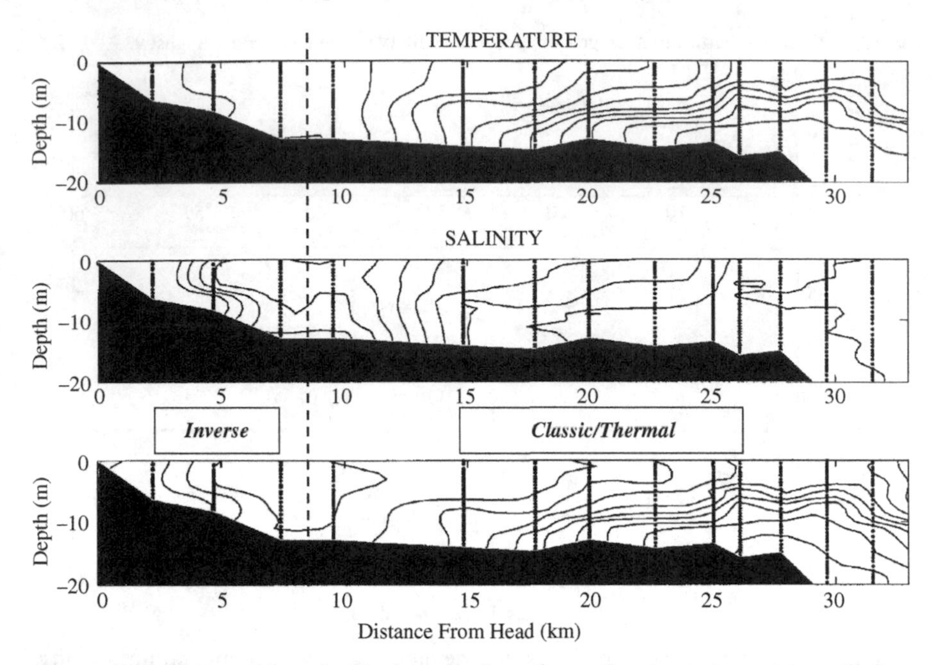

Figure 9.5. Observed longitudinal–vertical sections of water temperature, salinity and density (bottom panel) in San Diego Bay (California, USA) on 10 August 1993 (Largier *et al.*, 1996, 1997). Data obtained at dots (see vertical dotted lines). The mouth of the bay is on the right side of the section, at a distance of 26 km from the head. For temperature the contour interval is 1°C with a minimum of <14 °C at depth offshore of the mouth and a maximum of >25°C at the head of the estuary. For salinity the contour interval is 0.2 with a minimum of <33.2 offshore of the mouth and a maximum of >35.2 near the head. For density, the sigma-t contour interval is 0.2 kg/m^3 with a minimum of <23.2 about 8 km from the head, a maximum of >23.6 at the head and a maximum of >25.0 at depth offshore. An inverse density gradient is found in the innermost 8 km of the bay, with isolines at 5 km indicating the seaward flow of dense, warm, salty waters at depth (i.e., inverse circulation).

one-dimensional salt-flux model outlined above [equation (9.2)]. The importance of tidal exchange is typical in well-mixed estuaries and it is addressed in standard estuary texts (e.g., Officer, 1976; Dyer, 1997; Chapter 3 of this text).

Vertically sheared exchange flow has been observed, both in the thermal estuary and inverse estuary sections of low-inflow west-coast estuaries (e.g., San Diego Bay, Fig. 9.5). Inverse circulation has also been observed in other small to moderate estuary basins: Laguna Ojo de Liebre (Postma, 1965), Laguna de Guyamas (Valle-Levinson *et al.*, 2001), Mission Bay (Largier *et al.*, 1996), Shark Bay (Burling *et al.*, 1999), Elkhorn Slough (Nidzieko, 2009) and more. But, dynamical studies to date have only addressed larger hypersaline and inverse basins, such as Spencer Gulf (Nunes Vaz *et al.*, 1990) and the large marginal seas (e.g., Red Sea, Mediterranean

Sea, Gulf of California). There has been little study of inverse circulation dynamics in smaller basins, and of the role of inverse circulation in exporting excess salt from inverse/hypersaline estuaries. Ultimately, inverse circulation acts as a self-limiting feedback on hypersalinity in that the stronger the density excess in the estuary, the stronger the inverse circulation response and thus the more rapid the flushing of high-salinity water from the estuary (precluding further accumulation of salt and further increases in density) – see Fig. 9.2.

Following Officer (1976), a steady vertically sheared flow structure $u(z)$ can develop in response to a longitudinal density gradient $\partial_x\rho$ (with A_z for vertical eddy viscosity and z for vertical coordinate):

$$u(z) = \frac{g\partial_x\rho H^3}{48\rho A_z}\left[8(z/H)^3 - 9(z/H)^2 + 1\right]. \tag{9.6}$$

Maximum shear is given by $9g\partial_x\rho H^3/256\rho A_z$ (Chadwick *et al.*, 1996) and one can see that the circulation response will be larger for a stronger longitudinal density gradient, weaker vertical mixing, and deeper water. One thus expects to see inverse circulation in deep channels toward the head of the estuary, where the hypersalinity gradient is strongest and there is no density-compensating hyperthermal gradient. Further, tidally driven vertical mixing is weaker here, allowing stratified exchange flow to develop. Also, the longitudinal tidal diffusivity is weak and even weak buoyancy-driven inverse circulation may be important in the deeper channels of the inner basin (Largier *et al.*, 1996). This is where inverse circulation has been observed in San Diego and Mission Bays. Inverse circulation is expected to be strongest in basins with extensive shallows fringing a deeper channel – thus providing small average depth to enhance the volumetric importance of evaporative losses, but large channel depth to allow for a stratified exchange flow to develop (e.g., Laguna Ojo de Liebre; Postma, 1965). Largier *et al.* (1996) also cite unpublished reports of inverse circulation in tidal creeks through mangrove swamps in Mozambique, due to large evapotranspiration and muted tides in dense mangrove estuaries. The above discussion neglects the potential effect of wind on vertical mixing and longitudinal exchange. While wind does not appear to be a primary factor in Tomales Bay (Hearn and Largier, 1997), in more exposed and shallower basins, such as Laguna San Ignacio, wind is important and appears to preclude density-driven inverse circulation (Gutierrez de Velasco and Winant, 2004).

Considering tidal variability, Linden and Simpson (1988) conducted a laboratory experiment to represent fluctuations in vertical mixing, representing tidal stirring, and the disruption of this stratified exchange flow. They note that the time scale of the gravitational response is $(g\partial_x\rho/\rho)^{-1/2}$ and show that the buoyancy-driven longitudinal diffusivity is given by $K_b \sim g\partial_x\rho{\cdot}H^2\tau/16\rho$, where τ is the period of the turbulence modulation. Where $\partial_x\rho$ is large, gravitational response is quick and significant longitudinal exchange may occur in tidal pulses. In inverse estuaries, this gravitational adjustment will be assisted by strain-induced stratification during flood tides, suggesting that the strongest stratification and most effective stratified exchange flow will be seen at high tide (which is also when the water column is deepest). However, the longitudinal density gradient is typically weak in inverse estuaries and maxima (of order 10^{-4} kg/m^4) are found in the innermost basin where tidal currents and thus strain are weak. In addition to recent studies of the tidal modulation of stratification in inverse estuaries (Nidzieko, 2009), scaling estimates suggest that strain-induced periodic stratification will be weak.

In general, longitudinal exchange due to buoyancy effects is expected to be weak in most inverse estuaries – either due to weak hypersalinity gradients (in deeper basins) or to strong vertical mixing (in shallower basins). In the inner parts of San Diego Bay, where maximum inverse gradients of $\sim 0.5\times10^{-4}$ kg/m^4 are observed, empirical estimates of K_x are less than 10 m^2/s (and buoyancy-driven exchange must be less than this). This corresponds to a flushing time for this innermost inverse section of the bay of $L^2/K_x \sim 10^7$s $\sim 10^2$ days. This time scale is long enough for significant hypersalinity to develop and does not appear to limit the development of these inverse density gradients. Thus, in this and probably many other seasonal LIEs, the inverse structure is of secondary importance as a forcing term and really just a symptom of long-residence basins in dry climates. The evolution and demise of hypersaline and inverse conditions in these systems are on seasonal time scales and reflect the seasonal variation in hydrological forcing. This is discussed further in Section 9.6.

In the autumn there is an opportunity for an inverse circulation event, associated with rapid cooling of hypersaline waters. In southern California, heat fluxes reverse in autumn, months earlier than winter rains and inflow. Further, shallow-water temperatures can drop rapidly during an autumn cold-weather event, dropping several degrees ($\Delta_t\rho \sim 1$ kg/m^3) in a few days ($\ll$ flushing time), and increasing the density gradient much faster than other processes (evaporation or tidal diffusion). Observations from inner Mission Bay show such rapid cooling and development of inverse conditions in autumn 1992 (Fig. 9.6), as well as in October 2001 and 2002 (Fig. 9.7). In a basin with a deeper channel, an inverse-circulation flushing event may be expected following such a cooling event, but this was not evident in Mission Bay, where inner bay depths are less than 3 m. In 1992, the inverse density gradient of order 3×10^{-4} kg/m^4 persisted for a few weeks and eventually was terminated by the first winter rain in early December.

In special cases where the longitudinal density gradient dominates longitudinal exchange in LIEs (negligible tidal mixing), the longitudinal density gradient is controlled by the three terms in equation (9.5): heating, evaporation, and gravity-driven exchange flow. Stommel (1961) recognized the existence of alternative

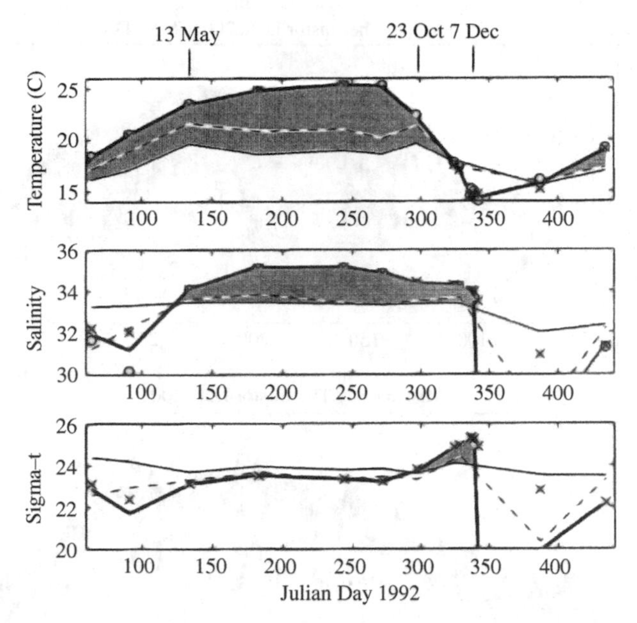

Figure 9.6. Observations of seasonal changes in water temperature, salinity and density (plotted as sigma-t for three stations in Mission Bay (California, USA) in 1992, as in Largier *et al.* (1996, 1997). The dark line is for the station near the head, while the fine line is for the station at the mouth, and the dashed line is for a station in Fiesta Bay (about halfway to the head). For much of the year the bay backwaters are warmer (shaded), while hypersalinity only develops around day 130 (early May). Cooling during October and November leads to a significant inverse gradient in density structure in November – prior to an inflow of low-salinity water around 7 December.

balances that represent alternative states for the longitudinal density structure. His box model has been adapted for analysis of shallow estuaries by Hearn (1998), who identified four states – the two stable states identified by Stommel (the familiar classical and inverse estuary states), as well as a "quasi-neutral" solution and an "intermittent" state. Hearn introduces the evaporative number Γ to denote the relative importance of evaporation, providing a comparison of the first two terms in equation (9.5) – thermal vs haline contributions to density. He also introduces an estuarine buoyancy number ε to compare the sum of the buoyancy flux terms (the first two terms) with the buoyancy-driven exchange term (the third term) and expresses this in terms of the ratio between basin length L and a buoyancy length L_b so that $\varepsilon = (L/L_b)^2$. This estuarine buoyancy number is included in a dimensionless exchange rate $r = \Delta\rho/\rho_e\varepsilon$, where $\Delta\rho$ is the estuary–ocean density difference and ρ_e is a reference density.

Hearn (1998) shows that one obtains a single density-driven solution for long basins, corresponding to inverse density gradient and inverse exchange flow ($r < 0$).

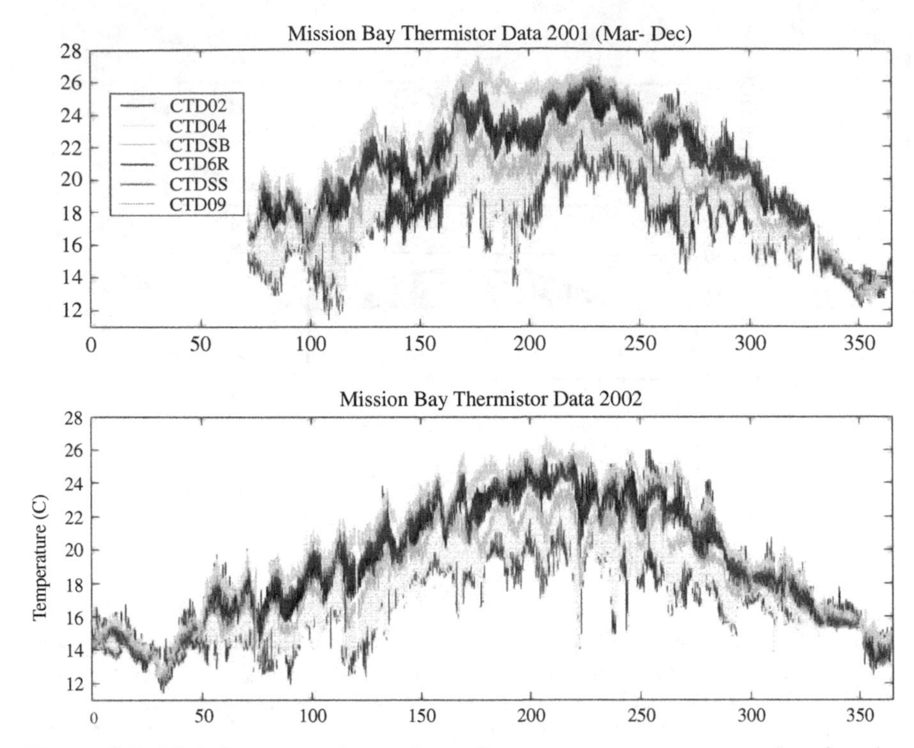

Figure 9.7. High-frequency observations of water temperature at several stations in Mission Bay (California, USA) for 2001 (upper panel) and 2002 (lower panel). Station CTD09 is near the head (about 9 km from the mouth). Station CTD02 is near the mouth (2 km inshore). The warm back-estuary waters cool strongly in October and November, leading to an inverse hypersaline period (cf. Fig. 9.6). At the end of November 2001, there is a sharp drop in temperature and the bay-wide thermal gradient reverses.

For shorter basins (smaller ε), the "Stommel bistability" occurs: there is a second stable steady solution corresponding to a thermal estuary with $r > 0$ as estuarine waters are warmer and less dense than ocean waters. The magnitude of r is greater for the thermal-estuary state, representing a scenario with faster estuary–ocean exchange that ensures cooler waters and a surface heat flux that dominates the surface buoyancy flux (heating effect exceeds evaporation effect). The alternative inverse-estuary state has slower exchange with a time scale that exceeds the time for ocean waters to attain thermal equilibrium temperatures; in this case, the buoyancy loss due to evaporation dominates the surface buoyancy flux. For the shortest basins, while the inverse estuary solution is possible, it is less likely to be achieved (few phase-plane trajectories lead to this stable state). For the same range of ε values where bistability is possible, a third unstable state exists mathematically, but this has no physical significance.

While theory shows that LIEs may exist in one of two alternate states, special conditions are required and it is unlikely that this bistability will be observed in most basins. First, given that many LIEs are seasonal, their trajectory toward the parameter space where bistability occurs must pass the thermal-estuary, classical-circulation state to get to the inverse-estuary state (i.e., the inverse-estuary state is not accessible from typical seasonal starting points). Second, natural systems experience much variability and ocean properties are not constant (e.g., Largier *et al.*, 2006). Further, perturbations toward the thermal-estuary state will have faster responses and thus briefer perturbations can be effective in bringing about that state than will perturbations toward the slower inverse-estuary state. Third, few natural systems have tide and wind forcing that is weak enough to allow ocean–estuary exchange to be controlled by the weak longitudinal density gradient that characterizes bistable parameter space; this is evident in the above-described observations from Californian LIEs. Fourth, the bistability results are obtained from a simple box model, in which mixing in the estuary is assumed to be complete. In reality, LIEs exhibit significant longitudinal structure that needs to be resolved by a multi-box model or a continuous one-dimensional model. In such a model, important longitudinal structure develops, including the separation of the thermal-estuary solution near the ocean and the inverse-estuary solution at greater distances from the ocean (if the basin is long enough). This is consistent with observed longitudinal structure – see the next section.

9.5. Longitudinal structure

Low-inflow estuaries exhibit a strong and characteristic longitudinal structure in water temperature, salinity, and density. Largier *et al.* (1996) identified four distinct zones, which can be explained through one-dimensional salt and heat balances (cf. Largier *et al.* 1997) – Fig. 9.8. Shortest residence times and properties closest to ocean values are found nearest to the mouth of the estuary, while longest residence times and most modified waters are found furthest from the mouth – expected where a weak non-zero freshwater inflow modifies the innermost waters of the LIE. Largier *et al.* (1996, 1997, 2000) describe LIEs with longitudinal exchange dominated by tidal diffusion, so that longitudinal diffusivity increases as the square of the distance from the landward end of the basin and residence times increase exponentially with distance from the mouth. However, even if the nature of the longitudinal exchange were different, one would still expect to observe least modified waters near the mouth and most modified waters near the head of the basin.

For the typical US west-coast scenario of a net evaporative basin, with cool ocean waters and a trickle inflow of freshwater, the four longitudinal zones are:

- *Marine zone.* Water is cool and salinity is similar to ocean. Where tidal mixing dominates, residence times are on the order of a day, and this zone extends about one tidal excursion from the mouth.
- *Thermal zone.* Water residence times are several days and warmer than the ocean. Given that surface buoyancy gain due to heating is greater than surface buoyancy loss due to

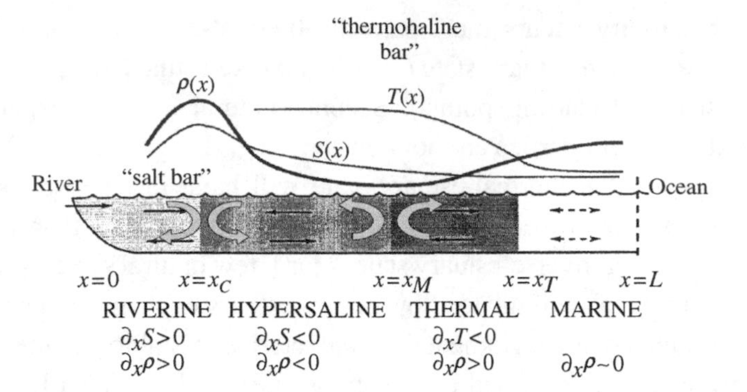

Figure 9.8. Schematic of four longitudinal zones typically found in a low-inflow estuary adjoining a cool ocean (Largier *et al.*, 1996): marine, thermal, hypersaline, and estuarine (or riverine). Gradients in water temperature, salinity and density are sketched. Near the head and near the mouth, one expects buoyancy-driven ouflow near-surface whereas the buoyancy structure tends to drive a seaward flow of dense hypersaline waters near-bottom. Reduced longitudinal exchange has been postulated at the "salt bar" where density is maximum and at the "thermohaline bar" where density is minimum.

evaporation, these waters are less dense than ocean waters and a thermal-estuary longitudinal gradient is observed. While transient thermal exchange flow is observed in Mission Bay and San Diego Bay (Chadwick *et al.*, 1996; Largier *et al.*, 1996, 1997), it is only in systems with weaker vertical mixing that thermal exchange flows may be important drivers of longitudinal exchange (e.g., Hamilton Harbor – Hamblin and Lawrence, 1990; Jervis Bay – Holloway *et al.*, 1992; Monteiro and Largier, 1999; Monismith *et al.*, 2006).

- *Hypersaline zone*. While there is weak hypersalinity in the "thermal zone", it is neither dynamically nor statistically significant. In the hypersaline zone, salinity increases due to evaporation are equal to or exceed the thermal influence on density. Residence times are typically a few weeks and the thermal buoyancy flux is zero as the surface water temperature is at the equilibrium value. With greater distances from the mouth, salinity increases further (but temperature is uniform) and an inverse density gradient may be observed in the inner hypersaline zone.
- *Estuarine zone*. In many LIEs there is a small inflow of freshwater at the head of the basin (but less than the evaporative loss of freshwater from the basin). Close enough to this inflow, salinities will decrease with increased distance from the ocean and the longitudinal density gradient is in the classical direction. This estuarine zone is expected at $x < R_r/EB$, where E is evaporation, R_r is river inflow, and B is basin width. Examples include van Diemens Gulf (Wolanski, 1986), Knysna Lagoon (Largier *et al.*, 2000), Tomales Bay (Abe, 2008), and the Norman, Normanby, and Marrett Rivers in north-eastern Australia (Ridd and Stieglitz, 2002).

The nature and extent (and existence) of these zones vary in response to external forcing. For example, the thermal zone will not occur unless the ocean water is cooler than equilibrium surface temperatures – thus thermal zones are absent from tropical systems (e.g., Wolanski, 1986; Smith, 1988; Savenije and Pages, 1992; Ribbe, 2006). Also, the estuarine zone will be absent when there is no freshwater inflow, e.g., Shark Bay (Burling *et al.*, 1999), Langebaan (Christie, 1981), San Diego Bay (Largier *et al.*, 1996), San Quintin (Camacho-Ibar *et al.*, 2003). And where the estuarine zone exists, it will expand and contract in response to changes in the river flow rate (e.g., Knysna Lagoon – Largier *et al.*, 2000). Likewise, the extent of the marine zone will expand and contract with fluctuations in tidal range. The hypersaline and thermal zones occupy the space between the marine and estuarine zones, expanding as the bounding zones contract. In these zones, the depth of the water relative to the surface buoyancy fluxes is a primary factor in the extent and intensity of the thermal and hypersaline zones.

Where a density maximum exists on the interface between the hypersaline and estuarine zones, Wolanski (1986) raised the possibility of a "salt bar", similar to the thermal bar associated with the 4°C density maximum in lakes. In the absence of other longitudinal and vertical mixing processes, a barrier to longitudinal exchange could develop on the interface of the counter-rotating circulations in the inverse hypersaline zone and the classical estuarine zone. A second hypothetical barrier could develop on the interface of the counter-rotating circulations in the inverse hypersaline zone and the classical thermal-estuary zone – forming a "thermohaline bar". While these density structures exist (Wolanksi 1986; Largier *et al.*, 1996), it is unlikely that other vertical and longitudinal mixing processes will be weak enough to allow this to happen, and evidence of the dynamical response is lacking.

9.6. Seasonal variability

Most LIE basins have flushing times of days to months and they exhibit marked seasonal variability – typically alternating between a classical state in the wet season and an inverse state in the dry season. Seasonal changes in the estuary–ocean density gradient follow the seasonal patterns in water temperature and salinity due to seasonally evolving heat and salt balances. The rate at which thermohaline patterns develop is also dependent on water depth in the basin. This seasonal cycle is illustrated for the wet-winter, dry-summer, cool-ocean Mission Bay (Figs 9.6 and 9.9):

- *Winter to spring*. In winter, the basin is hypothermal and hyposaline. As winter river inflow decreases in spring, any freshwater that is not flushed through density-driven estuarine circulation must be pumped out by tidal motions. This takes about a month following the reduction of inflows. At the same time, warmer spring weather heats the shallow waters.
- *Spring to summer*. In spring, estuarine waters are of similar salinity to the ocean, but warmer – representing a thermal estuary. However, longitudinal density gradients are

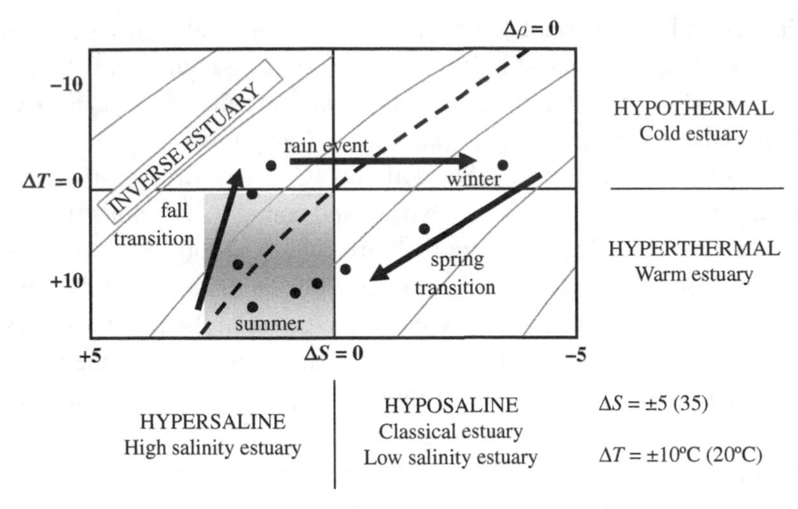

Figure 9.9. Illustration of a typical seasonal evolution of the thermohaline properties of water in a seasonally hypersaline estuary, like Mission Bay (California, USA). The background temperature–salinity–density axes are as in Fig. 9.1.

weak and estuarine residence times are long. During summer, the water warms a little more, to equilibrium values, and evaporation continues such that hypersalinity develops.

- *Summer to fall.* In late summer, the warm hypersaline estuary waters are of similar density to ocean waters. Surface cooling in fall reduces the heat excess of estuary waters, while retaining the hypersaline character, resulting in maximum inverse density gradients – and the possibility of inverse circulation events.
- *Fall to winter.* The inverse LIE in fall ends with the first winter rain, which shifts the estuary back to classical hyposaline conditions.

A similar seasonal cycle is observed in Tomales Bay (Fig. 9.10 and Largier *et al.*, 1997). This bay is at higher latitude and receives more rain than Mission Bay. The waters are not so warm, but the ocean is colder due to strong coastal upwelling (Largier *et al.*, 2006). During winter, Tomales Bay exhibits classical estuarine character, most notably following rain events. Nevertheless, the summer hypersaline period lasts for 3–4 months. Inverse conditions are rarely observed (Largier *et al.*, 1997).

Seasonal hypersalinity is only observed in estuaries with moderate residence times. Where estuaries exchange rapidly with the ocean, estuarine water properties will track ocean properties, with marginal thermohaline effects. With longer residence in shallow estuarine basins, waters will develop a distinct thermohaline character. However, where residence exceeds seasonal time scales, as in some coastal lagoons and coastal lakes (St Lucia Estuary, Lagoa de Araruama), or large basins (Mediterranean and Red Seas), the interannual variability in thermohaline character will dominate seasonal variability. Also, some basins in arid regions will not receive freshwater inflow every year, and LIE thermohaline conditions will persist for years.

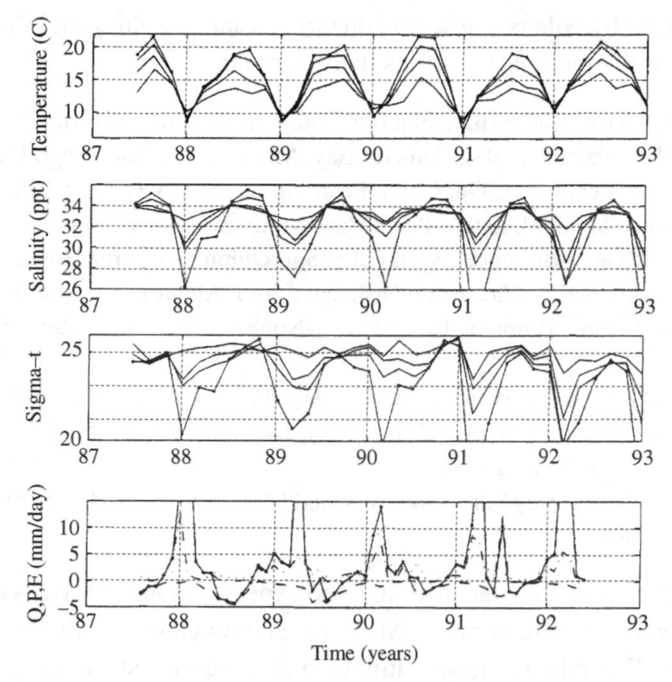

Figure 9.10. A 5-year record of bi-monthly water temperature, salinity, density and inflow/outflow in Tomales Bay (California, USA) – these data are also reported and analyzed in Largier *et al.* (1996, 1997). In the bottom panel, the solid line gives the freshwater input to the estuary: $R_r + P - E$. There is a recurrent seasonal pattern of warmer saltier waters in the back-bay during summer, but an absent or brief inverse gradient in the fall.

9.7. Examples of low-inflow estuaries

There are many examples of low-inflow estuaries, and many different types of estuaries that experience low-inflow periods. As suggested in the Introduction, it is quite possible that there are more estuaries worldwide with low-inflow periods than those without. However, the literature is still young and most LIE systems have not yet been described, or low-inflow periods have been neglected while describing wetter periods in a given estuary.

Here we review some of the systems in which hypersaline and/or inverse conditions have been observed – conditions that are hallmarks of LIEs. This is an incomplete review, not intended to be comprehensive, but rather intended to give some idea of how common these systems are and how widely they are found.

Temperate west coasts. Mid-latitude west coasts are characterized by Mediterranean or semiarid climate estuaries, with long dry summers. Hypersalinity is regularly observed at latitudes as high as 40°N (e.g., Humboldt Bay; Barnhart *et al.*, 1992) and 38°N (e.g., Tomales Bay – Largier *et al.*, 1997). In addition, these regions are characterized by cold ocean waters due to coastal upwelling and there are many observations of thermal estuary effects. For example:

- California (USA): southern Humboldt Bay, southern San Francisco Bay, Elkhorn Slough (Smith, 1971; Nidzieko, 2009), Tomales Bay, Mission Bay, San Diego Bay (Chadwick *et al.*, 1996; Largier *et al.*, 1996, 1997), Penasquitos Estuary (Zedler, 1983).
- Baja California (Mexico): Estero Punta Banda (Alvarez-Borrego *et al.*, 1977; Alvarez-Borrego and Alvarez-Borrego, 1982), Bahia San Quintin (Alvarez-Borrego *et al.*, 1975; Alvarez-Borrego, 1977; Chavez-de-Nishikawa and Alvarez-Borrego, 1974; Alvarez-Borrego and Alvarez-Borrego, 1982; Millan-Nuñez *et al.*, 1982; Camacho-Ibar *et al.*, 2003).
- South Africa: Langebaan (Christie, 1981; Day, 1981; Largier *et al.*, 1997), Milnerton Lagoon (Day, 1981), Kleinrivier (Day, 1981), Heuningnesrivier (Day, 1981), Kromme River (Scharler and Baird, 2003).
- Western Australia: Harvey Inlet (Lukatelich and McComb, 1986; Hearn, 1998; Hearn and Lukatelich, 1990).

These Mediterranean-climate LIEs are also expected along the coasts of Chile (e.g. Laguna Boyeruca), Spain, Portugal, Morocco, and much of the Mediterranean Sea.

Arid coasts. Extending to lower latitudes along west coasts, the climate becomes more arid, while coastal upwelling continues. The desert estuaries in these regions may not receive significant inflow every winter. They may still be seasonal, but they operate as LIEs with brief interludes of freshwater inflow and hyposaline conditions. Examples have been described in the following regions:

- Mexico – mainland west coast: Colorado River (Lavin *et al.*, 1998), Bahia de Guyamas (Valle-Levinson *et al.*, 2001), Caimanero-Huizache lagoon system (Moore and Slim, 1984).
- Mexico – Baja California peninsula: Laguna Guerrero Negro, Laguna Ojo de Liebre (Postma, 1965), Laguna San Ignacio (Winant and Gutierrez de Velasco, 2003; Gutierrez de Velasco and Winant 2004), Bahia Magdalena (Alvarez-Borrego *et al.*, 1975), Bahia Concepcion, Bahia Los Angeles.
- Iraq: Khowr Al-Zubair (Al-Ramadahn, 1988).
- Western Australia: Shark Bay (Smith and Atkinson, 1983; Burling *et al.*, 1999; Nahas *et al.*, 2005).

These arid region LIEs are also expected along the coast of Peru, northern Chile, Namibia (e.g., Walvis Bay), Morocco, Western Sahara, on the coasts of the eastern Mediterranean sea and the Arabian/Persian Gulf.

Subtropical coasts. Along many low-latitude east coasts, rainfall and river flow is confined to summers and there may be long dry periods during winter. There is also a tendency for multi-year wet or dry periods, with LIE conditions continuing for years and salinities of 70 or more being attained (e.g., Cassamance in

Senegal – Savenije and Pages, 1992 and St. Lucia in South Africa – Begg, 1978). These regions are characterized by warm ocean waters, in general, and thermal-estuary effects are not observed. The following regions exhibit LIEs:

- Gulf of Mexico: Laguna Madre of Texas (Smith, 1988), Laguna Madre de Tamaulipas, Corpus Christi Bay (Ritter and Montagna, 1999).
- South Africa: St Lucia Estuary (Begg, 1978).
- Brazil: Lagoa de Araruama (Kjerve *et al.*, 1996), Patos Lagoon (Fernandes *et al.*, 2005)
- Senegal: Saloum and Cassamance Rivers (Pages and Debenay, 1987; Savenije and Pages, 1992; Pages *et al.*, 1995).
- Sri Lanka: Puttalam Lagoon (Arrulanathan *et al.*, 1995).
- Australia: Hervey Bay (Ribbe, 2006), Norman, Normanby, and Marrett Rivers (Ridd and Stieglitz, 2002), Aligator River and Van Diemens Gulf (Wolanski, 1986).

Large systems. While the focus of this chapter is on smaller systems, there are some larger basins that can be considered as larger LIEs. Most notable are the studies of Spencer Gulf, in South Australia, where a seasonal inverse flushing event is observed with dense hypersaline waters draining downslope and seaward along the left boundary of the basin owing to Coriolis effects.

- Large seasonal systems: Spencer Gulf (Lennon *et al.*, 1987; Nunes and Lennon, 1987; Nunes Vaz *et al.*, 1990), St Vincent Gulf (de Silva Samarasinghe and Lennon, 1987; de Silva Samarasinghe, 1989), Golfo San Matias, northern Gulf of California (Lavin *et al.*, 1998), Gulf of Kachchh/Kutch (Vethamony *et al.*, 2007).
- Ice formation: Weddell Sea (Foster and Carmack, 1976), Ross Sea (MacAyeel, 1985), Beaufort Sea (Aagaard *et al.*, 1985). Also smaller basins, e.g., Peard Bay (Wiseman, 1979).
- Marginal seas (permanent LIE basins): Mediterranean Sea (Lacombe and Richez, 1982), Red Sea (Phillips, 1966; Tragou and Garrett, 1997), Gulf of California (Bray and Robles, 1991; Lavin *et al.*, 1998), Arabian/Persian Gulf (Banse, 1997), Great Australian Bight (Petrusevics *et al.*, 2009).

9.8. Summary

Low-inflow estuaries are found worldwide, occurring as transient states in wetter climates and as persistent states briefly interrupted following rain events in drier climates. In less arid areas, LIEs occur seasonally in estuaries with short-tailed inflow hydrographs (i.e., smaller watersheds). Along west coasts, or wherever coastal upwelling is important, the outer estuary is characterized by thermal estuary effects while the inner estuary exhibits hypersaline and inverse effects. In regions where ocean waters are warmer, thermal estuary effects are not found.

The intensity of hypersalinity is scaled by $E_{net}L^2/HK$, where K is a longitudinal mixing coefficient, E_{net} is the net evaporation rate, L is the basin length and H is the basin depth. Basins with large ocean–estuary exchange (i.e., large K/L^2) and/or weak evaporation will develop low levels of hypersalinity, although it will develop quickly. In contrast, basins with weaker exchange and stronger evaporation will develop high levels of hypersalinity, although it may take some time. The highest open-water hypersalinity levels are observed in long basins where dry periods have persisted for multiple years (e.g., Cassamance, Senegal and St Lucia Estuary, South Africa). Water depth is an important parameter as surface buoyancy fluxes have greater water column effects in shallower waters and because shallow waters reduce longitudinal exchange flows.

In this chapter, attention has been on the physical aspects of LIEs. However, these systems exhibit many common ecological characteristics – mostly related to the long residence times observed in these basins. Recent studies have related longitudinal patterns in fecal bacterial pollution (Largier *et al.*, 2003), nutrient levels (Smith *et al.*, 1991; Camacho-Ibar *et al.*, 2003), and phytoplankton concentrations (Kimbro *et al.*, 2009) to longitudinal patterns in residence and exchange rates. Also, it has been noted that the warmer waters in LIEs provide opportunities for invasive species introduced from regions with warmer ocean waters.

Low-inflow estuaries provide a view of the future state of estuaries subject to reduced inflows due to water extraction from inflowing rivers (e.g., Scharler and Baird, 2003), or due to droughts (e.g., Savenije and Pages, 1992), or changing hydrological climates. In these systems, the reduction of inflows will lead to reduced estuarine circulation and longer residence times, as observed in LIEs. Anthropogenic hypersaline conditions are already observed in the deltas of major rivers such as the Colorado River (Mexico – Lavin *et al.*, 1998) and the Indus River (Pakistan).

The challenge for those studying estuarine physics is twofold. First, there is a need to document LIE systems, carefully describing heat and salt balances as well as residence times and ecological linkages. Second, there is a great need for careful dynamical studies of exchange processes in small to moderate-sized LIEs and an assessment of when the weak longitudinal density gradients are important to the system. Or, are the intriguing "non-classical" thermal, hypersaline, and inverse gradients simply a symptom of estuarine state and not a feedback on LIE hydrodynamics?

References

Aagaard, K., J. H. Swift and E. C. Carmack (1985) Thermohaline circulation in the Arctic Mediterranean Seas. *J. Geophys. Res.* **90**, 4833–4846.

Abe, R. (2008) Temporal and spatial analysis of Lagunitas Creek Estuary. Research project, Bodega Marine Laboratory, 17 pp.

Al-Ramadahn, B. M. (1988) Residual fluxes of water in an estuarine lagoon. *Est. Coast. Shelf Sci.* **26**, 319–330.

Alvarez-Borrego, S., G. Ballesteros-Grijalvia and A. Chee-Barragan (1975) Estudio de algunas variables fisicoqumicas superficiales en Bahia San Quintın, en verano, otono e invierno. *Ciencias Marinas* **2**(2), 1–9.

Alvarez-Borrego, S., L. A. Galindo-Bect and A. Chee-Barragan (1975) Caracteristicas hidroquimicas de Bahia Magdalana, B.C.S. *Ciencias Marinas* **2**(2), 94–110.
Alvarez-Borrego S., M. J. Acosta-Ruíz and J. R. Lara-Lara (1977) Hidrología comparativa de Las Bocas de dos antiestuarios de Baja California *Ciencias Marinas* **4**(1), 1–11.
Alvarez-Borrego, J. and S. Alvarez-Borrego (1982) Temporal and spatial variability of temperature in two coastal lagoons. *CalCOFI Reports* **23**, 188–197.
Arrulanathan, K., E. M. S. Wiyeratne, L. Rydberg and U. Cederloef (1995) Water exchange in a hypersaline tropical estuary, the Puttalam Lagoon, Sri Lanka. *Ambio* **24**(7–8), 438–443.
Banse, K. (1997) Irregular flow of Persian (Arabian) Gulf waters to the Arabian Sea. *J. Mar. Res.* **55**, 1049–1067.
Barnhart, R. A., M. S. Boyd and S. E. Peguegrat (1992) *The Ecology of Humboldt Bay, California: An Estuary Profile*. US Fish & Wildlife Service, Arcata, CA, 121 pp.
Begg, G. (1978) The estuaries of Natal. *Natal Town and Regional Planning Report*, Vol. **41**.
Bray, N. A. (1988) Thermohaline Circulation in the Gulf of California. *J. Geophys. Res.* **93** (C5), 4993–5020.
Bray, N. A. and J. M. Robles (1991) Physical oceanography of the gulf of California. In J. P. Dauphin and B. R. T. Simoneit (eds), *The Gulf and Peninsular Province of the Californias*. AAPG Memoir 47, pp. 511–533.
Burling, M. C., G. N. Ivey and C. B. Pattiaratchi (1999) Convectively driven exchange in a shallow coastal embayment. *Cont. Shelf Res.* **19**, 1599–1616.
Camacho-Ibar, V. F., J. D. Carriquiry and S. V. Smith (2003) Non-conservative P and N fluxes and net ecosystem production in San Quintin Bay, Mexico. *Estuaries* **26**(5), 1220–1237.
Chadwick, D. B., J. L. Largier and R. T. Cheng (1996) The role of thermal stratification in tidal exchange at the mouth of San Diego Bay. In D. G. Aubrey and C. T. Friederichs (eds), *Buoyancy Effects on Coastal and Estuarine Dynamics*. Coastal and Estuarine Studies, Vol. 53, pp. 155–174.
Chavez-de-Nishikawa, A. G. and S. Alvarez-Borrego (1974) Hidrologia de Bahia San Quintin en inviemo y primavera. *Ciencias Marinas* **1**, 31–62.
Christie, N. D. (1981) Primary production in Langebaan Lagoon. In J. H. Day (ed.), *Estuarine Ecology*. Balkema, Rotterdam, pp. 101–115.
Day, J. H. (1981) *Estuarine Ecology* (With Particular Reference to Southern Africa). Balkema, Rotterdam.
deCastro, M., M. Gomez-Gesteira, I. Alvarez and R. Prego (2004) Negative estuarine circulation in the Ria of Pontevedra (NW Spain). *Est. Coast. Shelf Sci.* **60**, 301–312.
de Silva Samarasinghe, J. R. (1989) Transient salt-wedges in a tidal gulf: a criterion for their formation. *Est. Coast. Shelf Sci.* **28**, 129–148.
de Silva Samarasinghe, J. R. and G. W. Lennon (1987) Hypersalinity, flushing and transient salt-wedges in a tidal gulf – an inverse estuary. *Est. Coast. Shelf Sci.* **24**, 483–498.
Dyer, K. R. (1997) *Estuaries. A physical introduction*. John Wiley & Sons, New York.
Fernandes, E. H. L., K. R. Dyer and O. O. Moller (2005) Spatial gradients in the flow of Southern Patos lagoon. *J. Coast. Res.* **21**(4), 759–769.
Foster, T. D. and E. C. Carmack (1976) Frontal zone mixing and Antarctic bottom water formation in the southern Weddell Sea. *Deep-Sea Res.* **23**, 301–317.
Gutierrez de Velasco, G. and C. D. Winant (2004) Wind- and density-driven circulation in a well-mixed inverse estuary. *J. Phys. Oceanogr.* **34**, 1103–1116.
Hamblin, P. F. and G. A. Lawrence (1990) Exchange flows between Hamilton Harbor and Lake Ontario. In Proceedings of 1990 Annual Conference of the Canadian Society of Civil Engineering, Hamilton, Ont., pp. 140–148.

Hearn, C. J. (1998) Application of the Stommel model to shallow Mediterranean estuaries and their characterization. *J. Geophys. Res.* **103**, 10,391–10,404.
Hearn, C. J. and J. L. Largier (1997) The summer buoyancy dynamics of a shallow Mediterranean estuary and some effects of changing bathymetry: Tomales Bay, California. *Est. Coast. Shelf Sci.* **45**, 497–506.
Hearn, C. J. and R. J. Lukatelich (1990) Dynamics of Peel-Harvey Estuary, Southwest Australia. In R. T. Cheng (ed.), *Residual Currents and Long-term Transport*. Springer-Verlag, New York, pp. 431–450.
Holloway, P. E., G. Symonds and R. Numes-Vaz (1992) Water circulation in Jervis Bay. *Austr. J. Mar. Freshw. Sci.*
Kimbro, D. L., S. Largier and E. D. Grosholz (2009) Coastal cenographic processes influence the growth and size of a key estuarine species, the Olympia oyster. *Limnol. Oceanogr.* **54**, 1425–1437.
Kjerfve, B., C. A. F. Schettini, B. Kloppers, G. Lessa and H. O. Ferreira (1996) Hydrology and salt balance in a large hypersaline coastal lagoon: Lagoa de Araruama, Brazil. *Est. Coast. Shelf Sci.* **42**, 701–725.
Lacombe, H. and C. Richez (1982) The regime of the Straits of Gibraltar. In J. C. J. Nihoul (ed.), *Hydrodynamics of Semi-enclosed Seas*. Elsevier, Amsterdam, pp. 13–73.
Largier, J. L., C. J. Hearn and D. B. Chadwick (1996) Density structures in low-inflow "estuaries". In D. G. Aubrey and C. T. Friederichs (eds), *Buoyancy Effects on Coastal and Estuarine Dynamics*. Coastal and Estuarine Studies, Vol. 53, pp. 227–241.
Largier, J. L., S. V. Smith and J. T. Hollibaugh (1997) Seasonally hypersaline estuaries in Mediterranean-climate regions. *Est. Coast. Shelf Sci.* **45**, 789–797.
Largier, J. L., C. Attwood and J.-L. Harcourt-Baldwin (2000) The hydrographic character of Knysna Estuary. *Trans. Roy. Soc. South Africa* **55**(2), 107–122.
Largier, J., M. Carter, M. Roughan, D. Sutton, J. Helly, B. Lesh, T. Kacena, P. Atjai, L. Clarke, A. Lucas, P. Walsh and L. Carrillo (2003) *Mission Bay Contaminant Dispersion Study*. Scripps Institute of Oceanography, FIB, 77 pp.
Largier, J. L., C. A. Lawrence, M. Roughan, D. M. Kaplan, E. P. Dever, C. E. Dorman *et al.* (2006) WEST: A northern California study of the role of wind-driven transport in the productivity of coastal plankton communities. *Deep Sea Res.* II **53**(25–26), 2833–2849.
Lavin, M. F., V. M. Godinez and L. G. Alvarez (1998) Inverse-estuarine features of the upper Gulf of California. *Est. Coast. Shelf Sci.* **47**, 769–795.
Lennon, G. W., D. Bowers, R. A. Nunes, B. D. Scott, M. Ali, J. Boyle *et al.* (1987) Gravity currents and the release of salt from an inverse estuary. *Nature* **327**, 695–697.
Linden, P. F. and J. E. Simpson (1988) Modulated mixing and frontogenesis in shallow seas and estuaries. *Cont. Shelf Res.* **8**, 1107–1127.
Lukatelich, R. J. and A. J. McComb (1986) Nutrient levels and the development of diatom and bluegreen blooms in a shallow Australian estuary. *J. Plankton Res.* **8**, 597–618.
MacAyeel, D. R. (1985) Thermohaline circulation below the Ross Ice Shelf: a consequence of tidally induced vertical mixing and basal melting. *J. Geophys. Res.* **89**, 597–606.
Millan-Nuñ ez, R., S. Alvarez-Borrego and D. M. Nelson (1982) Effects of physical phenomena on the distribution of nutrients and phytoplankton productivity in a coastal lagoon. *Est. Coast. Shelf Sci.* **15**, 317–335.
Monismith, S. G., A. Genin, M. Reidenbach, G. Yahel and J.R. Koseff (2006) Thermally driven exchanges between a coral reef and the adjoining ocean. *J. Phys. Oceanogr.* **36**, 1332–1347.

Monsen, N. E., J. E. Cloern, L. V. Lucas and S. G. Monismith (2002) A comment on the use of flushing time, residence time and age as transport time scales. *Limnol. Oceanogr.* **47**, 1545–1553.

Monteiro, P. M. S. and J. L. Largier (1999) Thermal stratification in Saldanha Bay (South Africa) and subtidal, density-driven exchange with the coastal waters of the Benguela upwelling system. *Est. Coast. Shelf Sci.* **49**(6), 877–890.

Moore, N. H. and D. J. Slim (1984) The physical hydrology of a lagoon system on the Pacific coast of Mexico. *Est. Coast. Shelf Sci.* **19**(4), 413–426.

Nahas, E. L., C. B. Pattiaratchi and G. N. Ivey (2005) Processes controlling the position of frontal systems in Shark Bay, Western Australia. *Est. Coast. Shelf Sci.* **65**, 463–474.

Nidzieko, N. J. (2009) Dynamics of a seasonally low-inflow estuary: circulation and dispersion in Elkhorn Slough, California. Ph.D. Thesis, Stanford University, Stanford, CA.

Nunes, R. A. and G. W. Lennon (1987) Episodic stratification and gravity currents in a marine environment of modulated turbulence. *J. Geophys. Res.* **92**, 5465–5480.

Nunes Vaz, R. A., G. W. Lennon and D. G. Bowers (1990) Physical behavior of a large, negative or inverse estuary. *Cont. Shelf Res.* **10**(3), 277–304.

Officer, C. B. (1976) *Physical Oceanography of Estuaries (and Associated Coastal Waters)*. John Wiley & Sons, New York.

Pages, J. and J. P. Debenay (1987) Evolution saisonnaire de la salinite de la Cassamance: description et essai de modelisation. *Rev. d'Hydrol. Trop.* **20**, 203–217.

Pages, J., J. Lemoalle and B. Fritz (1995) Distribution of carbon in a tropical hypersaline estuary, the Cassamance (Senegal, west Africa). *Estuaries* **18**(3), 456–468.

Petrusevics, P., J. A. T. Bye, V. Fahlbusch, J. Hammat, D. R. Tippins and E. van Wijk Petrusevics (2009) High salinity winter outflow from a mega inverse-estuary – the Great Australian Bight. *Cont. Shelf Res.* **29**, 371–380.

Phillips, O. M. (1966) On turbulent convection currents and the circulation of the Red Sea. *Deep-Sea Res.* **13**, 1149–1160.

Postma, H. (1965) Water circulation and suspended matter in Baja California lagoons. *Neth. J. Sea Res.* **2**, 566–604.

Ribbe, J. (2006) A study into the export of saline water from Hervey Bay, Australia. *Est. Coast. Shelf Sci.* **66**, 550–558.

Ridd, P. V. and T. Stieglitz (2002) Dry season salinity changes in arid estuaries fringed by mangroves and saltflats. *ECSS* **54**, 1039–1049.

Ritter, C. and P. A. Montagna (1999) Seasonal hypoxia and models of benthis reponse in a Texas Bay. *Estuaries* **22**(1), 7–20.

Savenije, H. H. G. and J. Pages (1992) Hypersalinity: a dramatic change in the hydrology of Sahelian estuaries. *J. Hydrol.* **135**, 157–174.

Scharler, U. M. and D. Baird (2003) The influence of catchment management on salinity, nutrient stochiometry and phytoplankton biomass of Eastern Cape estuaries *S.A Est. Coast. Shelf Sci.* **56**, 735–748.

Schumann, E., J. Largier and J. Slinger (1998) Estuarine hydrodynamics. In B. R. Allanson and D. Baird (eds), *Estuaries of South Africa*. Cambridge University Press, Cambridge, pp. 27–52.

Schwartz, R. A. and J. Imberger (1988) Flushing behavior of a coastal marina. Environmental Dynamics, University of Western Australia, ED-88-259.

Smith, N. P. (1988) The Laguna Madre of Texas: hydrography of a hypersaline lagoon. In B. Kjerve (ed.), *Hydrodynamics of Estuaries, Volume II, Estuarine Case Studies*. CRC Press, Boca Raton, FL, pp. 21–40.

Smith, R. E. (1971) The hydrography of Elkhorn Slough, a shallow California coastal embayment. Moss Landing Marine Laboratories, Tech. Publ. 73–2.

Smith, S. V. and M. J. Atkinson (1983) Mass balance of carbon and phosphorus in Shark Bay, Western Australia. *Limnol. Oceanogr.* **28**(4), 625–639.

Smith, S. V., J. T. Hollibaugh, S. J. Dollar and S. Vink (1991) Tomales Bay metabolism: C–N–P stoichiometry and ecosystem heterotrophy at the land–sea interface. *Est. Coast. Shelf Sci.* **33**, 223–257.

Stommel, H. (1961) Thermohaline convection with two stable regimes of flow. *Tellus* **13**(2), 224–230.

Tragou, E. and C. Garrett (1997) The shallow thermohaline circulation of the Red Sea. *Deep-Sea Res. I* **44**(8), 1355–1376.

Valle-Levinson, A., J. A. Delgado and L. P. Atkinson (2001) Reversing water exchange patterns at the entrance to a semiarid coastal lagoon. *Est. Coast. Shelf Sci.* **53**, 825–838.

Vethamony P., M. T. Babu, M. V. Ramanamurty, A. K. Saran, A. Joseph, K. Sudheesh *et al.* (2007) Thermohaline structure of an inverse estuary – The Gulf of Kachchh: measurements and model simulations. *Marine Pollut. Bull.* **54**(6), 697–707.

Winant, C. D. and G. Guitteriez de Velasco (2003) Tidal dynamics and residual circulation in a well-mixed inverse estuary. *J. Phys. Oceanogr.* **33**, 1365–1379.

Wiseman, W. J. (1979) Hypersaline bottom water: Peard Bay, Alaska. *Estuaries* **2**(3), 189–193.

Wolanski, E. (1986) An evaporation-driven salinity maximum zone in Australian tropical estuaries. *Est. Coast. Shelf Sci.* **22**, 415–424.

Zedler, J. B. (1983) Freshwater impacts in normally hypersaline marshes. *Estuaries* **6**(4), 346–355.

10

Implications of estuarine transport for water quality

L.V. LUCAS
US Geological Survey

10.1. Introduction

In this chapter, some implications of estuarine transport for water quality are discussed. This is not an exhaustive review of all physical processes potentially important to water quality in estuaries. Rather, the focus is on (1) some fundamental relationships, concepts, and helpful idealizations (e.g., evolution equations for reactive scalars, transport time scales, scaling and non-dimensional numbers), (2) some common and often dominant physical processes in terms of their influence on estuarine water quality (e.g., stratification and turbulent mixing), and (3) some less prevalently discussed but probably widely important issues regarding high-frequency (i.e., intradaily) processes and their influence on water quality.

Here, "water quality" refers to the full range of suspended constituents (or "scalars", i.e., non-vector quantities) in an estuarine water column. These constituents may be dissolved or particulate, mineral, chemical, or biological, or they may represent physical properties of the water (e.g., temperature). The spatial distribution of a water quality constituent is influenced by the hydrodynamic environment in which it is suspended, but it may be additionally subject to motility, positive buoyancy, or negative buoyancy (e.g., some phytoplankton or zooplankton). Water quality scalars may be conservative (i.e., non-reactive, such as salt) or non-conservative (i.e., reactive and thereby potentially changing in concentration or form during transit; e.g., nitrogen, phosphorus, or phytoplankton). Hydrodynamic and transport processes are important not only because they "move stuff around" but also because, in the case of reactive scalars, those processes may expose the scalars to a range of environments, each of which may be associated with distinct rates of scalar transformation. Some water quality constituents are "active", meaning that they may produce feedbacks that influence the hydrodynamics, thus establishing a "two-way" physics–scalar relationship (e.g., salinity, temperature). Other water quality scalars may be "passive", meaning that they do not affect the hydrodynamics and thus maintain a "one-way" physics–scalar relationship.

"Water quality" encompasses quantities including (but not limited to) salinity (S), temperature (T), chlorophyll a (a proxy for phytoplankton biomass), dissolved or particulate nutrients or contaminants, suspended sediment (SS), suspended particulate matter (SPM), organic carbon (OC), and dissolved oxygen (DO). The most critical water quality constituent varies from system to system and depending on one's perspective. In some estuaries and bays, dissolved oxygen is of primary concern due to the observed stress of low DO on aquatic organisms such as fish; DO may in turn be closely related to other water quality constituents such as chlorophyll, nutrients, and active stratifying scalars such as salinity and temperature. In other systems, toxic contaminants, and perhaps the sediment particles to which they adsorb, may be of concern. Where drinking water quality is to be preserved, dissolved organic carbon (DOC) may be of utmost concern because some DOC fractions may interact with disinfectants during treatment to produce byproducts suspected to be carcinogenic. Water quality is often expressed in units of concentration (e.g., μg chl a/m^3 for phytoplankton biomass, or mg/m^3 for dissolved phosphorus), but of course there are exceptions (e.g., °C for temperature, or mSiemens/cm for specific conductance, a proxy for salinity).

10.2. Transport–reaction equations for water quality constituents

If we consider an infinitesimal cube-shaped control volume (Fig. 10.1) containing a potentially reactive water quality constituent of concentration C, we see that there are two ways for C within the control volume to change over time: (1) net transport of C into or out of the control volume through the control volume faces, and (2) internal sources or sinks for C within the control volume (i.e., "reactions"). The transport may have advective and/or diffusive components, and the source–sink processes may be biological (e.g., growth, predation, uptake) or chemical (e.g., decay, transformation to another chemical form).

The "conservation law" or "divergence" form (Hirsch, 1988) of the equation describing the transport and evolution of C combines advective and diffusive flux terms into one term, which is the divergence of the total flux:

$$\frac{\partial C}{\partial t} + \nabla \cdot \mathbf{F} = s \tag{10.1}$$

(∇ and bold variables are vectors). The internal source–sink ("reaction") term is s, and $\mathbf{F} = \mathbf{F}_x + \mathbf{F}_y + \mathbf{F}_z$ is the total flux vector. The flux vector in the k-direction (where k is either x, y, or z) and the expanded form of equation (10.1), respectively, are:

$$\mathbf{F}_k = \left(u_k C - K_k \frac{\partial C}{\partial k} \right) \mathbf{i}_k, \tag{10.2}$$

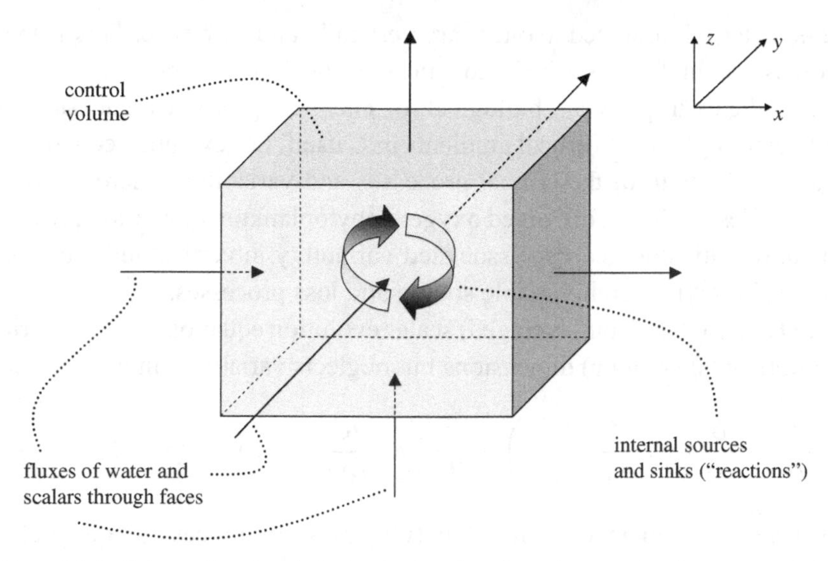

Figure 10.1. Schematic of an infinitesimal cube-shaped control volume containing a potentially reactive water quality constituent of concentration C. C inside the control volume can change due to transport through the control volume faces and/or due to source or sink ("reaction") processes within the control volume.

$$\frac{\partial C}{\partial t} = \frac{\partial}{\partial x}\left(K_x \frac{\partial C}{\partial x} - uC\right) + \frac{\partial}{\partial y}\left(K_y \frac{\partial C}{\partial y} - vC\right) + \frac{\partial}{\partial z}\left(K_z \frac{\partial C}{\partial z} - wC\right) + s(x, y, z, t), \tag{10.3}$$

where $\mathbf{i}_k$ is the unit vector in the k-direction, u_k is the velocity in the k-direction (a.k.a., "u" for the x-, "v" for the y-, and "w" for the z-directions), and K_k is the diffusivity or dispersivity for the k-direction in units of length squared over time. Equation (10.3) is the three-dimensional transport–reaction equation for a potentially reactive water quality constituent. If C is conservative, then $s = 0$. Usually in environmental applications, K_z refers to a turbulent diffusivity; depending on the horizontal dimensions of the control volume (i.e., numerical model grid spacing) over which equation (10.3) is solved, K_x and K_y may describe mixing due to turbulence or due to larger-scale dispersive processes.

One useful special case of the 3D scalar evolution equation (10.3) is represented by the one-dimensional vertical evolution equation:

$$\frac{\partial C}{\partial t} = \frac{\partial}{\partial z}\left(K_z \frac{\partial C}{\partial z}\right) - \frac{\partial}{\partial z}(wC) + s(z, t). \tag{10.4}$$

This equation is obtained by assuming horizontal fluxes do not vary in their respective directions (i.e., there is no net horizontal flux of C that could cause accumulation or depletion of C within the control volume). Additionally, the source–sink term is assumed to only vary vertically and with time. This 1D vertical equation is useful for describing aquatic environments characterized by strong vertical gradients in physical, biological, or chemical processes that could potentially generate significant vertical gradients in C itself. For example, equation (10.4) may capture the bulk of the critical processes and variability where non-uniform vertical profiles of C (e.g., dissolved oxygen, phytoplankton biomass) may arise due to vertical density stratification, associated variability in vertical turbulent mixing, and interactions with depth-variable source and loss processes.

The 2D horizontal (depth-averaged) scalar evolution equation captures variability in both horizontal (x and y) dimensions but neglects variability in the vertical (z):

$$\frac{\partial C}{\partial t} = \frac{\partial}{\partial x}\left(K_x \frac{\partial C}{\partial x} - UC\right) + \frac{\partial}{\partial y}\left(K_y \frac{\partial C}{\partial y} - VC\right) + s(x, y, t). \tag{10.5}$$

In this case, constituent concentration (C), transport processes (i.e., velocities, diffusivities), and source–sink terms may vary horizontally, but C is expected to be vertically well mixed (i.e., vertical fluxes of C are not expected to vary substantially within the water column and there are consequently no local zones of significant accumulation or depletion of C along the vertical dimension). U and V represent depth-averaged components of horizontal velocity. Although s (e.g., light-driven algal growth) may in reality vary with z, vertical mixing is assumed to be rapid enough compared to s to maintain vertical homogeneity in C. This equation may be appropriate for describing water quality in shallow water columns or those vigorously and thoroughly mixed by tides or wind.

In cases where the assumption of full mixing of C across the flow cross-section (i.e., across both the width and depth of flow) is justified, the 1D longitudinal (cross-sectionally averaged) scalar evolution equation may be useful:

$$\frac{\partial C}{\partial t} = \frac{\partial}{\partial x}\left(K_x \frac{\partial C}{\partial x} - UC\right) + s(x, t). \tag{10.6}$$

Equation (10.6) thus only describes longitudinal variability in C. Fluxes of C are assumed to be vertically and laterally uniform within the water column, as are internal sources and sinks; therefore, local zones of accumulation or depletion of C within a flow cross-section are not expected. Cross-sectional homogeneity of C, as assumed by equation (10.6), is most likely to occur in shallow, narrow systems such as streams and small rivers, although we may be compelled to make this approximation in other situations for simplicity's sake (see Section 10.3.4).

What do we do with equations such as those shown above? If spatial or temporal variability of driving processes or of system geometry can be simplified, then the scalar evolution equations may be simplified further and solved analytically, resulting in a closed-form solution for C (an example of such a simplification is given in Section 10.3.4). If the problem of interest is not so easily simplified, or if a detailed and realistic depiction of the aquatic system is the objective, then partial differential equations like equations (10.3)–(10.6) above may be solved numerically (i.e., numerical methods can be implemented within a computer model to estimate solutions for C stepwise across a model grid and at discrete points in time). Numerical solutions become increasingly necessary as spatial and temporal variability in internal sources and sinks, geometry, and boundary forcings increases.

10.3. Simplified horizontal transport, scaling, and non-dimensional numbers

10.3.1. Scaling

In engineering study and practice, "scaling" techniques are frequently used to estimate the magnitude or rate of processes, forces, or terms in equations; here we discuss the application of this tool to environmental and biological studies. Scaling allows rough comparison of magnitudes, and thus of the importance, of processes happening simultaneously. In fact, scaling can help in making decisions on the relative importance of individual terms within an equation, providing a semiquantitative justification for reducing complex equations such as equation (10.3) to simpler forms such as equations (10.4)–(10.6). Diverse scales such as time scales, length scales, velocity scales, force scales, and scales for biological or geochemical process rates can be estimated and compared.

Non-dimensional numbers are ratios of like scales or combinations of scales, combined in such a way that all dimensions cancel. Dimensionless numbers are frequently used to help delineate the combination of conditions necessary for a regime change in a system under study. A well-known non-dimensional number from fluid mechanics is the Reynolds number (Re), which represents the ratio of inertial forces to viscous forces and thus characterizes whether a flow will tend to be turbulent (large Re) or laminar (small Re). Another well-known dimensionless number from estuarine hydrodynamics is the Richardson number (Ri). Although there are several specific forms for Ri, it generally represents the ratio of the strength of vertical density stratification to the strength of turbulent mixing, thus characterizing whether an estuarine water column will tend to be stratified (large Ri) or vertically well mixed (small Ri).

A time scale characterizes the approximate amount of time required for a process to occur. As such, a time scale is the inverse of a rate. We can estimate time scales for transport across a defined domain or for a biological or geochemical process to cause a specified change in a quantity of interest. The smaller the time scale, the faster the process. Time scales for different processes can be compared by calculating time scale ratios to determine which process is fastest and therefore most important. If the time scales for two processes are similar in magnitude, then they are approximately equally fast and thus roughly equally important.

10.3.2. Transport time scales

Transport time scales such as "residence time", "flushing time", and "age" are often invoked to help explain variability of biological or chemical constituents in tidal and non-tidal aquatic ecosystems (Lucas *et al.*, 2009b). As first-order descriptions meant to represent the retention of water mass within defined boundaries (Monsen *et al.*, 2002), transport time scales are related to the time required for water and dissolved substances to pass through aquatic systems (Sheldon and Alber, 2002). Time scales for transport can be compared to time scales characterizing inputs or transformations of biological or chemical constituents in order to identify the dominant processes underlying variability in water quality. For example, transport time has been used to explain variability in thermal stratification, dissolved organic carbon, elemental ratios of heavy metals, algal biomass and primary production, dissolved nutrient concentrations, and isotopic composition (Monsen *et al.*, 2002). For reactive water quality constituents, a long transport time is seen as providing more time for biological or geochemical processes to occur, potentially allowing for significant modification of the mass of a constituent (e.g., algal biomass) present in an estuary, or possibly even modifying the dominant form of a constituent (e.g., dissolved vs particulate nitrogen).

Many estuarine researchers and resource managers have an intuitive understanding of the concepts of "residence time", "flushing time", and water "age". While these terms have distinct and precise definitions and methods of calculation, they are often confused and loosely and interchangeably used (Zimmerman, 1976; Takeoka, 1984; Sheldon and Alber, 2002). Because different practitioners, authors, and readers may employ different definitions for the same time scale term and because the various terms and estimation methods are not necessarily equivalent (Takeoka, 1984), it is important to present the definition and calculation method in discussions of transport time scales.

Flushing time

Flushing time is a bulk or integrative parameter describing the general exchange characteristics of a waterbody without identifying detailed underlying physical processes or their spatial distribution (Monsen *et al.*, 2002). Flushing time t_f (also known as "turnover time") may be estimated most simply as Vol/q, where Vol is the basin volume and q is the total volumetric flow rate of water in or out of the embayment (Monsen *et al.*, 2002; Sheldon and Alber, 2006). In river-dominated systems or time periods, q may be approximated by R (river flow); however, where gravitational circulation is present, q may deviate from R. Net (tidally averaged) flow may be used to represent q, but in that case this approach would not explicitly acknowledge the flushing potential of tidal dispersion processes (Monsen *et al.*, 2002). If one is concerned with the replacement of a scalar quantity, then t_f may be calculated as M/F, where M is the total mass of the scalar within the domain and F is the rate of scalar flux through the domain (Takeoka, 1984; Monsen *et al.*, 2002). The Vol/q and M/F flushing (or "turnover") times may be considered equivalent to average times of transit for water or scalars through a waterbody (Bolin and Rodhe, 1973; Takeoka, 1984; Sheldon and Alber, 2006).

Following Dyer (1973), Sheldon and Alber (2002) described flushing time as a replacement time specific to freshwater, such that t_f is calculated as the ratio of an initial freshwater volume within the estuary to the freshwater inflow rate:

$$t_f = \frac{Vol\left(\frac{S_{SW} - S_{avg}}{S_{SW}}\right)}{R}, \tag{10.7}$$

where S_{sw} and S_{avg} are the salinity of seawater and the average estuarine salinity, respectively. This time scale may be considered an average transit time for freshwater (or a constituent dissolved in it; Sheldon and Alber, 2002) from the head to the mouth of a region (Zimmerman, 1976). This method of calculating flushing time is termed the "freshwater fraction method" and is equivalent to the scalar M/F flushing time if the scalar is defined to be freshwater. Although equation (10.7) appears not to account for the flushing effects of seawater entering at the estuary mouth, Sheldon and Alber (2006) showed that, when additional flushing by ocean water is accounted for, the resultant flushing time estimate reduces to equation (10.7); the effects of flushing by ocean water are implicitly accounted for in equation (10.7) by S_{avg}. The freshwater fraction approach may be applied to a whole estuary or to a collection of several connected segments of an estuary for which the sum of segment-specific transit times equals the transit time for the entire estuary (Sheldon and Alber, 2002). Assumptions inherent to all of these flushing models are that the flushing agent (q or F) is constant, the system is at steady state, and the basin (or basin segment) is horizontally well mixed.

Even for the relatively simple Vol/q approach described above, the required quantities may not be known for the aquatic system of interest. In such a case, the "e-folding" flushing time may be calculated empirically with an exponential fit to a time series of outflow concentration $C_{out}(t)$ for a constituent mass that was instantaneously introduced to the system with initial concentration C_o (Monsen *et al.*, 2002):

$$C_{out}(t) = C_o \exp\left(-\frac{q}{Vol}t\right) = C_o \exp\left(-\frac{t}{t_f}\right). \tag{10.8}$$

This approach assumes instantaneous and complete mixing within the basin and, given its exponential form, implicitly assumes the introduced mass is never completely flushed. With this approach, at time $t = t_f$ only 63% ($1-e^{-1}$) of the initial mass has been flushed. Because the e-folding flushing time is an empirical approach, all processes helping the scalar to flush, including tidal dispersion, are implicitly included.

Another class of approaches for calculating estuarine flushing time is represented by "tidal prism" models, which explicitly acknowledge tides as a flushing agent. Although they may vary in detail, assumptions, and levels of complexity, these models in their most idealized form estimate turnover time of estuarine water as $(Vol)(t_{tide})/Vol_p$, where Vol is estuary volume, t_{tide} is tidal period, and Vol_p is tidal prism volume. Different authors have defined Vol to represent estuarine volume at low, high, or mid-tide, although the majority of those cited here recommend the latter. Tidal prism volume is usually defined as the difference between high- and low-tide estuary volumes, but assumptions vary amongst authors regarding whether this volume change is comprised purely of (what we shall call here) Vol_{fl}, the flood-tide volume added to the estuary from the coastal environment through the seaward boundary (e.g., van de Kreeke, 1983; Sanford *et al.*, 1992), or if Vol_p also includes freshwater that enters the estuary from upstream during flood tide (Sheldon and Alber, 2006). To maximize clarity here, and to explicitly account for the tidal (Vol_{fl}) and freshwater (Vol_{fw}) contributions to flushing, the simple form above is modified following Sheldon and Alber (2006):

$$t_f = \frac{(Vol)(t_{tide})}{Vol_{fl} + Vol_{fw}}, \tag{10.9}$$

where Vol_{fw} is the volume of freshwater entering the estuary during one tidal cycle. This formulation assumes that the estuary is well mixed and that all of the water flooding in through the seaward boundary is "new" to the estuary. Acknowledging that some fraction of the plume of water exiting the estuary on ebb tide may return on the subsequent flood, equation (10.9) may be adjusted with a "return flow factor" (b) following Sanford *et al.* (1992) and van de Kreeke (1983):

$$t_f = \frac{(Vol)(t_{tide})}{(1-b)Vol_{fl} + Vol_{fw}}. \tag{10.10}$$

If freshwater inflow is negligible, then Vol_{fw} disappears and equation (10.10) represents the case of tidal flushing alone. The return flow factor b represents the fate of the tidal prism when it is outside the embayment and is thus governed by interactions between the ebb flow emerging from the tidal embayment and the flow along the coast outside the embayment (Sanford *et al.*, 1992). A model for b derived by Sanford *et al.* (1992) depends on relative phases of effluent ebb and coastal currents, tidal period, tidal prism, estuary mouth geometry, and coastal current speed and water depth. The return flow factor may also be estimated using conservative tracers (Luketina, 1998). The effect of the return flow factor on flushing of scalars within an estuary can be considerable (Sanford *et al.*, 1992; Luketina, 1998). An expression similar to equation (10.10) was derived by Luketina (1998) for an unsteady representation of an instantaneous release within an estuary of a conservative substance other than salt.

Sheldon and Alber (2006) showed that, if salinity is treated as a tracer and if freshwater inflow, incomplete escape of ebb flow, and incomplete mixing of flood flow are accounted for, then flushing times estimated by the tidal prism and freshwater fraction approaches are identical. Those authors advised that freshwater fraction models are most appropriate for estuaries with high freshwater inflow (i.e., where there is a measurable salinity difference between the estuary and the ocean), while simple tidal prism models apply in well-mixed systems where freshwater inflow is low and gravitational circulation is weak (i.e., where the ocean–estuary salinity difference is negligible).

To provide an example of the variability in flushing time definitions and calculation methods, Monsen *et al.* (2002) applied a numerical model to a tidal lake, Mildred Island (California, USA). With the simple Vol/q approach and using a tidally averaged flow rate at the lake's main opening, t_f ranged from 31–50 days, while an e-folding flushing time was estimated to be about 8 days. The discrepancy between these two estimates stems from the lack of accounting for flushing by tidal dispersion with the simple Vol/q approach. Thus, where tidally driven oscillatory flow represents a potentially significant mixing process (e.g., where flow rate fluctuates significantly about its mean and geometry is complex), a simple Vol/q flushing estimate based on tidally averaged (Eulerian residual) flow may overestimate flushing time. In such a case, the Vol/q estimate may be improved with a tracer-based correction factor such as freshwater fraction, if tracer concentrations vary measurably between the embayment and adjacent outside waters; if they do not, then tidal prism or other approaches may be more appropriate. Following Dronkers and Zimmerman (1982), Monsen *et al.* (2002) also estimated t_f as M/F, where F was a scalar point mass loading rate within Mildred Island and M was a roughly steady-state aggregate scalar mass across the system; this t_f ranged from 8–9 days. Although consistent with the e-folding approach, this mass flux-based calculation demonstrated the real-world complications of changing river flows and management practices which make a true steady state difficult to achieve. Baek (2006) demonstrated, also with a numerical hydrodynamic model, that atmospheric forcing (i.e., wind-driven mixing and baroclinic circulation driven by atmospheric heating) can further reduce flushing times in this environment by 15–40%.

Residence time

Although the term "residence time" is often casually used to refer to quantities calculated using some flushing time approaches described above, we follow here a strict definition of residence time as the time taken by a water parcel or material element somewhere within the embayment to leave the embayment (Zimmerman, 1976; Dronkers and Zimmerman, 1982; Takeoka, 1984). Thus, residence time is location-specific since for any arbitrary start time water parcels at different locations within a waterbody will require different lengths of time to leave the embayment, depending on their proximity to openings and the local hydrodynamics (Takeoka, 1984; Sheldon and Alber, 2002). In tidal systems, initial tidal phase can also significantly impact residence time (Monsen *et al.*, 2002). Estimation of a true residence time often requires numerical simulations (Sheldon and Alber, 2002) or could be approximated with drogues or drifters in the field.

Monsen *et al.* (2002; see Fig. 10.2a, left panel) used a numerical model of river- and tide-driven hydrodynamics and numerical particle transport to calculate spatially variable residence time for the tidal lake, Mildred Island. The time for a particle to leave the lake for the first time was taken as the residence time associated with that particle's starting time and location. Residence times within Mildred Island ranged from <1 h to >168 h (168 h was the simulation length).

In tidal systems with oscillatory flow, water parcels and particles may visit, leave, and then revisit a defined region over time. If the particles or scalar constituents of interest are "reactive", with reaction rate (e.g., growth or loss rate) a time-varying function of location, then the strict definition of "residence time" (i.e., time for a particle within a waterbody to leave once) may not represent a very useful parameter for describing variability in the constituent. Monsen *et al.* (2002) presented the concept of "exposure time", the cumulative amount of time a particle or water parcel spends within a domain, integrating over repeated visits to the domain over multiple tidal cycles (Fig. 10.2a, right panel). This time scale may be relevant, for example, to phytoplankton biomass in a predominantly positive- or high-growth region with a low- or negative-growth region within it or adjacent to it. The long-term net growth of phytoplankton depends on the full range of growth–consumption conditions encountered along tidal trajectories, and spatial gradients in these conditions may be quite sharp (Thompson *et al.*, 2008; Lucas *et al.*, 2009a). Even if phytoplankton only spend a portion of each tidal cycle within a low- or negative-growth region, the cumulative amount of time exposed to that environment over many tidal cycles could influence the longer-term phytoplankton biomass trend (Lucas *et al.*, 1999b).

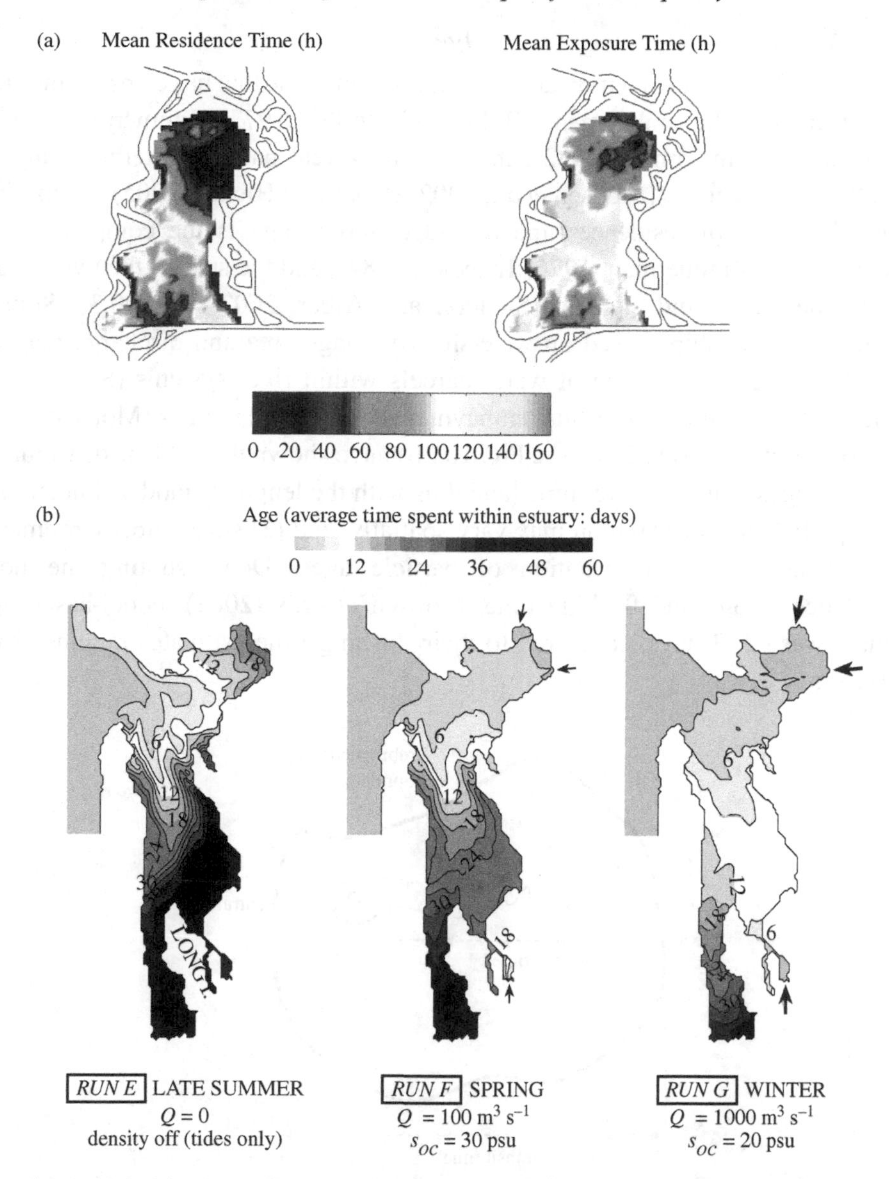

Figure 10.2. (a) Maps of mean residence time (left) and mean exposure time (right) calculated by a numerical model for Mildred Island, a tidal lake. The mean reflects the average value for each location over 24 simulations representing 24 different numerical particle release times staggered at 1-hour intervals. (Reproduced from Monsen *et al.*, 2002).

(b) Maps of water age calculated by a numerical model for Willapa Bay under three idealistic seasonal forcing scenarios amongst which river flow (Q), ocean salinity (s_{oc}), and inclusion of density effects varied. [Banas and Hickey, 2005. Copyright (2005) American Geophysical Union. Reproduced/modified by permission of AGU.]

Age

The age of a fluid element within an embayment is the time that has elapsed since the element entered the embayment (Bolin and Rodhe, 1973; Zimmerman, 1976; Dronkers and Zimmerman, 1982); thus, within a real natural waterbody, age is spatially heterogeneous (Monsen *et al.*, 2002; Banas and Hickey, 2005). According to the definition of residence time provided above, age is the complement to residence time (Zimmerman, 1976; Takeoka, 1984), and the sum of the two equals transit time for a fluid element (Sheldon and Alber, 2002; Fig. 10.3). Simple arithmetic relationships based on watershed drainage area and discharge may be derived to describe total age of water parcels within river systems (Søballe and Kimmel, 1987). For complex tidal embayments, numerical models (Monsen *et al.*, 2002; Banas and Hickey, 2005; see Fig. 10.2b) have shown that: (1) the distribution of particle ages may vary over time (and thus with the length of model simulation), (2) the particle age distribution may vary spatially, and (3) small subenvironments may contain a wide distribution of particle ages. Demonstrating the non-equivalence of age and flushing time, Monsen *et al.*'s (2002) model-based age estimate was ~1–3 days, compared to their flushing time estimates ranging from 8–50 days.

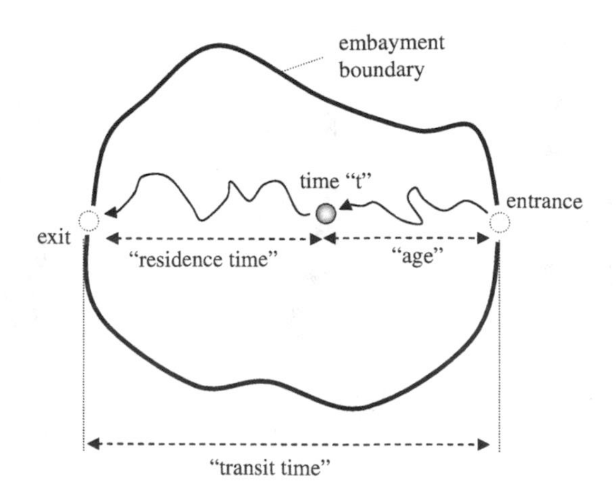

Figure 10.3. Two-dimensional schematic of an aquatic water body with two openings, one through which water parcels, scalar constituents, and/or particles enter and one through which they exit. The total time it takes a particle to pass through the embayment is the "transit time", which is the sum of water or particle "age" (the time since entering) and "residence time" (the time a parcel will remain in the embayment). "Age" and "residence time" are measured relative to an arbitrary time "*t*" and particle location within the water body.

10.3.3. Implications of transport time scales for water quality

In the previous section, we discussed several time scales characterizing horizontal transport, their definitions, how they are different, and various methods for calculating them. Now that we know some methods for estimating transport times, what do we do with them? Many scientists, engineers, and ecosystem managers use transport time scales to illuminate physical–chemical–biological interactions operating in their ecosystem, sometimes developing empirical "rules of behavior" between hydrodynamic processes and chemical or biological responses (e.g., by regressing constituent concentration against transport time). In some cases we compare transport time scales with time scales for growth, transformation, or loss processes potentially important in controlling a water quality constituent. Casting different kinds of processes into time scales places those processes in a common currency for ready comparison (Lucas *et al.*, 2009b). There are many examples of drawing on horizontal transport time scales to understand water quality. Here is a brief list:

- *Suspended particulate matter*: Uncles *et al.* (2002) surveyed 44 estuaries to show that for a given tidal range, long estuaries tend to have greater suspended particulate matter (SPM) concentrations within their high-turbidity regions than do short estuaries, due to the generally longer flushing times associated with physically longer estuaries (i.e., sediment is less likely to be lost from the estuary during freshets or ebb tides).
- *Dissolved nitrogen*: for several European estuaries, Middelburg and Nieuwenhuize (2000) calculated nitrogen turnover times (t_N, from dissolved nitrogen uptake rates) and compared them to water flushing times. Where $t_N > t_f$, nitrogen turnover processes are slow relative to transport and most of the nitrogen entering or produced in an estuary is expected to flush through it and get exported to the sea; whereas, if $t_N < t_f$, nitrogen turnover processes are relatively fast so nitrogen is efficiently recycled within the estuary. Nixon *et al.* (1996) showed that nutrient export from estuaries around the North Atlantic (as a fraction of inputs) is significantly and inversely correlated with log mean residence time.
- *Dissolved oxygen*: using a numerical model to calculate e-folding flushing times for the San Joaquin River, Monsen *et al.* (2007) showed that flow diversions away from the river's mainstem can significantly increase flushing times within a deep, dredged, frequently hypoxic portion of the river, shifting t_f from the order of days to weeks. Comparing (1) measured time series of river flow and dissolved oxygen, and (2) their calculated t_f values to BOD (biological oxygen demand) half-lives for that same region (~1–2 weeks; Volkmar and Dahlgren, 2006), it was concluded that diversion-influenced flushing times longer than a couple of weeks can lead to local oxygen depletions, a condition that can act as a barrier to upstream spawning migration of Chinook salmon.
- *Contaminants*: Baines *et al.* (2004) found that uptake of dissolved selenium by phytoplankton and bacteria, and thus the repackaging of Se into more bioavailable forms for upper trophic levels, was enhanced in a slowly flushed aquatic habitat relative to a nearby rapidly flowing channel habitat. On the other hand, the enhancement of phytoplankton

biomass over bacterial biomass in the slowly flushed environment caused selenium: carbon ratios in living particles, and thus the threat of toxicity to consumer organisms, to be lower than in the channel habitat.

- *Harmful algae*: Paerl and Huisman (2008) discussed how changes in the hydrologic cycle induced by global warming could enhance the dominance of cyanobacteria in aquatic systems. The sequence of elevated winter–spring rainfall (and associated increased nutrient discharges into water bodies) followed by extended summer drought periods (and increased residence times) can result in massive, harmful algal blooms.

Clearly, longer transport times provide more time for chemical or biological processes to act on a water parcel, dissolved constituent patch, or collection of suspended particles within an aquatic system, thus amplifying the effects of biogeochemical transformations and modifying the net downstream transport of biota, contaminants, or other reactive constituents (Søballe and Bachmann, 1984; Monsen *et al.*, 2007; Lucas *et al.*, 2009b).

In addition to the time scales for nutrient and BOD transformation mentioned above, time scales for other processes relevant to water quality can be estimated as (following Koseff *et al.*, 1993; Dame, 1996; Lucas *et al.*, 1998, 2009b; and others):

- *First-order growth*: $t_{growth} = 1 / \mu_{growth}$, where μ_{growth} is the specific (exponential) growth rate of a biological or chemical constituent (typical units for μ_{growth} are 1/day). This form of t_{growth} is known as an e-folding time scale for growth and is commonly used to represent the growth of phytoplankton. Similar forms could be applied to represent specific rates of exponential loss or decay.
- *Particle sinking*: $t_{sink} = \ell / w_s$, where ℓ is the relevant length scale (e.g., water column height, surface layer thickness) over which a particle is sinking, and w_s is particle sinking speed (typical units for w_s are m/day). This same length/velocity form may be used to characterize planktonic swimming (e.g., for zooplankton) or the rise of cyanobacteria in the water column associated with positive buoyancy.
- *Vertical turbulent mixing*: $t_{mix} = \ell^2 / K_z$, where K_z is vertical diffusivity (see Chapter 7 for more detail).
- *Pelagic grazing*: $t_{zoo} = 1 / ZP$, where ZP is the algal specific loss rate to zooplankton grazing (Lucas *et al.*, 2009b). For this time scale form to work, ZP must be cast in units of inverse time (e.g. 1/day).
- *Benthic grazing / advective loss through boundary*: $t_{ben} = \ell / BG$, where BG is benthic grazing rate, a potentially significant loss rate for phytoplankton by filter-feeding invertebrates living at the sediment–water interface. BG represents the volume of water filtered by organisms inhabiting a known bottom surface area within a known span of time; thus, units are typically m^3/m^2/day, which reduces to a (piston) velocity. Thus, the removal of phytoplankton from the water column by filter feeders feeding at a rate BG may be cast as an advective flux operating at the bottom boundary of the water column. ℓ is usually taken to be the water column height ($\ell = H$), thus invoking an implicit assumption that phytoplankton are well mixed throughout the water column. In this case, t_{ben} may be

thought of as the time it takes filter feeders to filter through the entire overlying water column (i.e., a water column "turnover" or "clearance" time). This time scale form could be applied to advective boundary fluxes for other water quality constituents.

Phytoplankton dynamics represents a prominent area of aquatic research and management for the application of horizontal transport time scales to unraveling variability in water quality. We know that concurrent hydrodynamic and biological (e.g., algal growth, grazing) processes can control the accumulation or depletion of phytoplankton biomass in aquatic systems. Examples of the use of transport time scales for characterizing and understanding phytoplankton dynamics are numerous. Koseff *et al.*, (1993) performed a dimensional analysis combining t_{ben}, t_{growth}, t_{mix}, and t_{sink} to delineate regimes of phytoplankton population growth and decay in a 1D vertical estuarine water column. Many authors have used comparisons of horizontal transport time scales (e.g., t_f) and t_{ben} to help identify processes regulating algal biomass in estuaries (Smaal and Prins, 1993; Dame, 1996; Strayer *et al.*, 1999). Although these sorts of analyses are perhaps most prominent in the area of phytoplankton dynamics, the time scale forms and overall approaches are adaptable to other water quality constituents.

10.3.4. A simple model of horizontal transport, sources, and sinks

Observations consistent with a positive relationship between transport time and phytoplankton biomass or production are common in the literature for rivers, lakes, reservoirs, estuaries, floodplains, and lagoons (Søballe and Kimmel, 1987; Howarth *et al.*, 2000; Schemel *et al.*, 2004; Torréton *et al.*, 2007; Paerl and Huisman, 2008). A positive phytoplankton–transport time (*P–T*) relationship means as transport time increases (or decreases), so does phytoplankton biomass or production (Lucas *et al.*, 2009b). The abundance of observations consistent with a positive *P–T* relationship may lead one to expect this kind of relationship in general. However, many other observations in aquatic systems reveal other types of *P–T* relationships, including negative, spatially variable, temporally variable, and non-monotonic relationships, as well as cases where no significant relationship is detected (Søballe and Bachmann, 1984; Basu and Pick, 1996; Walz and Welker, 1998; Paerl *et al.*, 2006; Strayer *et al.*, 2008).

To address the question of why we observe different relationships between phytoplankton biomass and transport time, Lucas *et al.* (2009b) presented a simple conceptual model (in mathematical and graphical form) of algal biomass in a generic aquatic system including parameters accounting for algal sources, losses, and advective transport. In this section, we summarize that model. Though originally derived to help explain the range of phytoplankton–transport time relationships observed in nature, the same relationship could be applied toward understanding other reactive, transported water quality constituents.

For suspended algae simultaneously growing and incurring loss through biological (e.g., pelagic and benthic grazing, senescence, disease) and physical processes (e.g., sedimentation) while being transported through an idealized advective steady-state system, the algal biomass concentration B at x distance downstream from the inlet was described by:

$$B(x) = B_{out} = B_{in} \exp\left(\frac{\mu_{growth} - \mu_{loss}}{u} x\right). \quad (10.11)$$

Equation (10.11) was derived from a time-varying one-dimensional longitudinal partial differential equation for streamwise advection and first-order growth and loss of phytoplankton [i.e., equation (10.6) without the diffusion term and with $[\mu_{growth} - \mu_{loss}]B$ substituted for "s"]. B_{in} is algal biomass concentration entering the system at the upstream boundary, μ_{growth} (1/day) is the algal specific growth rate, μ_{loss} (1/day) is the sum of specific loss rates due to processes operating between the upstream and downstream boundaries, and u is the characteristic velocity along the primary flow direction (for this idealized system, u is assumed unidirectional).

Equation (10.11) can be rewritten as a function of three time scales: t_{tran} (the transport time scale), t_{growth} (the algal growth time scale), and t_{loss} (the time scale for combined algal losses). The quantity t_{tran} is most simply represented with the Vol/q flushing time (see Section 10.3.2), which for the simple advective channel considered here reduces to x/u. t_{loss} is the reciprocal of the sum of specific loss rates (μ_{loss}), potentially including zooplankton grazing, the depth-averaged effect of grazing by benthic filter feeders, and algal sedimentation loss (see Section 10.3.3). Substituting the three time scales into equation (10.11) and rearranging a bit, Lucas *et al.* (2009b) derived the dimensionless expression for steady-state algal biomass:

$$\frac{\text{export}}{\text{import}} = B^*_{out} = \frac{B_{out}}{B_{in}} = \exp\left(\left[1 - \frac{1}{t^*_{loss}}\right] t^*_{tran}\right), \quad (10.12)$$

where $t^*_{loss} = t_{loss}/t_{growth}$ (representing the relative importance of growth and loss), $t^*_{tran} = t_{tran}/t_{growth}$ (representing the relative importance of transport and growth), and B^*_{out} is a non-dimensional number representing the biomass exported from the domain at the downstream boundary normalized by the biomass imported to the domain through the upstream boundary.

Figure 10.4 (reprised from Lucas *et al.*, 2009b) shows contours of B^*_{out} calculated from equation (10.12) and plotted as a function of t^*_{loss} and t^*_{tran}. Lucas *et al.*'s (2009b) conceptual model in mathematical [equation (10.12)] and graphical form (Fig. 10.4) shows why three basic P–T relationships may exist in an aquatic system. If loss dominates growth ($t^*_{loss} < 1$; lower part of Fig. 10.4), then biomass (B^*_{out})

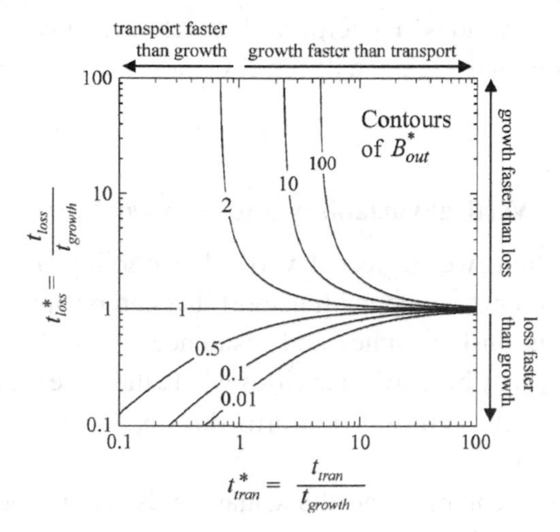

Figure 10.4. Contours of the dimensionless quantity B^*_{out}, the ratio of downstream (outgoing) to upstream (incoming) algal biomass, as a function of t^*_{loss} (time scale for algal loss normalized by the time scale for algal growth) and t^*_{tran} (time scale for transport normalized by the time scale for algal growth). This is the graphical form of Lucas *et al.*'s (2009b) conceptual model.

decreases as transport time (t^*_{tran}) increases. Whereas, if growth dominates loss ($t^*_{loss} > 1$; upper part of Fig. 10.4), then algal biomass increases with increasing transport time. If loss balances growth ($t^*_{loss} = 1$), then biomass does not change with changing transport time. Thus, the conceptual model (Lucas *et al.*, 2009b) shows that there are three possible *P–T* relationships (positive, negative, and none) and that the sign of the relationship depends on the growth–loss balance (t^*_{loss}).

Using data sets from around the world, Lucas *et al.* (2009b) showed that aquatic systems display the full range of t^*_{loss} (i.e., <1, >1, about 1) and thus of *P–T* relationships. They also showed that an individual system may exhibit a range of *P–T* relationships. The superposition of different relationships (e.g., in a regression of algal biomass vs transport time) may collectively portray an absence of a relationship. Subdividing *P–T* data into t^*_{loss} <1, >1, and ≈1 subgroups can help reveal multiple significant underlying relationships.

Complex hydrodynamic processes (e.g., tidally oscillatory flow and dispersion) could cause deviations from the conceptual model if it is used to predict algal biomass as a function of estimated transport time. So, if the simple conceptual model is used for prediction, care must be taken in choosing and calculating the relevant transport time scale. Although the conceptual model may not quantitatively predict algal biomass for systems that violate critical underlying assumptions (e.g., complex tidal systems), knowledge of the growth–loss balance (t^*_{loss}) is nonetheless informative for any aquatic environment, as it provides a useful index relating whether phytoplankton biomass within a defined region will tend to increase or decrease as transport time changes.

10.4. Vertical variability and transport processes

In the previous section, we neglected vertical variability and vertical transport processes, focusing on net horizontal transport that can be approximately captured by quantities such as flushing times and residence times. The approximation of vertical homogeneity may be justified in cases where the water quality constituent of interest is approximately vertically uniform due to, for example, thorough and vigorous vertical mixing.

But estuarine water columns and the scalar fields within them are not always vertically homogeneous. Estuaries by their very nature as meeting places of fresh and salty water are susceptible to the development of vertical gradients in density; consequently, gradients in other physical, biological, and chemical quantities and processes may develop or be reinforced. By whatever mechanism, if there is potential for the development of vertical gradients in important water quality parameters, then the nature, extent, and strength of vertical transport is also potentially important as it determines the degree of communication between vertically separated zones within the water column.

Vertical transport of water quality constituents in an estuary can be hydrodynamic (i.e., due to the motion of the fluid in which the constituent is suspended) or non-hydrodynamic (i.e., due to the motion of particles independent of the fluid motion). Vertical hydrodynamic transport in an estuary includes vertical turbulent mixing and vertical advection of water parcels. Vertical advection of water parcels may be associated with sharp topographic variations or secondary circulations such as those which sometimes develop around bends (see Chapter 5). Vertical turbulent mixing can be driven by wind stress at the water surface, interaction of the horizontal tidal current with the rough bottom boundary, unstable stratification, or velocity shear within the water column (see Chapter 7). Non-hydrodynamic transport may occur due to sinking of negatively buoyant particles (e.g., sediment, diatoms), rising of positively buoyant particles (e.g., some cyanobacteria), or the motility of suspended biota (e.g., swimming by some phytoplankton, zooplankton).

In this section, we will consider some of the predominant observed effects of vertical turbulent mixing and stratification on water quality in estuaries, the significance of the time scale of variability of stratification (e.g., tidal or longer), and the use of time scales to understand the combined effect of simultaneous diffusive mixing and non-hydrodynamic vertical advective processes.

10.4.1. Effects of stratification and turbulence

As detailed in Chapter 7, stable vertical density stratification, whether due to vertical salinity gradients (e.g., Pamlico River Estuary; Lin *et al.*, 2008), temperature gradients (e.g., Saldanha Bay; Monteiro and Largier, 1999), or both (e.g., Gulf of Mexico; Rabalais *et al.*, 2002), opposes vertical turbulent mixing. Stratification makes it more difficult for turbulence to mix a water column because mixing requires more energy in a stably stratified environment than in a homogeneous environment; hence, stratification quells mixing. On the other hand, turbulence erodes stratification and works to homogenize density; if turbulence successfully homogenizes density, then it may also homogenize other scalars.

The vertical mixing (K_z) profile can be particularly important if source and sink processes for scalars vary significantly in the vertical dimension (e.g., reaeration as a source of oxygen at the water surface vs oxygen consumption by bacteria in the lower water column; rapid photosynthesis and positive net algal growth in the well-lit surface layer vs net respiration and negative net algal growth in the dark lower water column). Sharp mixing gradients such as those caused by a pycnocline can help to isolate zones of scalar accumulation or depletion, resulting in sharp scalar concentration gradients. Rapid, thorough mixing, on the other hand, can transport scalars throughout the range of source–loss conditions existing within the water column, this range sometimes being well approximated by the depth-averaged condition (Lucas *et al.*, 1999a). Thus, vertical mixing processes can be important for water quality because they can connect the upper water column and the lower water column, thereby (1) exposing material from one zone to conditions in other zones, and (2) potentially (in the case of vigorous turbulent mixing) homogenizing the water column, allowing for approximations such as those made in the previous section.

Two water quality parameters commonly linked to stratification and vertical mixing processes are phytoplankton biomass (e.g., as chlorophyll *a*) and dissolved oxygen. These two parameters are linked to each other and are critical to the functioning and health of aquatic ecosystems. Phytoplankton are microscopic aquatic plants or "algae" living in the water column, thereby representing a critical, sometimes dominant (Sobczak *et al.*, 2002), component of fuel (organic carbon) for the base of the aquatic food web supporting upper trophic levels such as fish. Thus, phytoplankton are an important food source, but certain "blooms" of phytoplankton may be noxious or harmful, whether for the particular species that dominate (some of which may produce toxins harmful to aquatic organisms or to humans) or for the amount of biomass produced. If adequate nutrients and light are available, then dense phytoplankton blooms may develop within a water column; uneaten phytoplankton and fecal matter generated by the food web ultimately sink to the bottom, where they are decomposed by aerobic bacteria, potentially resulting in oxygen depletion (Rabalais *et al.*, 2002).

In many estuaries, this process is initiated or exacerbated by the development of density stratification, which can (1) isolate phytoplankton in the sunny upper water column where algal photosynthesis and growth are rapid and benthic grazers are remote, and (2) prevent replenishment of oxygen-poor bottom waters by reaeration processes occurring at the water surface. See Fig. 10.5 for observed linkages between

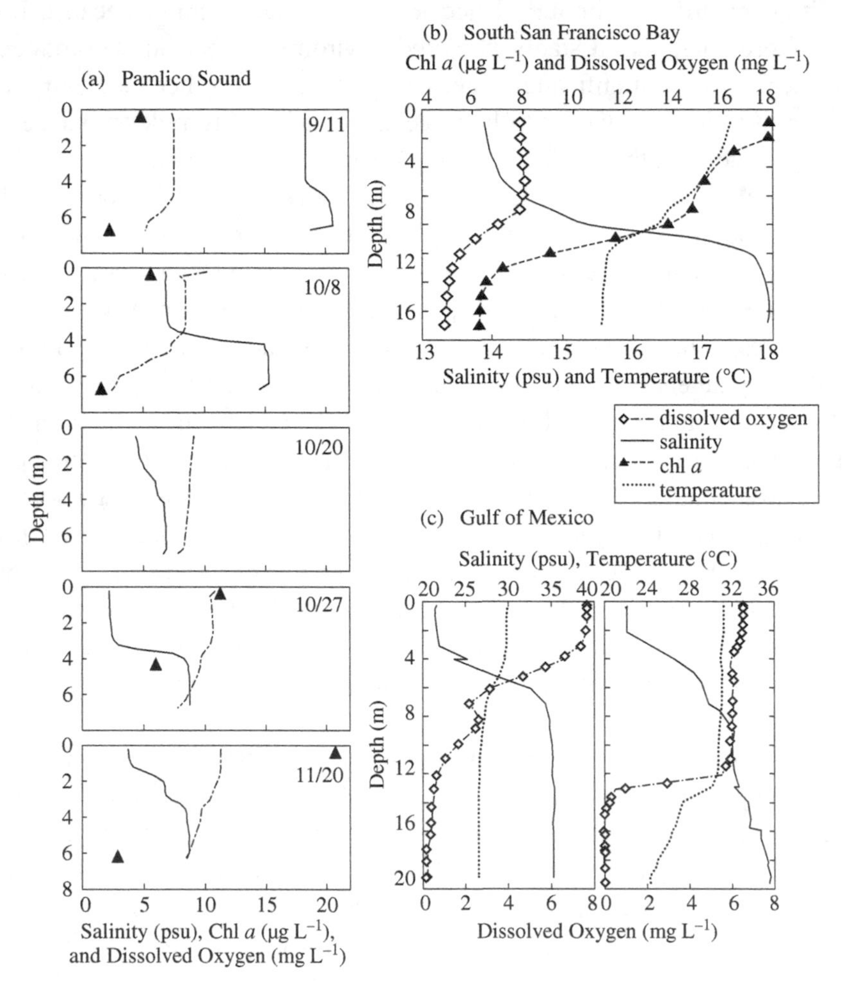

Figure 10.5. (a) Vertical distributions of salinity, dissolved oxygen, and chlorophyll *a* in Pamlico Sound in 1999. [Paerl *et al.*, 2001. Copyright (2001) National Academy of Sciences, USA. Reproduced/redrawn with kind permission of PNAS.]

(b) Vertical profiles of salinity, temperature, dissolved oxygen, and chlorophyll *a* at Station 28 in South San Francisco Bay on 9 May 2006 (http://sfbay.wr.usgs.gov/access/wqdata).

(c) Salinity, temperature, and dissolved oxygen profiles in the Gulf of Mexico. [Rabalais and Turner, 2006, figure 3, p. 230. Copyright (2006) Springer. Reproduced/redrawn with kind permission of Springer Science and Business Media and N. N. Rabalais. Data source: N. N. Rabalais, LUMCON.]

stratification and vertical variability in phytoplankton biomass (chlorophyll *a*) and dissolved oxygen (Paerl *et al.*, 2001; Rabalais and Turner, 2006; http://sfbay.wr.usgs.gov/access/wqdata).

Stratification can form and erode over a range of time scales (hours to months), but stratification that persists for at least days can have noticeable ecological and water quality consequences. In some systems, stratification is seen as a necessary ingredient for the development of hypoxia (low dissolved oxygen) in bottom waters (Rabalais *et al.*, 2002), while enhanced mixing energy provided by wind or tides may be sufficient to mix the water column and end a hypoxic episode (Bergondo *et al.*, 2005). Hypoxia or anoxia (no dissolved oxygen) can lead to stress, death, or emigration of benthic organisms (e.g., clams, mussels, oysters) and the demersal fish or crabs that consume them (Buzzelli *et al.*, 2002; Rabalais *et al.*, 2002), producing "dead zones" occupying areas as large as $O(100–10{,}000)$ km^2.

While stratification may provide a physical environment conducive to the development of hypoxia, the phytoplankton driving that process require adequate nutrients for cell growth and division. Most commonly limiting to estuarine phytoplankton are nitrogen and phosphorus, which are supplied by rivers, urban areas, agricultural runoff, and atmospheric deposition. Thus, nutrient inputs to estuaries are frequently correlated with river flow (Swaney *et al.*, 2008). While phytoplankton growth and biomass can be limited by a dearth of nutrients, they are not necessarily high when nutrient concentrations are high because the response of a coastal ecosystem to nutrient loading may be mediated by a suite of other processes (Cloern, 2001). Along with light limitation by turbidity, flushing (discussed in the previous section), and grazing by pelagic and benthic consumers, the coupling between stratification and vertical mixing, as discussed above, represents a key interaction that can determine how coastal ecosystem water quality will respond to nutrient inputs. "Eutrophication", which has been strictly defined as an increase in the supply rate of organic matter (Nixon, 1995), may more broadly be considered as the full set of biogeochemical and ecological responses, either direct or indirect, to anthropogenic fertilization of ecosystems at the land–sea interface (Cloern, 2001).

Density stratification and vertical gradients in mixing, along with the oxygen and algal biomass gradients that may accompany them, are heavily implicated in transformations of other water quality parameters such as toxic contaminants. For example, mercury methylation (the Hg conversion process producing methyl mercury, the Hg compound predominating in fish tissues) has been observed to occur in the oxic–anoxic transition zone within the pycnocline of a stratified estuary (Mason *et al.*, 1993). Stratification and vertical mixing can also influence the remineralization and vertical distributions of nutrients in systems where the sediments represent a significant nutrient source (Nagy *et al.*, 2002).

10.4.2. Significance of the time scale of stratification

In Chapter 4, different types and time scales of stratification were discussed. Gravitational circulation may produce persistent salinity stratification (i.e., lasting for days or longer) when a strong longitudinal salinity gradient coincides with weak tidal mixing, such as during a neap tide in the presence of significant freshwater inflow. The "estuarine Richardson number" (see Fischer *et al.*, 1979) and "horizontal Richardson number" (see Monismith *et al.*, 1996) provide tools for estimating values of R, $\partial S/\partial x$, u or other parameters that, for a given set of conditions, would allow the water column to persistently stratify. Tidal straining of approximately vertical isopycnals can cause shorter-lived, tidal time scale stratification (usually on ebb tides) when the longitudinal salinity gradient is weaker and/or tidally driven vertical turbulent mixing is stronger than in the persistent stratification case. This tidally variable stratification is called "SIPS" (strain induced periodic stratification; Simpson *et al.*, 1990) and, in some estuaries, may be the kind of stratification that prevails during high-energy spring tides when a longitudinal salinity gradient is present (Monismith *et al.*, 1996; Lucas *et al.*, 1998). Intradaily (thermal) stratification may also be caused by diurnal heating and cooling cycles.

What is the significance of intradaily or intratidal variations in stratification for estuarine water quality? Numerical model results have suggested that, while algal bloom development in an isolated deep estuarine channel may be significantly influenced by persistent stratification, tidally periodic stratification does not increase the likelihood of a bloom beyond that of a constantly unstratified water column (Lucas *et al.*, 1998).

The effect of periodic short-lived stratification may vary, however, between systems. Joordens *et al.* (2001) observed in the Rhine outflow region of the Dutch coast that periods of stratification during the tidal cycle were characterized by cell retention in the lower water column due to sinking and worse conditions for phytoplankton growth; whereas, during well-mixed periods of the tidal cycle, the sinking phytoplankton (and possibly nutrients) are mixed up into the upper water column where light is plentiful and photosynthesis can proceed. Taylor and Stephens (1993) showed that diurnal variations in turbulent mixing and mixed layer depth can be critical in setting bloom timing in the thermally stratified ocean. Tidally periodic stratification has been shown to enhance the particle trapping tendencies of flood-dominated estuaries (those that fill faster than they empty) by allowing particles to sink during stratified (weakly mixed) ebb tides but resuspending particles during destratified (more strongly mixed) floods when up-estuary transport occurs (Chant and Stoner, 2001). This process could directly or indirectly affect estuarine biota and water quality through transport and trapping of larvae acting as passive particles or through tidal variability in food availability, turbidity levels, or ecological parameters that are functions of the turbidity field.

Although the specific effects of hourly scale variations in stratification and mixing on ecology and water quality may be site- and time-specific, these few examples demonstrate that high-frequency variations in the physical environment can yield long-term net effects different from a single persistent condition. For this reason, knowing whether stratification is persistent or periodic over time scales of hours is critical to our understanding of how an estuarine ecosystem works. Stratification is usually not measured continuously (i.e., with moored instrumentation); rather, vertical density profiles typically are measured relatively infrequently. Temporally coarse monitoring (e.g., with period $\geq$1 tidal period) of stratification in an estuary could be misleading because it could indicate that stratification is present at the moment of observation but not indicate whether the stratification persists through the tidal cycle or through the day–night period. Therefore, if the ecosystem and water quality effects of stratification are of interest, then short-term high-frequency (hourly scale or finer) measurements of stratification may prove valuable in showing (1) whether hourly scale periodic stratification occurs in that estuary and (2) the conditions under which we should expect hourly scale periodic vs persistent stratification to form.

10.4.3. Interactions of sinking with vertical mixing

In Section 10.4.2, the importance of vertical turbulent mixing for sinking particles was briefly discussed. Depending on its strength, mixing can counteract the sinking of negatively buoyant particles, and the relative strengths of sinking vs mixing can have ecological implications. For example, a weak mixing environment may not be the most advantageous for diatoms (heavy, silica-coated phytoplankton) if the water column is deep or turbid; in this case, sinking could cause them to advect downward out of the euphotic zone and into the lower water column where light limitation causes respiration to dominate over photosynthesis and where benthic consumers may be present. The balance between turbulent mixing and sinking can be represented by the turbulent Peclet number, the ratio of the mixing time scale to the sinking time scale (Lucas *et al.*, 1998):

$$Pe_t = \frac{t_{mix}}{t_{sink}} = \frac{\ell^2/K_z}{\ell/w_s} = \frac{\ell w_s}{K_z}, \qquad (10.13)$$

where ℓ is the vertical length scale of interest (e.g., surface layer depth or water column height; see also Section 10.3.3). If $Pe_t \gg 1$, sinking dominates vertical transport and mixing is slow by comparison. This case may be conceptualized by a cloud of quickly sinking particles, with turbulent mixing causing modest diffusion about the sinking center of mass. If $Pe_t \ll 1$, turbulent mixing dominates vertical transport and sinking is slow by comparison. This case may be conceptualized by a cloud of slowly sinking particles that are frequently and vigorously transported upward by large, energetic turbulent eddies, resulting in a uniform vertical particle distribution (Lucas *et al.*, 1998). If $Pe_t \approx 1$, then mixing and sinking are about equally important.

In addition to the case of sinking phytoplankton, equation (10.13) can be used to assess the relative influence of mixing and advection for other sinking particles such as sediment or for certain phytoplankton or zooplankton that rise or swim. For example, consider zooplankton swimming at a speed of w_{swim} = 0.01 m/s in an unstratified 10 m deep water column with depth-averaged current speed of about 0.1 m/s. To find out whether the swimming zooplankton can overcome turbulent mixing to successfully migrate from the bottom to the top of the water column, we can calculate Pe_t. Using the $K_z = 0.067Hu_*$ (Fischer *et al.*, 1979) estimate for turbulent mixing coefficient in a unidirectional flow, where u_* (shear velocity) is taken as 10% of the depth-averaged current speed, we obtain $Pe_t = Hw_{swim}/K_z = 14.9$. Since Pe_t indicates that swimming is an order of magnitude faster than turbulent mixing, the zooplankton should be able to overcome mixing and successfully travel to the top of the water column.

10.5. Hourly scale variability

The time scales of variability associated with potentially important estuarine processes range from seconds to decades (Fig. 10.6a). Within the lower end of that range, we know that important physical and biogeochemical processes in many coastal regions vary with diel cycles of solar radiation or wind and with the diurnal or semidiurnal cycles of flood/ebb currents and tidal shallowing and deepening of the water column. However, logistical and technological constraints have historically limited our ability to observe the finer-scale variability of estuarine processes (Lucas *et al.*, 2006), thereby often constraining our conceptual models of estuarine function to incorporate process variability occurring over time scales longer than a day or a tidal cycle (e.g., the effects of persistent stratification or river flow on biogeochemical quantities).

Over the past couple of decades, continual advances in oceanographic instrumentation have allowed for the measurement of water quality parameters and of underlying physical processes at ever higher frequencies (e.g., Fig. 10.6b). These technological improvements are opening increasingly more resolved windows through which we can examine the variability of physical, biological, and chemical variables, thereby allowing us to entertain and explore the potential long-term significance of processes and interactions occurring over time scales finer than a

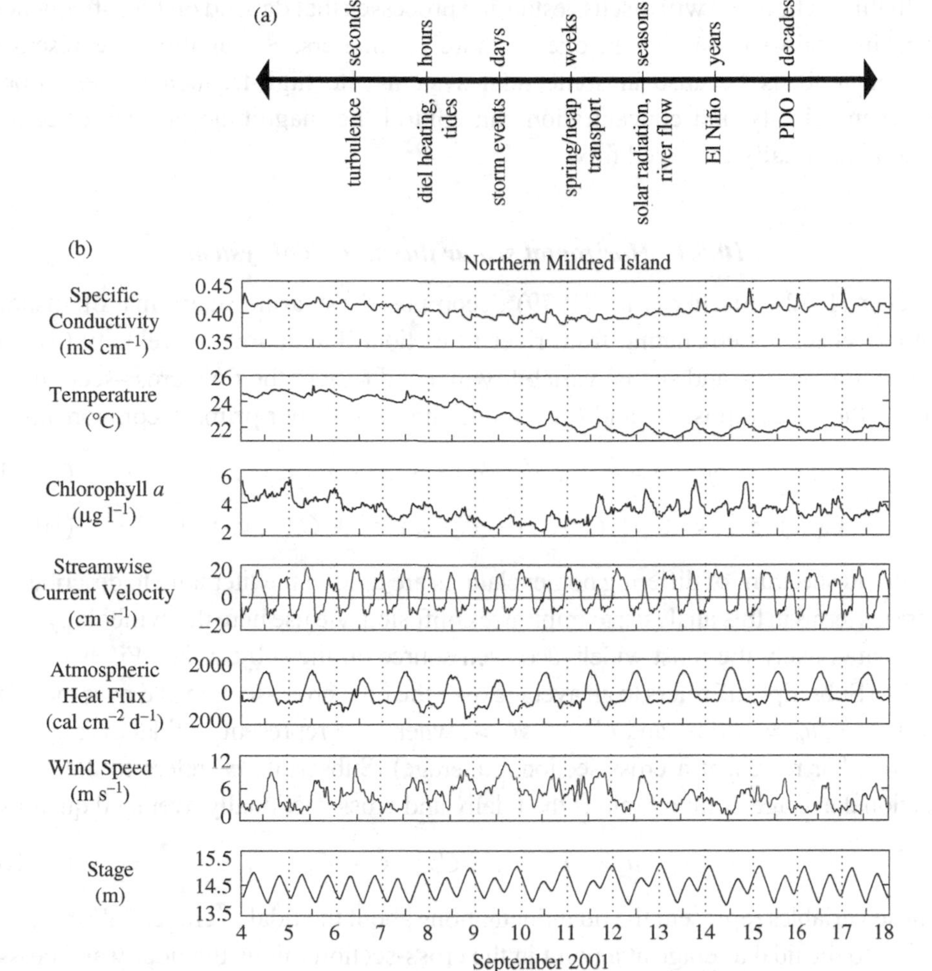

Figure 10.6. (a) Schematic describing the range of time scales of variability characterizing processes influencing estuaries.

(b) High-frequency measurements of water quality and underlying physical forcings in northern Mildred Island, California. (Reproduced/modified from Lucas *et al.*, 2006.)

day. Parallel advancements in computing power provide us with another continually improving toolset (numerical modeling) with which to explore highly resolved physical–biological–chemical interactions in a detailed way. These tools allow for the discovery of higher-frequency variability that may not have previously been known to exist and can initiate the development of new conceptual models which include more process variability occurring over shorter time scales.

In this section, we will discuss estuarine processes that depend on high-frequency (i.e., intradaily) variability in one or more parameters. Scalar fluxes represent a particular focus because in some tidal systems the high-frequency interactions between velocity and concentration can control the magnitude or even direction of total net tidally averaged flux.

10.5.1. Horizontal scalar flux in a tidal system

Fischer (1972), Fischer *et al.* (1979) decomposed estuarine scalar mass transport into mechanisms emanating from river flow, tidal fluctuations in velocity (u) and concentration (C), and spatial variability in u and C over the flow cross-section. To do so, they decomposed u and C each into the sum of four primary components:

$$u(x, y, z, t) = u_a + u_c(x, t) + u_s(x, y, z) + u'(x, y, z, t), \tag{10.14}$$

$$C(x, y, z, t) = C_a + C_c(x, t) + C_s(x, y, z) + C'(x, y, z, t). \tag{10.15}$$

(A broad spectrum of differing nomenclatures are used in the literature to describe the various terms in this analysis; to minimize confusion, we use here the symbology used in what is likely the most widely accessed source on the subject, i.e., Fischer *et al.*, 1979.) Subscript "*a*" refers to an average over the flow cross-section and over the tidal cycle (i.e., $u_a = <\bar{u}>$, and $C_a = <\bar{C}>$, where $<>$ represents a tidal average and the overbar represent a cross-sectional average). Subscript "*c*" refers to the cross-sectional average at time t minus the tidally and cross-sectionally averaged quantity:

$$u_c = \bar{u} - u_a, \quad C_c = \bar{C} - C_a. \tag{10.16}$$

These variables represent the tidal fluctuations about the tidal average. Subscript "*s*" refers to the tidal average at a point in the cross-section minus the tidally and cross-sectionally averaged quantity:

$$u_s = \langle u \rangle - u_a, \quad C_s = \langle C \rangle - C_a. \tag{10.17}$$

These variables represent tidally averaged spatial deviations about the cross-sectional mean. The primed terms are the terms that are left over after the averaged terms above are subtracted from the observed variables; u' is described as representing unsteady shear flow (Fram *et al.*, 2007; Martin *et al.*, 2007). The total tidally averaged transport of constituent C through a cross-section of area A is as follows, after substitution and multiplication of the above velocity and concentration components:

$$\underset{(1)}{\dot{M} = A\langle\overline{uC}\rangle} = \underset{(2)}{RC_a} + \underset{(3)}{A\langle u_c C_c\rangle} + \underset{(4)}{A\overline{u_s C_s}} + \underset{(5)}{A\langle\overline{u'C'}\rangle}. \tag{10.18}$$

Term (1) is the total tidally averaged flux over the cross-section (in mass per unit time). Term (2), the "advective" flux, is driven by the river flow *R*. Terms (3)–(5) are the "dispersive" flux terms arising from different mechanisms. Term (3), which is the time average of the tidal correlation of the cross-sectional averages (i.e., driven by temporal deviations from the mean), is considered to represent "tidal pumping" and "tidal trapping" mechanisms and is frequently a major dispersive mechanism (Martin *et al.*, 2007). Term (4) is the "steady" or "residual" circulation term driven by time-averaged spatial profiles of *u* and *C*; this term includes mechanisms such as gravitational circulation. In cases where thorough mixing over the cross-section and minimal shear (e.g., negligible gravitational circulation) may be assumed, then this term may be neglected. Term (5) represents "unsteady" or "oscillatory" shear flow and short-time scale random motions. To obtain equation (10.18), it is assumed that tidal fluctuations in *A* are small enough to assume it is constant. See Fischer (1972), Fischer *et al.* (1979) for further discussion of other neglected and presumably small terms, and for the suggested further decomposition of u_s, C_s, u', and C' into transverse and vertical variations (Fram *et al.* 2007 provides an example of this with real observational data). The relative magnitudes of the terms in equation (10.18) may vary between estuaries and also temporally for an individual system, depending on variability in freshwater input, tidal range, excursion and velocity, geometry, strength of vertical mixing, and distributions of source–loss processes that can set up spatial gradients in constituents of interest.

Often we can estimate the advective term (2) relatively easily by obtaining some estimate of river flow and mean constituent concentration. The dispersive terms are more difficult to estimate because they require much more temporally and/or spatially detailed measurements, usually through the deployment of moored and/or boat-mounted instrumentation. What might we miss if do not make the detailed measurements and instead assume advective flux dominates over the dispersive terms [i.e., if we assume term (1) = term (2)]?

For locations and constituents with significant concentration gradients and large tidal excursions, the tidal pumping component of dispersion [term (3)] can dominate advective and other dispersive components, in some cases causing the direction of total net flux to be opposite the direction of the advective flux (Lucas *et al.*, 2006; Fram *et al.*, 2007; Martin *et al.*, 2007; see Fig. 10.7). A non-zero tidal pumping flux requires a spatial gradient in *C* because it results from a difference between flood and ebb concentrations (i.e., if the distribution of *C* is uniform, then at a location *C* is the same on ebb and flood and there is consequently no net tidal pumping flux). Tidal pumping flux can vary seasonally due to changes in concentration gradient; for example, chlorophyll *a* gradients may shift in magnitude or direction depending on the seasonally shifting locations of algal blooms (e.g., estuary vs ocean), potentially reversing the direction of dispersive and even total chlorophyll *a* flux (Martin *et al.*,

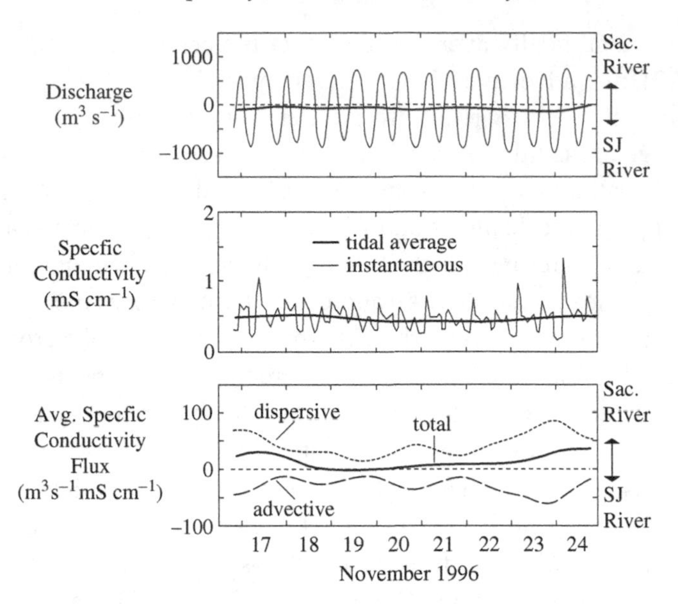

Figure 10.7. Measurements of instantaneous and tidally averaged discharge (top panel) and specific conductivity (middle panel), and tidally averaged dispersive, advective, and total specific conductivity flux (bottom panel) for Threemile Slough, California (a tidal channel connecting the Sacramento and San Joaquin Rivers), in November 1996 (C. A. Ruhl unpublished data; California Department of Water Resources California Data Exchange Center public communication). Here, dispersive and total conductivity flux were directed toward the Sacramento River, which is the opposite direction of discharge and advective flux. (Reproduced from Lucas *et al.*, 2006.)

2007). This flux can also vary over the spring–neap cycle, as tidal excursion length typically increases with the stronger currents of spring tides. Advective flux [term (2)] is almost always down-estuary; however, because the tidal pumping flux is always down-gradient, a single set of tidal conditions will produce dispersive fluxes of opposite directions (up-estuary vs down-estuary) for constituents with oppositely oriented concentration gradients (Gardner and Kjerfve, 2006). For example, Sylaios *et al.* (2006) found that average tidal pumping fluxes of nitrogen was directed into a small Mediterranean lagoon, while average tidal pumping flux of phosphorus and chlorophyll *a* were directed out of the lagoon. Tidal pumping flux may also evolve seasonally in response to changes in the relative phasing between velocity and concentration (Fram *et al.*, 2007).

Biological or geochemical processes can interact with physical processes to modify net fluxes into or out of a tidal embayment. For example, interaction of diurnal phytoplankton production and biomass with semidiurnal tidal currents can result in fortnightly (spring–neap-like) pulses in dispersive algal biomass export (Lucas *et al.*, 2006). Grazing by benthic bivalves can remove algal biomass entering an embayment on flood tides, causing net biomass flux to be directed into the embayment (Banas *et al.*, 2007). Photosynthesis that exceeds consumption of dissolved oxygen within a tidal system can cause outward dispersive DO flux at the mouth (Gardner *et al.*, 2006).

10.5.2. Other high-frequency interactions with water quality

The tidally periodic stratification dynamics discussed in Section 10.4.2 and the high-frequency components of scalar flux described in Section 10.5.1 represent two important categories of hourly scale physical interactions with water quality. There are many other examples of tidally or atmospherically driven intradaily physical influences on estuarine water quality. For example, Rocha (1998) found that the tidally driven drying–inundation cycle of intertidal sediments drove a rapid convective turnover of sediment pore water and large export of ammonium into the water column. Tidal cycles of resuspension and deposition of suspended particulate matter (SPM) can cause exchanges between dissolved and particulate organic carbon pools in the water column due to changes in mineral surface to water ratios affecting sorption processes (Middelburg and Herman, 2007). Tidally shallowing and deepening water column height has been shown with a numerical model to interact with the diel cycle of solar radiation and benthic grazing to control weekly to monthly scale trends in phytoplankton biomass accumulation or depletion (Lucas and Cloern, 2002). Desmit *et al.*, (2005) showed that large modeling errors may be incurred if key parameters for a phytoplankton productivity model (e.g., light attenuation coefficient, incident solar radiation) are represented with time-averaged, as opposed to time-varying, values.

10.5.3. Implications of high-frequency variability for sampling

If water quality constituents vary with large amplitude over time scales shorter than the sampling period, then temporally coarse sampling could provide misleading information. An example is provided in Fig. 10.8a, which shows that temporally coarse sampling (in this case, a hypothetical daily noon-time measurement scheme) of a diurnally varying water quality constituent could entirely miss the peaks and underestimate the mean. Practical and logistical limitations often dictate that high-resolution sampling cannot be performed in a sustained fashion. However, knowledge of *whether* and *how* a key scalar varies in a system can help in optimizing the match between sampling goals and sampling design. Figure 10.8b presents a subsampling exercise that started with a real 10-min time series of chlorophyll *a* in a tidal environment (Lucas *et al.*, 2006); the observed variability was shown to result from a combination of semidiurnal tidal advection, diurnal wind-induced

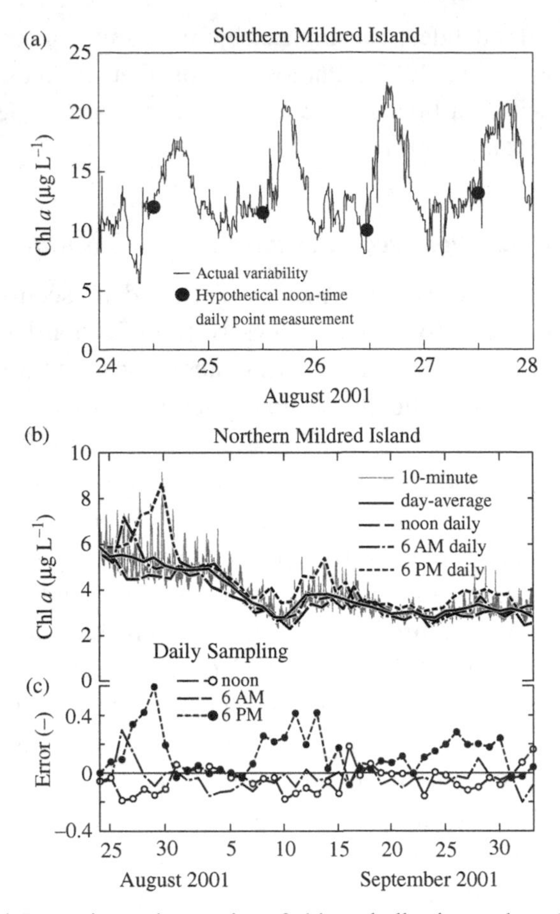

Figure 10.8. (a) Ten-minute time series of chlorophyll *a* in southern Mildred Island, California (L. V. Lucas and T. S. Schraga unpublished data), showing large-amplitude diurnal oscillations in concentration that may be missed or misrepresented by coarse daily measurements.

(b) Three chlorophyll *a* subsampling schemes (daily sampling at noon, 6AM, and 6PM) compared with day-averaged chlorophyll *a* and the original 10-min time series from northern Mildred Island. (c) Error for each subsampling scheme, calculated relative to the day average. [Panels (b) and (c) reproduced/modified from Lucas *et al.*, 2006.)

mixing, and diurnal algal growth. Three daily sampling schemes were applied to that time series to reveal what information coarser (in this case daily) sampling of chlorophyll in that environment would have provided. If we assume the sampling goal is to represent daily mean chlorophyll, then we see that the 6PM sampling is the worst, producing "fake blooms" due to the fact that chlorophyll in the southern source region of this water body is highest at the end of the daily photoperiod. The 6AM and noon-time sampling provide acceptable representations of the mean. Even if high-frequency sampling is not sustainable, short-duration high-frequency reconnaissance studies can help us: (1) identify whether large-amplitude, high-frequency variability in key water quality variables exists in our system; (2) determine the time of day or tidal phase that optimizes the match between sampling scheme and the sampling goal (e.g., minimum dissolved oxygen, maximum temperature, average salinity); and (3) allow for estimation of error (Fig. 10.8c) associated with individual sampling schemes (Lucas *et al.*, 2006).

10.6. Summary

In this chapter, fundamental transport equations were reviewed, and the concepts of scaling, transport time scales, and physical–biological–chemical time scale comparison were discussed. One of the most important but widely variable features of estuarine physics – density stratification – was described in relation to reactive and ecologically significant quantities such as phytoplankton biomass and dissolved oxygen. Physical processes varying over hourly time scales and their implications for long-term trends, time-averaged quantities (e.g., net scalar fluxes), and lower-frequency monitoring approaches were discussed. There are many more key connections between hydrodynamic processes and estuarine water quality than were covered in this chapter. For example, estuarine fronts, upwelling in the adjacent coastal ocean, residual circulation features such as eddies, lateral transport, internal waves, and modifications to ambient hydrodynamics by aquatic vegetation represent just a few more classes of physical processes known to influence variability in water quality on local, regional, or system-level scales.

Acknowledgments

Many thanks to Stephen Baines, Jim Cloern, Rochelle Labiosa, and Joan Sheldon for their helpful comments and suggestions and Jeanne DiLeo for her artistic brilliance. My gratitude to Neil Banas, Barbara Hickey, Nancy Monsen, Hans Paerl, Nancy Rabalais, and Eugene Turner for granting me permission to redraw and/or reprint their previously published figures.

References

Baek, S. (2006) *The role of atmospheric forcing in determining transport in a shallow tidal lagoon*. University of California, Berkeley, CA.

Baines, S. B., N. S. Fisher, M. A. Doblin, G. A. Cutter, L. Cutter and B. E. Cole (2004) Light dependence of selenium uptake by phytoplankton and implications for predicting selenium incorporation into food-webs. *Limnol. Oceanogr.* **49**, 566–578.

Banas, N. S. and B. M. Hickey (2005) Mapping exchange and residence time in a model of Willapa Bay, Washington, a branching, macrotidal estuary. *J. Geophys. Res.* **110**, C11011.

Banas, N. S., B. M. Hickey, J. A. Newton and J. L. Ruesink (2007) Tidal exchange, bivalve grazing, and patterns of primary production in Willapa Bay, Washington, USA. *Mar. Ecol. Progr. Ser.* **341**, 123–139.

Basu, B. K. and F. R. Pick (1996) Factors regulating phytoplankton and zooplankton biomass in temperate rivers. *Limnol. Oceanogr.* **41**, 1572–1577.

Bergondo, D. L., D. R. Kester, H. E. Stoffel and W. L. Woods (2005) Time-series observations during the low sub-surface oxygen events in Narragansett Bay during summer 2001. *Mar. Chem.* **97**, 90–103.

Bolin, B. and H. Rodhe (1973) A note on the concepts of age distribution and transit time in natural reservoirs. *Tellus* **25**, 58–62.

Buzzelli, C. P., R. A. Luettich, Jr., S. P. Powers, C. H. Peterson, J. E. McNinch, J. L. Pinckney and H. W. Paerl (2002) Estimating the spatial extent of bottom-water hypoxia and habitat degradation in a shallow estuary. *Mar. Ecol. Progr. Ser.* **230**, 103–112.

Chant, R. J. and A. W. Stoner (2001) Particle trapping in a stratified flood-dominated estuary. *J. Mar. Res.* **59**, 29–51.

Cloern, J. E. (2001) Our evolving conceptual model of the coastal eutrophication problem. *Mar. Ecol. Progr. Ser.* **210**, 223–253.

Dame, R. F. (1996) *Ecology of Marine Bivalves: An Ecosystem Approach*. CRC Press, Boca Raton, FL.

Desmit, X., J. P. Vanderborght, P. Regnier and R. Wollast (2005) Control of phytoplankton production by physical forcing in a strongly tidal, well-mixed estuary. *Biogeosciences* **2**, 205–218.

Dronkers, J. and J. T. F. Zimmerman (1982) Some principles of mixing in tidal lagoons. In Proceedings of the International Symposium on Coastal Lagoons, Bordeaux, France. *Oceanologica Acta*.

Dyer, K. R. (1973) *Estuaries: A Physical Introduction*. John Wiley & Sons, Chichester.

Fischer, H. B. (1972) Mass transport mechanisms in partially stratified estuaries. *J. Fluid Mech.* **53**, 671–687.

Fischer, H. B., E. J. List, R. C. Y. Koh, J. Imberger and N. H. Brooks (1979) *Mixing in Inland and Coastal Waters*. Academic Press, New York.

Fram, J. P., M. A. Martin and M. T. Stacey (2007) Dispersive fluxes between the coastal ocean and a semienclosed estuarine basin. *J. Phys. Oceanogr.* **37**, 1645–1660.

Gardner, L. R. and B. Kjerfve (2006) Tidal fluxes of nutrients and suspended sediments at the North Inlet–Winyah Bay National Estuarine Research Reserve. *Est. Coast. Shelf Sci.* **70**, 682–692.

Gardner, L. R., B. Kjerfve and D. M. Petrecca (2006) Tidal fluxes of dissolved oxygen at the North Inlet–Winyah Bay National Estuarine Research Reserve. *Est. Coast. Shelf Sci.* **67**, 450–460.

Hirsch, C. (1988) *Numerical Computation of Internal and External Flows, Volume 1, Fundamentals of Numerical Discretization*. John Wiley & Sons, New York.

Howarth, R. W., D. P. Swaney, T. J. Butler and R. Marino (2000) Climatic control on eutrophication of the Hudson River Estuary. *Ecosystems* **3**, 210–215.

Joordens, J. C. A., A. J. Souza and A. Visser (2001) The influence of tidal straining and wind on suspended matter and phytoplankton distribution in the Rhine outflow region. *Cont. Shelf Res.* **21**, 301–325.

Koseff, J. R., J. K. Holen, S. G. Monismith and J. E. Cloern (1993) Coupled effects of vertical mixing and benthic grazing on phytoplankton populations in shallow, turbid estuaries. *J. Mar. Res.* **51**, 843–868.

Lin, J., H. Xu, C. Cudaback and D. Wang (2008) Inter-annual variability of hypoxic conditions in a shallow estuary. *J. Mar. Syst.* **73**(1–2), 169–184.

Lucas, L. V. and J. E. Cloern (2002) Effects of tidal shallowing and deepening on phytoplankton production dynamics: a modeling study. *Estuaries* **25**, 497–507.
Lucas, L. V., J. E. Cloern, J. R. Koseff, S. G. Monismith and J. K. Thompson (1998) Does the Sverdrup critical depth model explain bloom dynamics in estuaries? *J. Mar. Res.* **56**, 375–415.
Lucas, L. V., J. R. Koseff, J. E. Cloern, S. G. Monismith and J. K. Thompson (1999a) Processes governing phytoplankton blooms in estuaries. I: The local production–loss balance. *Mar. Ecol. Progr. Ser.* **187**, 1–15.
Lucas, L. V., J. R. Koseff, S. G. Monismith, J. E. Cloern and J. K. Thompson (1999b) Processes governing phytoplankton blooms in estuaries. II: The role of horizontal transport. *Mar. Ecol. Progr. Ser.* **187**, 17–30.
Lucas, L. V., D. M. Sereno, J. R. Burau, T. S. Schraga, C. B. Lopez, M. T. Stacey *et al.* (2006) Intradaily variability of water quality in a shallow tidal lagoon: mechanisms and implications. *Est. Coasts* **29**, 711–730.
Lucas, L. V., J. R. Koseff, S. G. Monismith and J. K. Thompson (2009a) Shallow water processes govern system-wide phytoplankton bloom dynamics: a modeling study. *J. Mar. Syst.* **75**, 70–86.
Lucas, L. V., J. K. Thompson and L. R. Brown (2009b) Why are diverse relationships observed between phytoplankton biomass and transport time? *Limnol. Oceanogr.* **54**, 381–390.
Luketina, D. (1998) Simple tidal prism models revisited. *Est. Coast. Shelf Sci.* **46**, 77–84.
Martin, M. A., J. P. Fram and M. T. Stacey (2007) Seasonal chlorophyll *a* fluxes between the coastal Pacific Ocean and San Francisco Bay. *Mar. Ecol. Progr. Ser.* **337**, 51–61.
Mason, R. P., W. F. Fitzgerald, J. Hurley, A. K. Hanson, P. L. Donaghay and J. M. Sieburth (1993) Mercury biogeochemical cycling in a stratified estuary. *Limnol. Oceanogr.* **38**, 1227–1241.
Middelburg, J. J. and P. M. J. Herman (2007) Organic matter processing in tidal estuaries. *Mar. Chem.* **106**, 127–147.
Middelburg, J. J. and J. Nieuwenhuize (2000) Uptake of dissolved inorganic nitrogen in turbid, tidal estuaries. *Mar. Ecol. Progr. Ser.* **192**, 79–88.
Monismith, S. G., J. R. Burau and M. T. Stacey (1996) Stratification dynamics and gravitational circulation in Northern San Francisco Bay. In J. T. Hollibaugh (ed.), San Francisco Bay: The Ecosystem. Pacific Division of the American Association for the Advancement of Science, pp. 123–153.
Monsen, N. E., J. E. Cloern, L. V. Lucas and S. G. Monismith (2002) A comment on the use of flushing time, residence time, and age as transport time scales. *Limnol. Oceanogr.* **47**, 1545–1553.
Monsen, N. E., J. E. Cloern and J. R. Burau (2007) Effects of flow diversions on water and habitat quality: examples from California's highly manipulated Sacramento–San Joaquin Delta. *San Francisco Est. Watershed Sci.* **5**, Article 2.
Monteiro, P. M. S. and J. L. Largier (1999) Thermal stratification in Saldanha Bay (South Africa) and subtidal, density-driven exchange with the coastal waters of the Benguela Upwelling System. *Est. Coast. Shelf Sci.* **49**, 877–890.
Nagy, G. J., M. Gomez-Erache, C. H. Lopez and A. C. Perdomo (2002) Distribution patterns of nutrients and symptoms of eutrophication in the Rio de la Plata River Estuary System. *Hydrobiologia* **475/476**, 125–139.
Nixon, S. W. (1995) Coastal marine eutrophication: a definition, social causes, and future concerns. *Ophelia* **41**, 199–219.
Nixon, S. W., J. W. Ammerman, L. P. Atkinson, V. M. Berounsky, G. Billen, W. C. Boicourt *et al.* (1996) The fate of nitrogen and phosphorus at the land–sea margin of the North Atlantic Ocean. *Biogeochem.* **35**, 141–180.

Paerl, H. W. and J. Huisman (2008) Blooms like it hot. *Science* **320**, 57–58.
Paerl, H. W., J. D. Bales, L. W. Ausley, C. P. Buzzelli, L. B. Crowder, L. A. Eby *et al.* (2001) Ecosystem impacts of three sequential hurricanes (Dennis, Floyd, and Irene) on the United States' largest lagoonal estuary, Pamlico Sound, NC. *PNAS* **98**, 5655–5660.
Paerl, H. W., L. M. Valdes, B. L. Peierls, J. E. Adolf and L. W. Harding (2006) Anthropogenic and climatic influences on the eutrophication of large estuarine ecosystems. *Limnol. Oceanogr.* **51**, 448–462.
Rabalais, N. N. and R. E. Turner (2006) Oxygen depletion in the Gulf of Mexico adjacent to the Mississippi River. In L. N. Neretin (ed.), *Past and Present Water Column Anoxia*. Springer-Verlag, New York, pp. 225–245.
Rabalais, N. N., R. E. Turner and W. J. Wiseman (2002) Gulf of Mexico hypoxia, A. K. A. "The Dead Zone". *Annu. Rev. Ecol. Syst.* **33**, 235–263.
Rocha, C. (1998) Rhythmic ammonium regeneration and flushing in intertidal sediments of the Sado estuary. *Limnol. Oceanogr.* **43**, 823–831.
Sanford, L. P., W. C. Boicourt and S. R. Rives (1992) Model for estimating tidal flushing of small embayments. *J. Waterw. Port Coast. Ocean Eng.* **118**, 635–654.
Schemel, L. E., T. R. Sommer, A. B. Müller-Solger and W. C. Harrell (2004) Hydrologic variability, water chemistry, and phytoplankton biomass in a large floodplain of the Sacramento River, CA, U.S.A. *Hydrobiologia* **513**, 129–139.
Sheldon, J. E. and M. Alber (2002) A comparison of residence time calculations using simple compartment models of the Altamaha River Estuary, Georgia. *Estuaries* **25**, 1304–1317.
Sheldon, J. E. and M. Alber (2006) The calculation of estuarine turnover times using freshwater fraction and tidal prism models: a critical evaluation. *Est. Coasts* **29**, 133–146.
Simpson, J. H., J. Brown, J. Matthews and G. Allen (1990) Tidal straining, density currents, and stirring in the control of estuarine stratification. *Estuaries* **13**, 125–132.
Smaal, A. C. and T. C. Prins (1993) The uptake of organic matter and the release of inorganic nutrients by bivalve suspension feeder beds. In R. F. Dame (ed.), *Bivalve Filter Feeders in Estuarine and Coastal Ecosystem Processes*. Springer-Verlag, New York, pp. 271–298.
Søballe, D. M. and R. W. Bachmann (1984) Influence of reservoir transit on riverine algal transport and abundance. *Can. J. Fish. Aquat. Sci.* **41**, 1803–1813.
Søballe, D. M. and B. L. Kimmel (1987) A large-scale comparison of factors influencing phytoplankton abundance in rivers, lakes, and impoundments. *Ecology* **68**, 1943–1954.
Sobczak, W. V., J. E. Cloern, A. D. Jassby and A. B. Muller-Solger (2002) Bioavailability of organic matter in a highly disturbed estuary: the role of detrital and algal resources. *PNAS* **99**, 8101–8105.
Strayer, D. L., N. F. Caraco, J. J. Cole, S. Findlay and M. L. Pace (1999) Transformation of freshwater ecosystem by bivalves. *Bioscience* **49**, 19–27.
Strayer, D. L., M. L. Pace, N. F. Caraco, J. J. Cole and S. Findlay (2008) Hydrology and grazing jointly control a large-river food web. *Ecology* **89**, 12–18.
Swaney, D. P., D. Scavia, R. W. Howarth and R. Marino (2008) Estuarine classification and response to nitrogen loading: insights from simple ecological models. *Est. Coast. Shelf Sci.* **77**, 253–263.
Sylaios, G. K., V. A. Tsihrintzis, C. Akratos and K. Haralambidou (2006) Quantification of water, salt, and nutrient exchange processes at the mouth of a Mediterranean coastal lagoon. *Environ. Monit. Assess.* **119**, 275–301.

Takeoka, H. (1984) Fundamental concepts of exchange and transport time scales in a coastal sea. *Cont. Shelf Res.* **3**, 311–326.

Taylor A. H. and J. A. Stephens (1993) Diurnal variations of convective mixing and the spring bloom of phytoplankton. *Deep-Sea Res. II* **40**, 389–408.

Thompson, J. K., J. R. Koseff, S. G. Monismith and L. V. Lucas (2008) Shallow water processes govern system-wide phytoplankton bloom dynamics: a field study. *J. Mar. Syst.* **74**, 153–166.

Torréton, J.-P., E. Rochelle-Newall, A. Jouon, V. Faure, S. Jacquet and P. Douillet (2007) Correspondence between the distribution of hydrodynamic time parameters and the distribution of biological and chemical variables in a semi-enclosed coral reef lagoon. *Est. Coast. Shelf Sci.* **74**, 766–776.

Uncles, R. J., J. A. Stephens and R. E. Smith (2002) The dependence of estuarine turbidity on tidal intrusion length, tidal range, and residence time. *Cont. Shelf Res.* **22**, 1835–1856.

van de Kreeke, J. (1983) Residence time: application to small boat basins. *J. Waterw. Port Coast. Ocean Eng.* **109**, 416–428.

Volkmar, E. C. and R. A. Dahlgren (2006) Biological oxygen demand dynamics in the Lower San Joaquin River, California. *Environ. Sci. Technol.* **40**, 5653–5660.

Walz, N. and M. Welker (1998) Plankton development in a rapidly flushed lake in the River Spree system (Neuendorfer See, Northeast Germany). *J. Plankton Res.* **20**, 2071–2087.

Zimmerman, J. T. F. (1976) Mixing and flushing of tidal embayments in the western Dutch Wadden Sea. Part I: Distribution of salinity and calculation of mixing time scales. *Neth. J. Sea Res.* **10**, 149–191.

Index

Printed in the United States
By Bookmasters